Mimetic trio documented by Springer and Smith–Vaniz (1972a): above *Meiacanthus nigrolineatus* (left) and *Ecsenius gravieri* (right), below female *Plagiotremus townsendi;* all from Gulf of Aqaba. Photographs by Victor G. Springer; frontispiece courtesy of Herbert R. Axelrod, TFH Publications, Inc.

THE SABER-TOOTHED BLENNIES, TRIBE NEMOPHINI (PISCES: BLENNIIDAE)

WILLIAM F. SMITH-VANIZ
Department of Ichthyology
Academy of Natural Sciences of Philadelphia

MONOGRAPH 19
THE ACADEMY OF NATURAL SCIENCES OF PHILADELPHIA
19th AND THE PARKWAY
PHILADELPHIA, PENNSYLVANIA 19103

MONOGRAPH 19
THE ACADEMY OF NATURAL SCIENCES
OF PHILADELPHIA
1976

Printed by the George M. Carpenter Fund for the
Encouragement of Original Scientific Research

Published December 17, 1976
Fulton Press Inc., Lancaster, Pennsylvania 17603

Dedicated to my parents

Mr. and Mrs. Julius T. Smith-Vaniz

for encouraging my interest in natural history

Cover Illustration:

Radiograph of *Petroscirtes breviceps*

TABLE OF CONTENTS

ABSTRACT

The blenniid tribe Nemophini is revised based on the examination of slightly more than 1,500 collections. A total of five genera (including one genus with two subgenera and two genera each with three subgenera) and 41 species (including three species each with two subspecies) are recognized. The genera recognized are: *Petroscirtes, Aspidontus, Xiphasia, Meiacanthus* and *Plagiotremus.* Descriptions are given for all taxa, including two new subgenera, 13 new species and one new subspecies, and identification keys are provided for all taxa except subspecies. Illustrations and distributional maps are given for all species and subspecies. Sexual dimorphism, geographic variation and distribution, biology, nomenclature, evolutionary trends and relationships are discussed. More than a third of the species of Nemophini appear to be involved in mimetic relationships. Possible reasons for such interactions are discussed and several leads for additional study are suggested.

Meiacanthus is the only fish genus known to possess toxic buccal glands. The three subgenera of *Meiacanthus* can be distinguished solely on the basis of gland morphology. *Plagiotremus* possesses unique trophic specializations associated with an obligatory, predacious scale-eating habit. Short ligaments extend from the anterodorsal margin of the dentary to a bony spur positioned posteromesially on each dentary incisor. This ligamentous arrangement allows the incisors a certain amount of movement, both laterally and anteroposteriorly, which presumably facilitates scraping and snagging of scales.

Nine nominal genus-group names and 28 species names are relegated to synonymy. A neotype is designated for *Petroscirtes variabilis* Cantor and lectotypes are selected for *Salarias viperidens* De Vis, *Petroscirtes fluctuans* Weber, *Petroscirtes oualanensis* Günther, *Petroskirtes rhinorhynchos* Bleeker and *Petroskirtes tapeinosoma* Bleeker.

INTRODUCTION

The tribe Nemophini, or saber-toothed blennies, comprises the most specialized group of blenniid fishes. The collective common name refers to the large dentary canines, which are used only in defense or intraspecific battling. The structural and behavioral diversity found in the Nemophini is remarkable for such a small group of closely related fishes, so much so that some workers have assigned the genera of this tribe to three different families. The adults of all species are semipelagic and, unlike most blennies, spend much of their time hovering above the substrate or actively swimming in the water column. This behavior is facilitated by a well-developed swimbladder that is absent only in the two species of *Xiphasia,* which are unusual among the Nemophini in having extremely elongate bodies and apparently nocturnal habits. Except for a single eastern Pacific species, the tribe is restricted to the Indo-Pacific region. The adults of most species are restricted to depths of 1-15 meters on or in the vicinity of coral reefs, although one species typically occurs in estuarine or freshwater habitats. Many of the species of Nemophini are brilliantly colored and adapt well to aquaria. Several of the species involved in mimetic relationships have made excellent subjects for ethological studies.

None of the genera of Nemophini has been treated adequately systematically. Knowledge of the tribe has been confined largely to original descriptions and faunal studies which were based on relatively few specimens and resulted in many errors, both of nomenclature and identification. The objectives of this study are to: (1) define the genera and species of Nemophini; (2) identify evolutionary trends and possible relationships within the group; (3) elucidate structural and behavioral adaptations; and (4) thus, hopefully, stimulate other workers to study the biology and behavior of these fascinating fishes.

Since this study is primarily revisionary, one may ask how well it agrees with the previous literature and how complete it is in terms of species sampled. Previously, 17 nominal genus-group names and 76 nominal species names were available in the Nemophini. I recognize 7 of these generic group names and 28 of these species names. In addition, I describe two new subgenera, 13 new species and one new subspecies, for a total of 5 genera (including one genus with two subgenera and two genera each with three subgenera) and 41 species (including three species each with two subspecies) in the tribe Nemophini.

This study is based on examination of more than 4,000 specimens comprising slightly more than 1,500 collections. The availability of specimens has increased significantly in recent years as a result of increased use of SCUBA and ichthyocides as a means of collection. Indeed, more than a third of the collections and specimens utilized in this study were obtained during the past decade. Many species of the Nemophini have restricted distributions and appear to be the product of recent speciation. Judging from the fact that seven species are presently known from single collections, it is reasonable to expect that more undescribed species await discovery.

ACKNOWLEDGMENTS

This monograph represents the major portion of a doctoral dissertation submitted to the faculty of the University of Miami. I wish to express my deepest gratitude to Dr. C. Richard Robins who, as my major professor at the University of Miami, stimulated and considerably broadened my perspective of ichthyology and Dr. Victor G. Springer, National Museum of Natural History, who suggested this research and has been a constant source of help and inspiration throughout this study. Grateful thanks are extended to James E. Böhlke, Academy of Natural Sciences of Philadelphia, C. Richard Robins, Victor G. Springer and Jon C. Staiger, University of Miami, for reading the entire manuscript and for their helpful suggestions and critical comments. I have also profited from discussions with Stanley H. Weitzman, National Museum of Natural History, which helped me to more fully understand the principles of phylogenetic systematics.

This study would not have been possible without the generous cooperation of a large number of people. The following individuals facilitated specimen examination, either through hospitality at their institutions or by providing information on, or loans of, specimens (abbreviations refer to institutions and are explained in the methods and materials section): C. Lavett Smith, AMNH; John R. Paxton and Douglas F. Hoese, AMS; James E. Böhlke, ANSP; John E. Randall, BPBM; Alwyne C. Wheeler and Peter J. Whitehead, BMNH; William N. Eschmeyer and Lillian Dempster, CAS; Phillip C. Heemstra and I. S. R. Munro, CSIRO; P. Kailola, DASF; Robert K. Johnson and Loren P. Woods, FMNH; the late H. Steinitz, HUJ; Norma Chirichigno F., IMP; W. Nellen, IMUK; Christine Karrer, ISZZ; Robert J. Lavenberg and Camm C. Swift, LACM; Myvanwy M. Dick, MCZ; M. L. Bauchot, MNHN; Wolfgang Klausewitz, NFIS; Paul Struhsaker, NFMS; J. Moreland, NMNZ; Paul Kähsbauer, NMV; P. Fourmanoir, Centre ORSTOM, Noumea, New Caledonia; R. J. McKay, QMB, and formerly WAM; M. Boeseman, RMNH; Margaret M. Smith and Thomas H. Fraser (formerly), RUSI; P. A. Hulley, SAM; Carl L. Hubbs and Richard H. Rosenblatt, SIO; Warren C. Freihofer, SU; Boyd W. Walker and John Bleck, UCLA; Harry T. Kami, UG; Reeve M. Bailey, UMMZ; George S. Losey, Jr., University of Hawaii; Tokiharu Abe, University of Tokyo; Victor G. Springer, USNM; Lev Fishelson, UTAI; Arthur D. Welander, UW; Walter Fasbender, Vanderbilt Museum Commission; Gerald R. Allen and J. B. Hutchins, WAM; Han Nijssen and I. J. H. Isbrücker, ZMA; A. G. K. Menon, ZSIC; W. Ladiges, ZSZM; Yoshiaki Tominaga, ZUMT.

I am especially grateful to M. Boeseman for his assistance in searching the collections in Leiden for types and in helping me to correctly identify and select Bleeker holotypes and lectotypes.

The following individuals made special efforts to collect specimens for this study, many of which were received as gifts: Gerald R. Allen, Western Australian Museum; Bruce A. Carlson, University of the South Pacific, Fiji; Hitoshi Ida, Kitasato University, Japan; and Victor G. Springer. Alfred L. Wolf of Philadelphia, with the assistance of Rodney Jonklaas, succeeded in the difficult task of collecting and shipping live blennies from the Maldive Islands that I was particularly interested in examining.

John E. Randall generously made available excellent color transparencies of numerous species, provided notes on life colors and behavior, and kept me informed of recent collections that he had made from critical areas. Valuable color photographs and transparencies were also provided by Gerald R. Allen, M. Boeseman, Bruce A. Carlson, William P. Davis, Douglas F. Hoese, Hitoshi Ida, Barry C. Russell and Victor G. Springer. William H. Hulet, Marine Biomedical Institute, Galveston, Texas, provided the scanning electron micrographs shown in Figure 43. I thank the Director, Rijksmuseum van Natuurlijke Historie, Leiden, for permission to publish the photographs of Bleeker's unpublished illustrations of *P. solorensis* and *thepassi.*

Alan R. Emery, Royal Ontario Museum, provided information on the behavior of *Petroscirtes thepassi;* and Lev Fishelson sent me a copy of his manuscript (since published) on the reproduction and behavior of *Meiacanthus nigrolineatus.*

Eugenia B. Böhlke, ANSP, provided many of the radiographs utilized in this study. C. E. Dawson, Gulf Coast Research Laboratory, facilitated statistical analysis of some of my data and William R. Taylor, USNM, answered questions on nomenclature. Mary H. Fuges illustrated the specimens shown in Figures 104, 112 and 137. The nearly illegible first draft of the manuscript was carefully typed by Christina G. Fuiman and the final draft was expertly typed by Debbie Bauman. Finally, I wish to thank my wife, Esther, for her encouragement and many hours devoted to tabulating data and proofreading.

Most of this work was done at the Rosenstiel School of Marine and Atmospheric Sciences, University of Miami and the Academy of Natural Sciences of Philadelphia. I am grateful for the working space and facilities provided by these institutions which made it possible for me to accomplish this task. Partial support for visits to museums was provided by a Grant-in-Aid of Research from the Society of the Sigma Xi.

HISTORICAL RESUMÉ OF TRIBE NEMOPHINI

The oldest suprageneric name available for the tribe Nemophini was proposed by Kaup (1858: 169). He stated that his new genus *Nemophis* (= *Xiphasia*) formed "a distinct group among the riband-shaped fishes, and is perhaps the type of a different family, which we may call Nemophidae." In his paper describing *Xiphasia trachypareia,* Bleeker (1864) stated that it was a blennioid related to *Petroscirtes* and that it might also be allied to Kaup's *Nemophis.* Gill (1865) also considered his new genus *Plagiotremus* to be related to *Nemophis* and classified it as a subfamily (Plagiotrematinae) within the Nemophidae. Regan (1912: 267) reaffirmed the blennioid affinities of *Xiphasia* and pointed out similarities between it and related genera. In his provisional synopsis of the genera of Blenniidae, Norman (1943) correctly recognized the intimate relationship of the genera treated in this study, but did not formally name the group. Springer's (1968) major contribution to blenniid classification firmly established the Nemophini as a natural unit. He defined the group osteologically, recognizing it as a subfamily, and listed all nominal genera referable to it. In their paper describing the monotypic tribe Phenablenniini, Springer and Smith-Vaniz (1972b) compared in detail the blenniid tribes and discussed their relationships. Because *Phenablennius* was somewhat intermediate between the Blenniinae and Nemophinae, especially with respect to the vestigial basisphenoid, they chose not to recognize any subfamilies in the Blenniidae.

METHODS AND MATERIALS

Measurements. — Those proportional measurements that proved to be most useful in defining species usually involved sexually dimorphic characters and are presented as a series of scatter diagrams. The lengths of paired structures (i.e., outer caudal-fin rays and pelvic fins) frequently were unequal, and the length of the longest such structure was measured to reduce the variation presumably resulting from mechanical injury. Measurements were made point-to-point with needlepoint dial calipers and recorded to the nearest 0.1 millimeter. Unless otherwise stated, all proportional measurements given in the descriptions are expressed as a percentage of standard length. Measurements used in this study are defined below.

Standard length (SL): Distance from bony mid-tip of snout to midbase of the caudal fin.

Preanal length: Distance from bony mid-tip of snout to origin of the anal fin.

Least body depth: Vertical measurement from origin of the anal fin to the base of dorsal fin.

Dorsal-spine lengths: Length of dorsal-fin spines measured from anterobasal point of insertion to the distal tip.

Pelvic-fin length: Length of longest segmented ray measured from anterobasal point of insertion to the distal tip.

Caudal-fin ray length: Length of longest segmented ray measured from anterobasal point of insertion to the distal tip.

Meristic characters. — Numbers of dorsal-, anal- and caudal-fin rays, vertebrae, pleural and epipleural ribs were usually determined from radiographs.

In the dorsal and anal fins each spine and segmented ray is typically supported by a separate pterygiophore, except for the last spine and first ray (Fig. 97) which share a common pterygiophore. Occasionally a pterygiophore lacked an external element, usually marked by a conspicuous gap between two external elements. When such an anomalous condition existed, the external element count was increased by one for each such pterygiophore.

Anal-fin spines are frequently difficult to discern externally. With rare exception the anal fin always has two spines.

Pelvic fins are typically present in all species, except *Plagiotremus celebesensis* and *iosodon* (see species account). The pelvic fins invariably consist of a spine (not visible externally) and three simple rays, the innermost not always apparent.

All counts of caudal-fin elements are reported as a two-part formula, a plus sign separating the count for elements associated with the complex dorsal hypural fan (recorded first) from those of the complex ventral hypural fan. Caudal-fin elements include the spinelike dorsal and ventral procurrent rays and principal (segmented) rays. The distinction between principal and procurrent rays is somewhat arbitrary. For enumeration purposes, principal rays are defined as segmented rays (typically robust) that are attached firmly to the hypural bones. In *Xiphasia* there are 10 (rarely 11) principal rays, 11 in the other genera of Nemophini. In some genera, especially *Meiacanthus,* the posteriormost dorsal and ventral "procurrent" caudal-fin rays may be segmented; frequently only a single, distal segmental joint (easily overlooked) is present. Because of the ease in overlooking such structures, only rays with two or more segmental joints are included in counts of "segmented rays."

Vertebral counts distinguish precaudal from

caudal vertebrae. When combined counts are given, precaudal vertebrae are recorded first. The first caudal vertebra includes the first centrum with a well-developed haemal spine. Sometimes the last precaudal vertebra bears a moderate haemal spine, in which case the distal tip of the spine is positioned in advance of the first anal-fin pterygiophore and is noticeably shorter than the haemal spine of the next vertebra posteriorly. Occasionally the determination of the first caudal vertebra was somewhat subjective. However, the ranges of precaudal and caudal vertebrae recorded for individual species are based on unequivocal specimens. The terminal caudal vertebra includes the hypural-bearing centrum. Infrequently specimens exhibited partial fusion of adjacent vertebrae; in such cases the neural spines are usually apparent and considered to equal the number of vertebrae involved in the fusion.

Epipleural rib counts include all vertebrae bearing an epipleural rib, regardless of whether a pair or only one rib was present, or whether on the right or left side. The anterior epipleural ribs were not apparent on radiographs, but invariably began on the first vertebra in cleared and stained specimens. Occasional vertebrae lacking epipleural ribs that occur between vertebrae bearing epipleural ribs were counted as if they had these ribs.

Teeth. — All species of Nemophini have a large, recurved canine tooth posteriorly on each dentary. A somewhat smaller canine is also present posteriorly on each premaxillary in *Petroscirtes, Aspidontus* (typically lost in adult *A. taeniatus,* see Table 12), *Xiphasia* and *Meiacanthus.* Two canines on a side occasionally occur together in juveniles and (less frequently) adults of all genera.

Plagiotremus lacks premaxillary canines, but in juveniles the posteriormost incisors are canine-like in appearance. These teeth (one or two on each premaxillary) usually are separated from the other incisors by a diastema and have relatively sharp tips instead of being broadly rounded. Unlike canines they are not recurved or noticeably larger than adjacent incisors. Occasional specimens exhibit incisors that are intermediate in appearance between the two extremes.

The number of incisors increases with increase in standard length in all species of Nemophini. Scatter diagrams showing the relationship of incisors to standard length are given for all species of *Aspidontus* (Fig. 30), *Xiphasia* (Fig. 37), *Plagiotremus* (Figs. 74-77), and for selected species of *Petroscirtes* (Fig. 14) and *Meiacanthus* (Fig. 45). Tooth counts given in the descriptions are of the total number of incisors for either both dentaries or both premaxillaries. The range records the minimum and maximum counts obtained, but are not necessarily those of the smallest and largest specimen examined (size limits indicated).

Gill openings. — In adults of all Nemophini the gill opening is restricted. In *Petroscirtes* and *Meiacanthus* it is restricted to the area dorsal to the dorsalmost pectoral-fin ray. In the other genera the ventral margin of the gill opening may be opposite or well below the dorsalmost pectoral-fin ray. The relative extent of the gill opening may vary considerably but tends to be more restricted in adults of *Aspidontus* (see Table 13) and *Xiphasia.* The position of the ventral margin of the gill opening relative to a particular fin ray is often a matter of judgment and allowance of plus or minus one or two rays should be assumed.

Cephalic sensory pores. — Pore series recognized in this study are shown in Figure 1. The pore series or units recognized are in some cases arbitrary, but were adopted because they facilitated a precise description of pore distributions. Illustrations of typical pore patterns for representative species of each genus are given in the systematic accounts. Table 1 gives a summary of selected cephalic pore counts for the genera of Nemophini. Cephalic pores are often difficult to see and are easily overlooked. A fine jet of compressed air and strong illumination and magnification greatly facilitate examination. In some species pores may be hidden by fleshy cirri. Fortunately such cirri are associated only with cephalic pores, individual pores always occurring near the base of a cirrus. Pore series illustrated in Fig. 1 are defined below. Those pore series that may not be readily apparent on specimens are defined as follows:

Mandibular (MD) series: Pores on the lower jaw beginning with pore nearest the symphysis of the jaw and terminating with the pore positioned immediately posterior to the dentary bone. In those species of *Plagiotremus* that have 3 MD pores the posterior pore is frequently reduced and difficult to detect.

Preopercular (POP) series: Pores connected by

TABLE 1. — Summary of selected cephalic pore counts in genera of Nemophini.[1]

	Mandibulo-preoperculo-temporal				Circumorbital		Interorbital
	MD	POP	LT	ST	IFO	SO	IO
Petroscirtes	3	6	3	3–5	7–9	2	4
Aspidontus	3	6	3	3	6–7	2	4
Xiphasia	3	6	3	2–3	7–8	2	4
Meiacanthus	2–3	6	2–3	3–4	7–9	2	4
Plagiotremus	2–3	5	3	3	6–7	1–2	2 or 4

[1] Refer to Figure 1 and text for methodology of pore counts. Abbreviations: MD, mandibular; POP, preopercular; LT, lateraltemporal; ST, supratemporal; IO, interorbital. Except for pores in the ST and IO series, counts are bilateral (one side only).

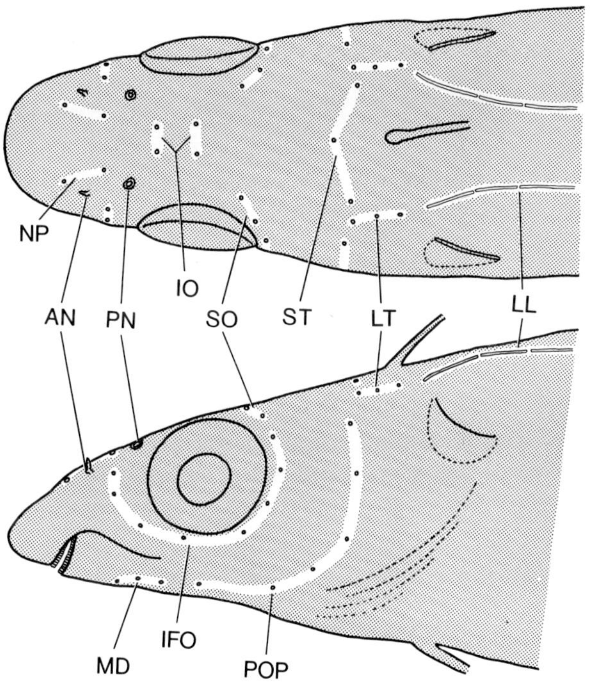

Figure 1. — Semidiagrammatic illustration of cephalic sensory pore series in *Petroscirtes thepassi*. Abbreviations: NP, nasal pores; AN, anterior nostril; PN, posterior nostril; IFO, infraorbital (includes only pores associated exclusively with infraorbital bones); SO, supraorbital; IO, interorbital (both sides included in count; species of Nemophini have either two or four IO pores); MD, mandibular; POP, preopercular; LT, lateral temporal; ST, supratemporal (both sides included in count; species of Nemophini have 3-5 ST pores); LL, lateral-line tubes.

the canal of the preopercular bone, including the terminal pore at each end of the bone. In the illustrations showing pore patterns of representative species, the dorsalmost POP pore is indicated by an arrow.

Lateral temporal (LT) series: In all species, except *Meiacanthus ditrema,* there are 3 LT pores (unilaterally). They are aligned lengthwise, horizontally and constitute the lateralmost pores above the POP series. In *M. ditrema* there are only 2 LT pores (Fig. 40) and they constitute the two posteriormost pores on either side of the head.

Supratemporal (ST) series: Pores on the dorsum of the head between the LT series. There may be 3-5 pores in this series. A formula is used to describe individual pore patterns (i.e., 1-1-2 indicates that there are 4 pores, one on the left side, two on the right side and one in front of the dorsal fin at the mid-point of the supratemporal canal).

Supraorbital (SO) series: Pores located near the posterodorsal rim of orbit that are associated exclusively with the frontal bone. In the illustrations showing pore patterns of representative species the ventralmost SO pore is indicated by an arrow.

Cirri. — A fleshy flap or cirrus is associated with the anterior pore of each dentary in most species of Nemophini, a sexually dimorphic character in all genera except *Petroscirtes.* Additional cirri are variously associated with other cephalic pores in *Petroscirtes* (all species except *marginatus*) and in large males of *Meiacanthus anema* (Fig. 51). Cirri are frequently minute and can be easily overlooked. A simple orbital cirrus may also occur on the fleshy dorsum of the eye in some species of *Petroscirtes.* In several species the orbital cirrus is well developed in juveniles but in large adults is often minute or absent. Cirri are defined based on position as follows:

Orbital cirrus: Cirrus located on the dorsum of each eye.

Postorbital cirri: Cirri asociated with the supraorbital pores.

Symphyseal mandibular cirrus: Cirrus associated with the anteriormost pore of each dentary.

Preopercular cirri: Cirri associated with the preopercular pores.

Predorsal cirrus: Cirrus associated with the pore located at the mid-point of the supratemporal canal.

Nuchal cirri: Cirri associated with the supratemporal pores exclusive of the predorsal cirrus.

Posttemporal cirrus: Cirrus associated with the posteriormost pore in the lateral temporal series (typically present in all species of *Petroscirtes,* except *marginatus*).

Swim bladder. — All genera of Nemophini have swim bladders (Fig. 89) except *Xiphasia.* The swim bladder is a very delicate structure in blenniids and is easily overlooked unless a careful dissection is made. In old museum specimens or fresh material that was not fixed in formalin immediately after capture, the swim bladder tends to collapse and is virtually impossible to detect.

Larvae. — The smallest individuals of each species of Nemophini examined in this study are listed in Table 2. Except for some proportional differences (larger head and eyes, etc.), and the position of the dorsal fin in *Xiphasia,* these small specimens closely resemble the adults.

Weber (1913, Fig. 114) and de Beaufort (1951, Fig. 53) published the same figure of a planktonic larva identified as *Petroscirtes* spec. Springer and Gomon (1975) reproduced the same illustration and convincingly demonstrated (p. 46) that it was based on a member of the tribe Omobranchini. The illustrated specimen agrees with their definition of "larvae" in having a well-developed preopercular projection or spine; it also has a large area of dark melanophores on the pectoral fin, which was also characteristic of most of the larvae of *Omobranchus* that they examined. The smallest nonlarva reported in their study was 12.7 mm SL; the largest larva (of 7 species) ranged from 12.6-18.1 mm. Although specimens between 7.2-18.0 mm SL were available for 19 species of Nemophini, none possessed the "larval" characters of *Omobranchus.* If the larvae of the Nemophini do indeed lack preopercular spines, this may be another character useful in defining the tribe. This possibility is uncertain, however, as the younger stages of several species of Nemophini are unavailable (note especially *Aspidontus*).

Osteology. — Osteological terminology follows that of Springer (1968), except that the circumorbital bones are here called infraorbitals. Knob-like processes occur on each side of the base of the neural arches of the third to fifth through tenth vertebrae in all genera except *Xiphasia.* Strong ligaments extend from these processes to the proximal ends of the associated epipleurals. Since I know of no specific name for these structures, they are here called "neural arch processes."

Material used for osteological study was cleared and stained using the enzyme digestion method of Taylor (1967a, 1967b). All osteological drawings were made with the aid of a camera lucida. When bones from the right side of specimens were illustrated, the drawings were reversed so that all bone illustrations would be similarly oriented. In a few instances representative species of other blenniid tribes were examined for the presence or absence of an osteological character. These specimens are not listed below, but include only material previously cited in Springer (1968). In the case of a few rare species of Nemophini the number of infraorbital bones was determined from partial dissection of non-cleared and stained specimens.

Cleared and stained specimens of Nemophini are listed below alphabetically by genus and species. Numbers in parentheses refer to standard lengths in millimeters.

Aspidontus dussumieri: ANSP 128041, male (53.4); ANSP 128044, male (35.5); USNM 204549, female (100.3).

Aspidontus taeniatus: ANSP 122373, male (48.2); ANSP 128567, 2 females (48.2-57.4); ANSP 128568, male (47.5); USNM 142228, male (90.4); USNM 199432, male (50.1); USNM 200339, female (67.5); USNM 201562, male (63.9).

Meiacanthus anema: ANSP 128535, 4 males (38.3-40.6), female (29.5).

Meiacanthus atrodorsalis: ANSP 124872, male (59.1); ANSP 124873, male (42.4).

Meiacanthus ditrema: ANSP 128491, male (26.2), female (27.4); USNM 195706, male (37.1).

Meiacanthus grammistes: ANSP 128536, 3 males (26.0-46.2), female (42.9).

Meiacanthus lineatus: USNM 201363, 2 males (46.0-60.4), female (50.0).

Meiacanthus mossambicus: USNM 201484, female (53.3).

Meiacanthus nigrolineatus: USNM 200601, male (37.1).

Meiacanthus procne: USNM 205291, female (40.0).

Meiacanthus smithi: USNM 199422, female (46.2).

Petroscirtes ancylodon: USNM 147975, female (75.0); USNM 167027, 2 males (57.9-89.5).

Petroscirtes breviceps: ANSP 128559, 2 males (56.3-71.3).

Petroscirtes fallax: USNM 201800, male (42.6), female (44.0).

Petroscirtes lupus: ANSP 124405, male (62.0), female (64.3).

Petroscirtes mitratus: ANSP 128560, male (39.8), 2 females (32.7-41.7); USNM 199433, male (53.7); USNM 200567, female (38.6); USNM 200599, male (51.6).

Petroscirtes thepassi: USNM 205296, male (46.6), female (54.1).

Petroscirtes variabilis: USNM 138250, 2 females (49.9-54.3).

Petroscirtes xestus: ANSP 124874, male (45.9); USNM 115531, male (46.0).

Plagiotremus azaleus: USNM 200392, 3 males (44.4-48.2); USNM 256392-F6, 3 females (36.9-40.6).

Plagiotremus celebesensis: CAS SU67356, juv. (54.9).

Plagiotremus iosodon: USNM 205372, female (39.9).

Plagiotremus phenax: ANSP 128561, 2 females (35.9-38.2).

Plagiotremus laudandus: ANSP 126230, male (49.3), female (39.6); USNM 201413, female (31.4).

Plagiotremus rhinorhynchos: ANSP 124875, female (72.4); ANSP 124876, male (67.8); ANSP 111015, 2 males (52.9-57.5).

Plagiotremus tapeinosoma: ANSP 124877, male (32.5); ANSP 124878, male (60.5), female (38.9); USNM 195704, male (71.8), female (68.9); USNM 200338, male (70.4).

Plagiotremus townsendi: USNM 205379, male (31.6), female (37.5).

Xiphasia matsubarai: ANSP 128537, 2 males (195-208); USNM 200562, female (224).

Xiphasia setifer: ANSP 128699, 2 females (267-282); USNM 205292, female (374).

Color pattern. — Descriptions are deliberately brief and stress diagnostic features of the color pattern. In the descriptions "bars" and "bands" refer to vertical markings; "stripes" refer to horizontal markings.

Good specimen illustrations are an essential part of any revisionary study of blenniid fishes. Economic considerations dictated the use of photographs of museum specimens (often not in the best condition) as the primary method of illustration. In the individual species accounts references are given to previously published illustrations of good quality, especially color photographs.

Synonymies. — Only references to original descriptions are given in the synonymies for previously described species.

Material examined. — In order to conserve space in the lists of specimens examined, all SL measurements have been rounded to the nearest mm except in the case of holotypes or lectotypes. The number of specimens and the size range in mm SL are given in parentheses.

Only general localities are given for most collections of common species. All available locality data are listed for rare or new species. USNM collections with the only collection data "Dutch East Indies-Longley" are presumed to have come from the Banda Sea as suggested by Springer (1971: 42). The abbreviation "I." denotes an isolated island or component island of a group of islands generally recognized as a natural unit; the abbreviation "Is." refers to a group of islands (i.e., Marshall Is.). Distinction is not always made between atolls and islands. Depth measurements have been converted to meters (m) where practical. Collectors and date of capture are usually given only for new species.

No sequence of listing collection localities in the material examined sections or in the tables showing geographic variation is entirely satisfactory. However, in an effort to be consistent, presentation is from west to east and clockwise in the central Pacific, roughly as follows: Persion Gulf — Red Sea — E.

Africa — Madagascar — Maldive Is. — Sri Lanka — Andaman Is. — Cocos-Keeling Is. — Indonesia — W. Australia — S. China Sea — Philippine Is. — Japan — Mariana Is. — Caroline Is. — Marshall Is. — Gilbert Is. — Line Is. — Marquesas Is. — Society Is. — Samoa Is. — Tonga Is. — Fiji Is. — New Hebrides Is. — Solomon Is. — New Guinea — E. Australia — New Caledonia — Lord Howe I.

Institutional abbreviations. — The following abbreviations are used to denote the location of specimens cited in this study and the affiliations of individuals who facilitated loan requests.

AMNH	American Museum of Natural History, New York City
AMS	Australian Museum, Sydney (catalog numbers usually with the prefixes I, IA and IB)
ANSP	Academy of Natural Sciences of Philadelphia
BMNH	British Museum (Natural History), London
BPBM	Bernice P. Bishop Museum, Honolulu
CAS	California Academy of Sciences, San Francisco (catalog numbers with the prefix GVF refer to George Vanderbilt Foundation register numbers)
CSU	California State University, San Diego
CSIRO	Commonwealth Scientific and Industrial Research Organization, Division of Fisheries and Oceanography, Cronulla, N. S. W., Australia (catalog numbers with prefix A or B)
DASF	Fisheries Office, Department of Agriculture, Stock and Fisheries, Kanedobu, Papua (catalog numbers typically with prefix FO)
FFUM	Faculty of Fisheries, Prefectural University of Mie
FMNH	Field Museum of Natural History, Chicago
HA	Private collection of H. Aoyagi (deceased) — now in possession of wife
HUJ	Department of Zoology, Hebrew University, Jerusalem (catalog numbers frequently with the prefixes F, E and SLR)
IMP	Instituto del Mar del Peru
IMUK	Institut für Meereskunde an der Universität Kiel
ISZZ	Institut für Spezielle Zoologie und Zoologisches Museum, Berlin (East)
LACM	Los Angeles County Museum
MCZ	Museum of Comparative Zoology, Harvard University, Cambridge, Massachusetts
MNHN	Museum National d'Histoire Naturelle, Paris (catalog numbers frequently with the prefix A)
NFIS	Natur-Museum und Forschungs-Institut Senckenberg, Frankfurt
NMFS	National Marine Fishery Service, Honolulu (collections mostly transferred to University of Hawaii)
NMNZ	National Museum of New Zealand, Wellington
NMV	Naturhistorisches Museum, Vienna
QMB	Queensland Museum, Brisbane (catalog numbers frequently with the prefix I)
RMNH	Rijksmuseum van Natuurlijke Historie, Leiden
RUSI	Rhodes University, J. L. B. Smith Institute of Ichthyology, Grahamstown, South Africa
SAM	South African Museum, Capetown
SIO	Scripps Institution of Oceanography, University of California, La Jolla
SU	Stanford University, Division of Systematic Zoology (collection transferred to CAS)
UCLA	University of California at Los Angeles
UG	University of Guam, Agana
UMMZ	University of Michigan, Museum of Zoology, Ann Arbor
USNM	National Museum of Natural History, Smithsonian Institution, Washington, D.C.
UTAI	University of Tel Aviv, Israel (catalog numbers with the prefix NS)
UW	University of Washington, Seattle
WAM	Western Australian Museum, Perth (catalog numbers with the prefix P)
ZIAS	Zoological Institute, Academia Sinica, Shanghai
ZITU	Zoological Institute, University of Tokyo
ZMA	Zoologisch Museum, Universiteit van Amsterdam
ZSIC	Zoological Survey of India, Calcutta (catalog numbers with the prefix F)
ZSZM	Zoologisches Staatsinstitut und Zoologisches Museum, Hamburg

ZUMT Department of Zoology, University Museum, University of Tokyo

While this monograph was in press I examined important collections of Nemophini from St. Brandon Shoals (Cargados Carajos) and the Agalega Islands obtained during a recent Smithsonian expedition sponsored by Mr. Lewis H. Strauss. All of these collections are cited in the material examined sections but, except for *Meiacanthus fraseri,* have not been indicated on the distributional maps.

RECOGNITION OF GENERA, SPECIES AND SUBSPECIES

Genera. — Probably no taxon is as misused or unstable as the genus. This is because the genus is a subjective category for which no objective definition can be given, other than that it comprises a monophyletic grouping that is separated from similar units by a decided gap. All too often ichthyologists have been guilty of emphasizing differences rather than similarities with the result that the genus is frequently meaningful to a relatively few workers who are thoroughly familiar with the group. Bolin (1947) discussed the history of generic splitting in the marine cottids of California and gave an eloquent plea for a more conservative approach employing the use of other categories (i.e., subgenera) to call attention to evolutionary lines of intermediate magnitude. In keeping with the above philosophy, my concept of generic limits is broader than that of some ichthyologists and I use subgenera to a correspondingly greater extent.

Recognition of taxa above the species level and interpretation of relationships is, as far as practical, based on the principles of phylogenetic analysis. The philosophy and methodology of this type of analysis has been described in detail by Hennig (1966) and summarized by Kavanaugh (1972) and Griffiths (1972: 5-28). Overall similarity is rejected as a basis for establishing kinship. Common possession of primitive characters (symplesiomorphy) neither indicates nor precludes a close relationship. The only basis for assuming that two or more taxa constitute a monophyletic group is the common possession of derived characters (synapomorphy). This premise remains valid regardless of whether the synapomorphy is expressed identically in all the included taxa, or whether it is present in different derived conditions.

Species. — Despite the wide acceptance of the "biological species" concept, it is difficult to apply. The majority of the species recognized in this study are allopatric, and for some the only reliable distinguishing character is coloration. In their discussion of endemic Hawaiian fishes Gosline and Brock (1960: 24-25), have clearly stated the problem inherent in taxonomic recognition of allopatric forms. In this study I have adopted the following general guidelines for taxonomic recognition of allopatric forms. For a more detailed discussion refer to the individual species accounts for those examples cited below. (1) When I was unable to make a reasonably definitive decision (e.g., the two endemic Hawaiian species of *Plagiotremus*) the status quo was maintained in the interest of stability of nomenclature. (2) When two closely related allopatric forms could be consistently (and only) recognized on the basis of qualitative differences in color pattern and the distribution of one or both forms coincided with zoogeographic patterns exhibited by other organisms (Thorne, 1963: 332), they were considered to be different species. Although this decision is nothing more than a subjective choice, it can be consistently applied and is perhaps most compatible with number one. (3) When two forms differed only in quantitative aspects of their color patterns and had broadly disjunct distributions which probably reflected uneven collecting efforts (e.g., the two color forms of *Meiacanthus nigrolineatus* or *M. lineatus*) they were treated as conspecific species. (4) When one of two forms with real allopatric distributions differed only in having a xanthic color pattern (e.g., the Fijian population of *Meiacanthus atrodorsalis* or *Plagiotremus laudandus*) they were recognized as subspecifically distinct. (5) When two sympatric forms differed only in quantitative aspects of their color patterns and were suspected or known to be mimicking different organisms of corresponding coloration (e.g., the two Marquesan color forms of *Aspidontus taeniatus*) they were treated as conspecific species.

Subspecies. — I use subspecific names only when a practical purpose is served by recognition of the names and the zone of intergradation is relatively narrow (or absent in the case of allopatric populations). I recognize subspecies in three species: *Aspidontus taeniatus, Meiacanthus atrodorsalis* and *Plagiotremus laudandus*. Refer to individual species accounts for discussion.

SEXUAL DIMORPHISM AND SEX RATIO

Primary Sexual Dimorphism. — The primary sexual dimorphism found in the Nemophini is typical of most, if not all, blenniids. In both sexes the urogenital opening is posterior to the anus. In females the first anal-fin spine is greatly reduced and concealed by a hood or flap of skin. The relatively large pocketlike urogenital opening is readily exposed when the skin is depressed at the anterior base of the flap. In males the urogenital opening is a minute orifice at the tip of a short tube. In adult males both anal-fin spines are usually distinct and always more apparent than in females. The primary sexual dimorphism of *Petroscirtes breviceps* has been illustrated by Tomiyama (1951).

Secondary Sexual Dimorphism. — At least some species of all genera of Nemophini exhibit secondary sexual dimorphism. The symphyseal mandibular cirri, an important diagnostic character of the Nemophini, are sexually dimorphic in all genera except *Petroscirtes*. In *Meiacanthus,* sexually dimorphic characters are especially useful in evaluating interspecific relationships and as an aid to identification.

1. Anal-fin spines: In nuptial males of *Petroscirtes* and at least a few species of *Meiacanthus* the tips of the anal-fin spines are swollen and padlike, and the epidermis is often rugose. Smith (1974) described these structures in a number of blenniids and discussed their possible function. The extent of the rugosity and the size of the pads was found to be seasonally variable and maximally developed at the height of the reproductive season. The degree of rugosity is also interspecifically variable. In an important paper not cited by Smith, Blüm (1972) described these "pads" in *Blennius sphinx* Valenciennes noting that the epidermis of the spine tips contains glands with excretory ducts and pores. The highly rugose anal-fin spines of nuptial males of *Petroscirtes mitratus* and *fallax* are shown in Figure 2.

2. Modified Tips of Segmented Dorsal-fin Rays: The tips of the segmented dorsal-fin rays of large nuptial males of all genera of Nemophini (except possibly *Xiphasia,* see below) become expanded and slightly rugose, somewhat similar to that of the anal-fin spines. Springer and Gomon (1975) noted the same condition for nuptial males of *Omobranchus.* In at least a few large males of *Petroscirtes* and *Meiacanthus* the tips of the more posterior dorsal-fin spines are similarly modified.

3. Anal-fin Tabs: Small, lappet-like structures develop subdistally to the tips of the segmented rays of the anal fin in nuptial males of all genera of Nemophini, with the possible exception of *Xiphasia.* Judging from the large number of adult males that lack these structures, it seems likely that they persist for a relatively short period of time during or near the height of reproductive activity. I have not observed the structures in *Xiphasia,* but this may simply be a reflection of the relatively small number of adult males examined. I have made a cursory survey for the presence of these structures in the Blenniidae and found them to be present in a large and diverse number of genera and species. As far as known all males of the Blenniidae care for the eggs until hatching and it is possible that the anal-fin "tabs" are in some way related to that function. These structures have not been observed in any other teleosts and I suspect that they will prove to be an important diagnostic character of the Blenniidae.

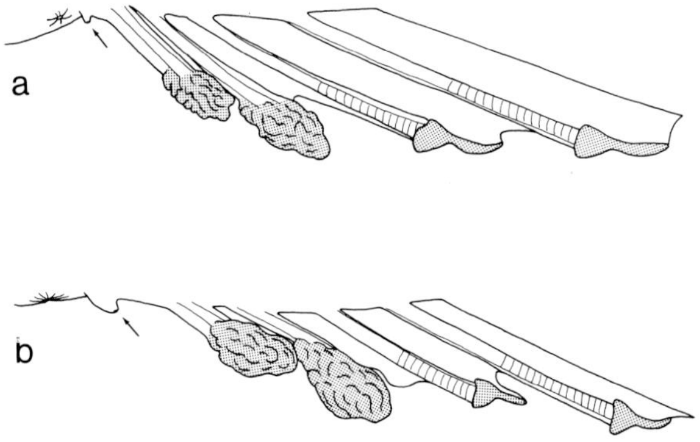

Figure 2. — Anal fin of nuptial males of two species of *Petroscirtes* showing nearly identical modification of spines and tips of segmented rays: a, *mitratus,* USNM 199433, 56.8 mm; b, *fallax,* USNM 201798, 69.8 mm. Arrow points to urogenital papilla.

In the Phenablenniini (Smith-Vaniz, 1975: Fig. 3) and at least some species of the Omobranchiini the "tabs" apparently are bilateral structures and thus differ from those of the Nemophini.

4. Symphyseal Mandibular Cirri: The presence of symphyseal mandibular cirri distinguishes the Nemophini from the other blenniid tribes. In *Petroscirtes* these cirri are present in the juveniles (found in specimens of *P. mitratus* as small as 12.4 mm) as well as adults of both sexes. (Apparently secondarily absent in *P. marginatus* for which only a single female known.) In the other genera, however, they are absent in juveniles, sexually immature males and most females. Although these cirri may occasionally be present in large females, they are seldom as well developed as in males of comparable size. In most species the occurrence of mandibular cirri in males coincides with the expression of other sexually dimorphic characters. The smallest male *Xiphasia setifer* with mandibular cirri was 321 mm SL; no males of *X. matsubarai* were found to have cirri but the largest male was only 265 mm.

5. Color Pattern: The life coloration of one or both sexes of many species of Nemophini is unknown, so information on sexual dichromatism is lacking. Mature males of *Plagiotremus townsendi, phenax* and at least some populations of *tapeinosoma* have an orange spot in the dorsal fin that females lack. The adult males of *Meiacanthus anema* are darker and lack the conspicuous throat markings of females and immature males. In at least some populations of *M. ditrema* the dorsal fin of males differs from females in having a dark submarginal stripe. In most blenniids it is the male that has the most conspicuous color pattern but adult females of *Aspidontus dussumieri,* which differ from males in having an intense black spot anteriorly in the dorsal fin, are an exception.

6. Fin Spine and Ray Lengths: *Petroscirtes* is unique within the Nemophini in having the first or second dorsal-fin spine of adults noticeably longer than the others. In most species these spines are longest in males. The failure of some authors to appreciate the amount of variation, due to sexual dimorphism, in the length of the first two dorsal-fin spines of *P. mitratus* resulted in the naming of several junior synonyms.

The pelvic fins are longer in the males of at least some species of all genera except *Xiphasia,* and the same is true for the caudal fin for all genera except *Aspidontus.* In most species only the outermost segmented rays of the caudal fin are elongate, but in males of *Xiphasia setifer* it is the innermost pair of rays that are filamentous. In some species of *Meicanthus* all of the inner rays are slightly elongated and the interradial membranes of the caudal fin are deeply incised (Fig. 3a). In *M. vittatus* and *geminatus* only the males have incised caudal-fin membranes. The caudal fin is incised in both sexes of *M. grammistes* but the condition is more pronounced in males.

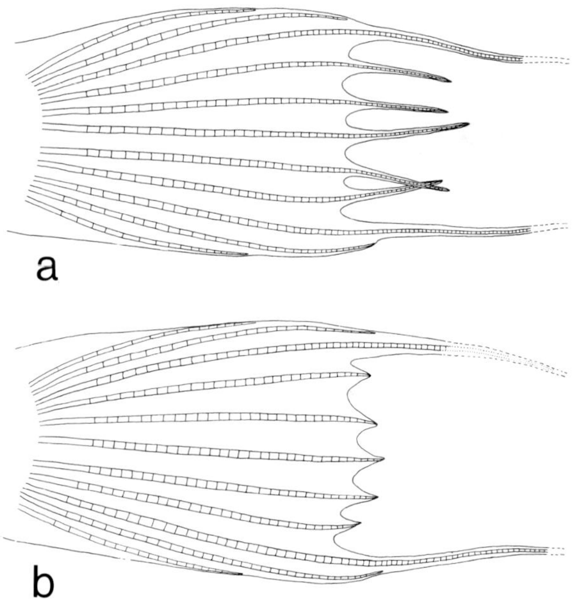

Figure 3.—Caudal fins of nuptial males of two species of *Meiacanthus* showing differences in degree to which inner rays are exserted beyond the margin of the interradial membranes: a, *vittatus,* ANSP 126900, 43.7 mm SL; b, *lineatus,* USNM 201363, 42.8 mm.

7. Flap on First Doral-fin Spine: Large males of some species of *Petroscirtes* have a fleshy flap on the anterior margin of the first dorsal-fin spine.

8. Standard Length: The largest male is larger than the largest female in 26 of the 33 species for which both sexes were available (Table 2). Length-frequency histograms (Figs. 4-8) are also given for most species for which at least 5 adults of each sex were available. The disparity between the num-

TABLE 2. — Size distribution of species of tribe Nemophini: smallest individual, number examined (excluding specimens under 20 mm SL) and maximum size.

Species	smallest individual	Females		Males	
		N	maximum size	N	maximum size
Petroscirtes					
mitratus	7.2	130	53.9	164	60.9
xestus	9.3	11	33.8	45	56.8
springeri	—	—	—	1	49.7
breviceps	9.7	71	83.5	118	110.0
fallax	11.0	11	56.3	32	73.2
marginatus	—	1	39.2	—	—
variabilis	11.6	70	105.4	132	124.7
thepassi	21.5	25	57.3	18	58.7
ancylodon	12.2	28	78.6	47	91.5
lupus	11.9	36	99.1	75	110.2
Aspidontus					
taeniatus	33.3	73	92.4	144	99.5
dussumieri	21.4	31	100.3	60	98.3
Xiphasia					
setifer	16.3	30	460.0	26	532.0
matsubarai	140.0	8	298.5	5	265.0
Meiacanthus					
anema	13.4	118	45.7	181	71.0
ditrema	19.0	15	34.5	22	39.8
lineatus	10.9	26	55.3	36	67.6
grammistes	18.0	89	84.2	69	87.3
kamoharai[1]	12.3	4	79	6	87
vittatus	12.3	10	43.3	8	45.5
geminatus	25.7	—	—	4	61.3
nigrolineatus	11.4	41	57.0	61	65.7
mossambicus	17.5	10	69.2	12	62.5
fraseri	33.8	16	61.0	11	54.3
smithi	11.7	17	56.1	15	64.4
atrodorsalis	12.5	74	64.4	99	67.9
procne	15.9	6	42.8	7	56.0
reticulatus	—	1	70.5	—	—
bundoon	—	1	46.3	4	63.7
phaeus	—	1	55.8	—	—
Plagiotremus					
spilistius[2]	—	2	140.3	1	164.3
celebesensis[3]	54.3	—	—	—	—
iosodon[3]	38.0	—	—	—	—
laudandus	22.0	31	57.9	33	52.0
phenax	21.0	18	42.4	11	42.9
townsendi	18.1	26	37.5	33	35.4
rhinorhynchos	18.9	92	70.3	135	95.5
ewaensis	39.4	13	56.4	15	88.3
tapeinosoma	29.3	129	92.4	171	121.2
goslinei	32.7	17	53.3	16	41.6
azaleus	14.2	121	77.8	136	71.6

[1] Maximum size for both sexes based on Tomiyama (1956b).

[2] Sze-Chung (1957) reported the "lengths" (presumably SL) for five specimens of each sex as follows: females 122.3-153 mm; males 133-174 mm.

[3] Only juvenile specimens have been collected.

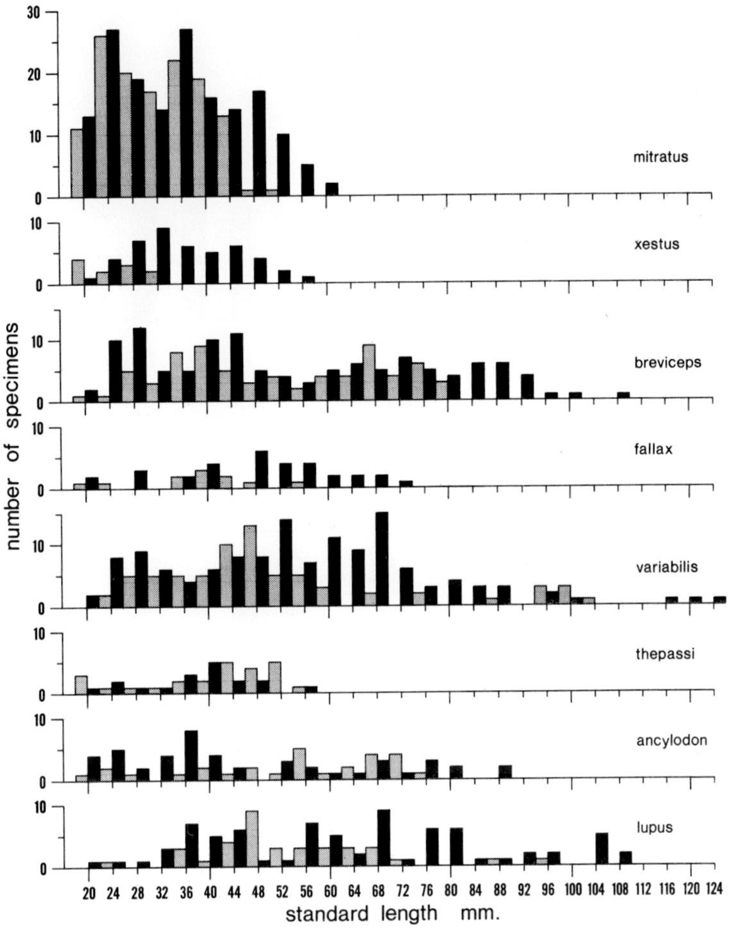

Figure 4. — Size-frequency histograms for selected species of *Petroscirtes*. Lower class limits rounded to nearest 0.5 mm; black bars = males; shaded bars = females.

ber of large males to females is especially great in *Petroscirtes mitratus* and *xestus* (Fig. 4) and *Meiacanthus anema* (Fig. 7). Selection pressure in a promiscuous mating system, which is probably characteristic of at least *Petroscirtes* and *Meiacanthus* (see general section of biology for *Meiacanthus*), would favor increase in body size in males owing to the advantage of large size in courtship display and intermale dominance relations.

9. Meristic Characters: No sexual dimorphism in meristic characters was noted.

Sex Ratio. — The ratio of males to females is unexpectedly high for a number of species (Table 2). Sexes of blenniids are relatively easy to distinguish externally, especially the large individuals. If the explanation for the uneven ratios is incorrect sex determination of immature individuals, one might expect the ratio of males to females to be higher at smaller sizes; inspection of the length-frequency histograms (Figs. 4-8) does not generally support such an interpretation. (This may perhaps partially explain the unusually high ratio in *Aspidontus,* however, because sex is difficult to determine in this genus and there does appear to be a proportionally greater number of males at the smaller sizes.)

An alternative interpretation for the uneven sex

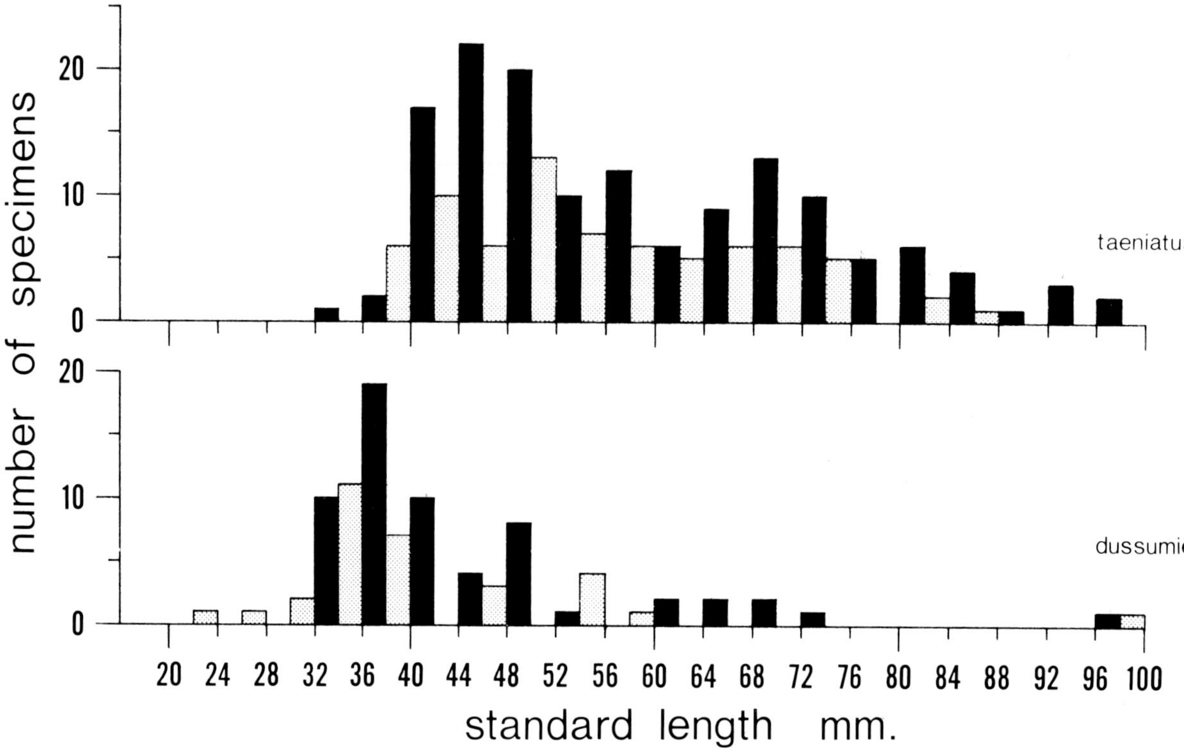

Figure 5. — Size-frequency histograms for species of *Aspidontus.* Explanation as in Figure 4.

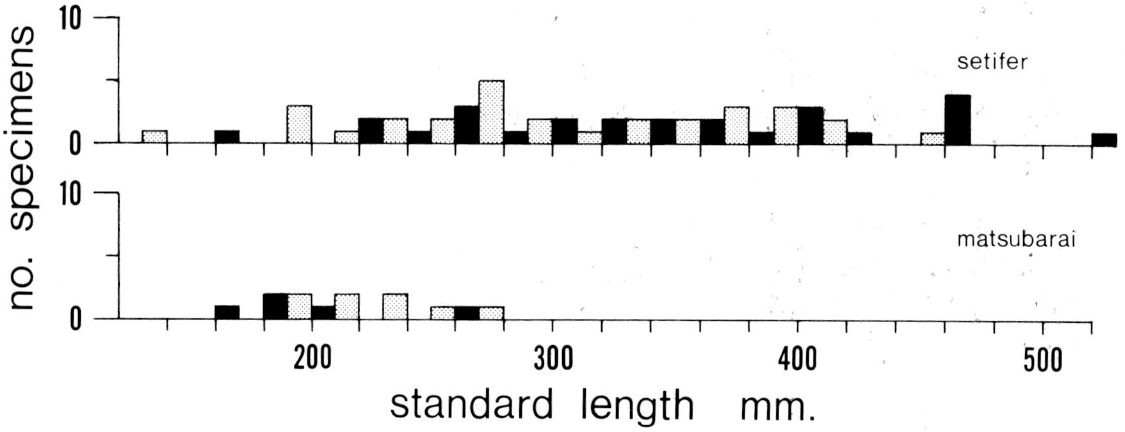

Figure 6. — Size-frequency histograms for species of *Xiphasia.* Explanation as in Figure 4.

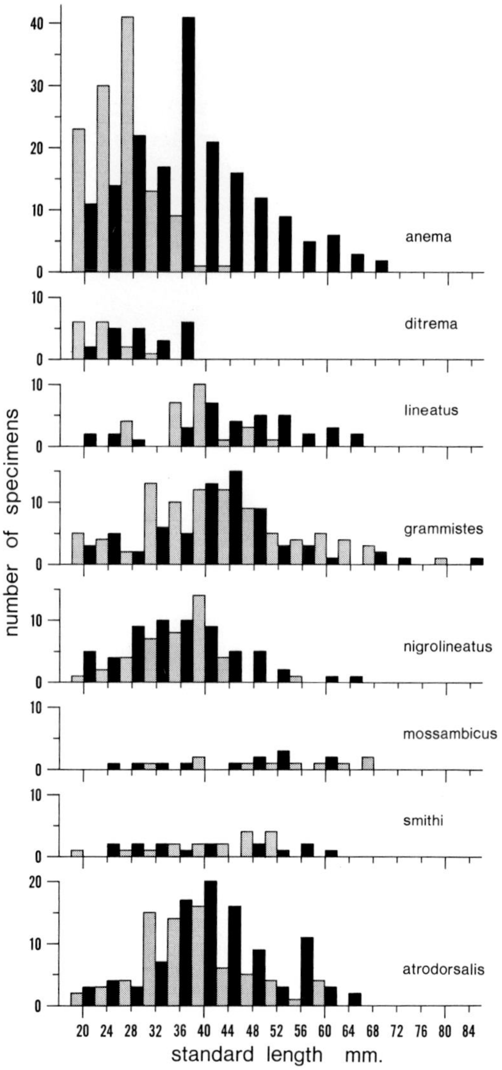

Figure 7. — Size-frequency histograms for selected species of *Meiacanthus*. Explanation as in Figure 4.

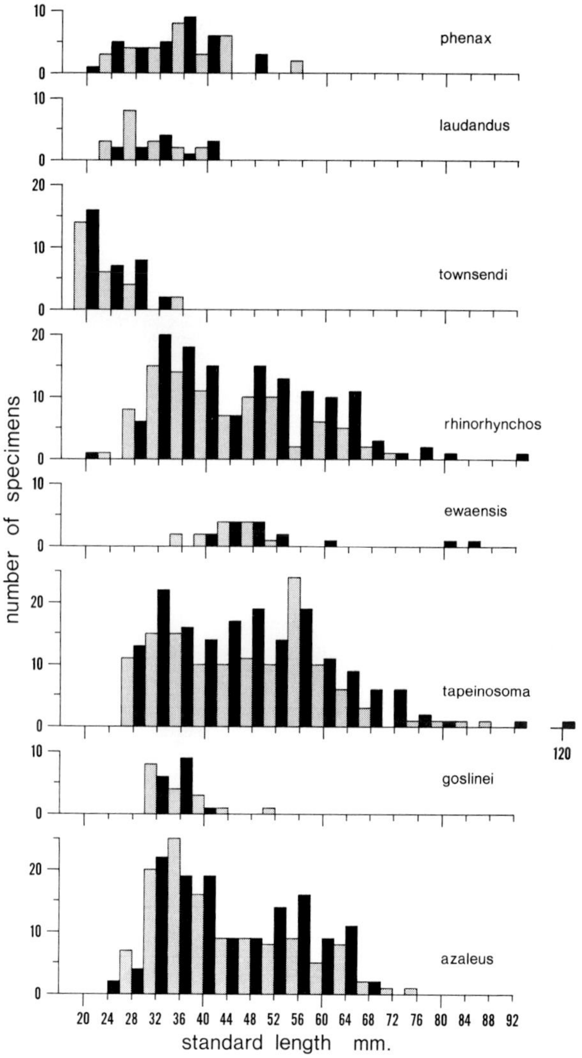

Figure 8. — Size-frequency histograms for selected species of *Plagiotremus*. Explanation as in Figure 4.

ratios might be differential niche utilization or behavior resulting in biased sampling for one sex. This possibility deserves additional study. Williams (1954) has documented differential vertical distribution of the sexes in the blennioid fish *Gibbonsia elegans* (Cooper). Differential niche overlap might also explain the functional significance of differences in number of incisoriform teeth between the sexes in some species of *Ecsenius* (Springer, 1971: 5) and *Omobranchus* (Springer and Gomon, 1975: 16). Sexual dimorphism coupled with differential niche utilization has been documented for birds (Selander, 1966) and the lizard *Anolis* (Schoener, 1967). Although no statistical analyses were made in the present study, tooth counts were plotted separately for both sexes of several species and no obvious differences were noted.

MIMICRY

Springer and Smith-Vaniz (1972a) gave a detailed account of a mimetic complex involving *Meiacanthus nigrolineatus, Plagiotremus townsendi* and *Ecsenius gravieri* (Pellegrin), and summarized all other known or suspected mimetic relationships involving blenniid fishes. Several additional cases of mimicry are reported here and some of our previous statements are modified in the light of additional observations and collections. Many of the presumed cases of mimicry involving blenniids (Fig. 9) are not well documented and would make excellent subjects for future investigation.

Mimicry is the result of complex evolutionary processes and, as with many biological phenomena, it is difficult to provide precise definitions that fit all mimetic situations. Care especially must be taken not to confuse cases of convergent evolution for mimicry. The distinction between camouflage and mimicry may also be unclear. For example, when one species mingles with a school of another species of similar appearance and thus gains an advantage by appearing inconspicuous to potential predators, is this a case of camouflage or mimicry? What if under similar circumstances the advantage gained from the association is inconspicuousness to potential prey? Mimicry is here defined as the similarity in external appearance of two species of animals, at least one of which derives an advantage from the similarity. The mimetic interactions summarized in Figure 9 probably involve three distinct types of mimicry.

Batesian mimicry is the resemblance of a palatable (broad sense) species to an unpalatable (broad sense) species. Excluding *Plagiotremus,* all species that mimic *Meiacanthus* are believed to be Batesian mimics. The toxic buccal glands of *Meiacanthus* (see general section on biology) account for its popularity as a model. There is a widespread concept which holds that a Batesian mimic must be relatively rare compared to its model if the mimetic association is to persist. Springer and Smith-Vaniz (1972a: 13) listed a number of hypothetical situations which would permit a Batesian mimic to occur in high numbers relative to its model. The following three situations could certainly be valid with respect to *Meiacanthus:*

1. Where the models are unharmed in the process of being tested by the predators. This permits a model to be tested repeatedly, thus increasing the effective number of models without increasing their actual number.

2. Where negative experiences of the predators have a stronger, more lasting effect than positive experiences.

3. Where because of behavioral differences the mimic is encountered less often than the model or is more difficult to catch.

Müllerian mimicry is the resemblance of two unpalatable (broad sense) species to each other. In this type of mimicry each species (mimic) reinforces the other. Elements of Müllerian mimicry may come into play in the interactions between species of *Plagiotremus* and *Meiacanthus*. Feeding experiments (Springer and Smith-Vaniz, 1972a: 21) and unpublished data (Springer, pers. comm.) clearly revealed that *Plagiotremus townsendi* (both dead and alive individuals) is rejected as food by some predators. Although more study is needed, it is likely that some (if not all) species of *Plagiotremus* have a noxious quality that make them unacceptable as prey to some predators. This would explain how those species that do not mimic a *Meiacanthus* escape heavy predation while spending most of the daylight hours hovering in the water column.

Aggressive mimicry is the resemblance of a "pre-

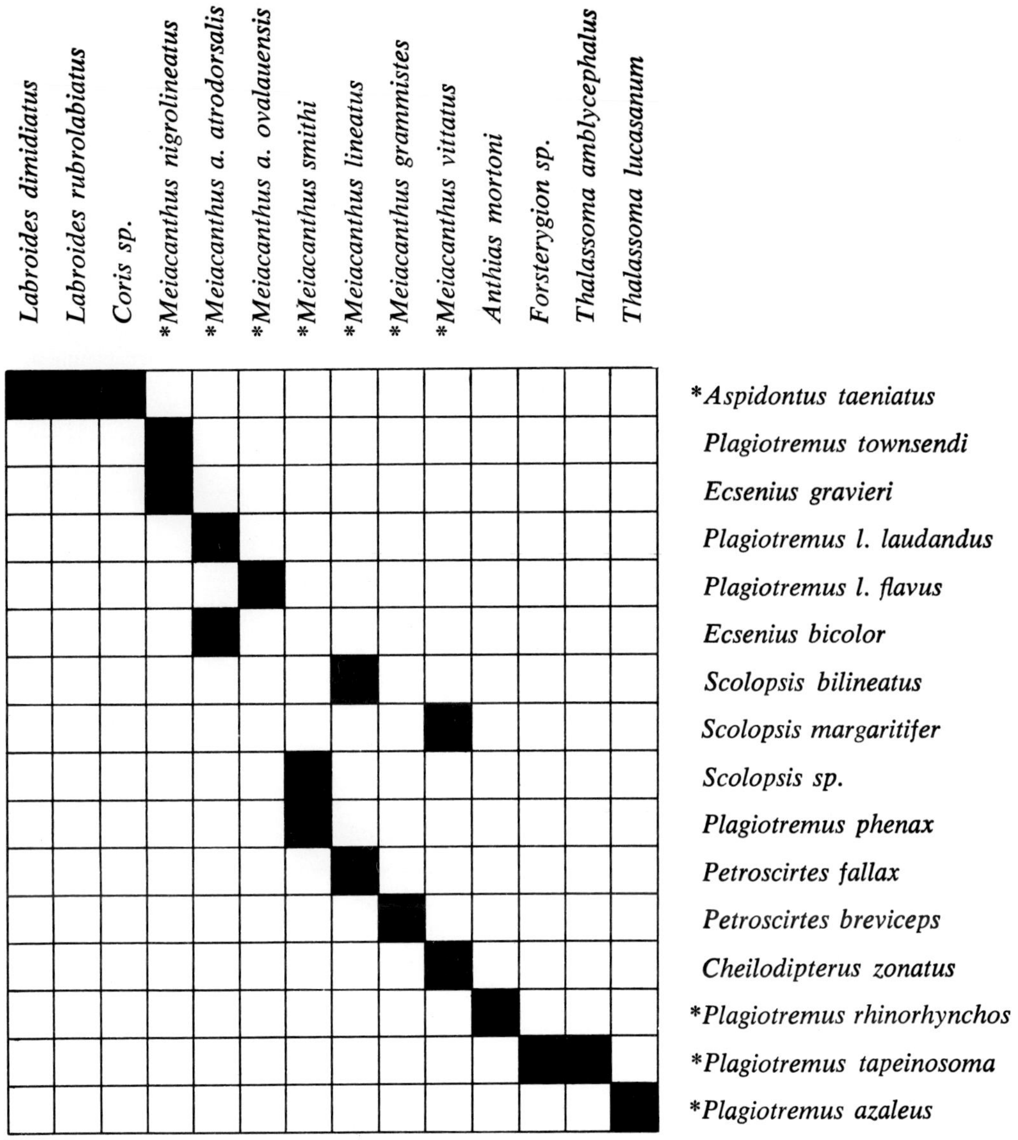

Figure 9. — Documented or suspected mimetic interactions involving species of Nemophini. Presumed models are at top of figure; refer to accounts of species with asterisk for discussion. (See also species accounts for discussion of possible aggressive mimetic interactions involving *Plagiotremus rhinorhynchos* — *Labroides dimidiatus* and *Plagiotremus tapeinosoma* — *Trachinops taeniatus*.)

datory" or aggressive species to a nonaggressive or useful species. The similarity functions as a disguise that enables the aggressive species to gain access to a prey (the third species) that confuses it with the nonaggressive species. *Aspidontus taeniatus* is a classical example of an aggressive mimic. The adults are remarkably similar, both in appearance and swimming behavior, to the beneficial cleaner wrasse *Labroides dimidiatus* (Valenciennes). This disguise enables the blenny to approach within striking distance of unsuspecting prey from which it bites off fin rays and attached membranes. Detailed accounts of this mimetic association have been published by Eibl-Eibesfeldt (1959), Randall and Randall (1960), Wickler (1960; 1961; 1963; 1968) and Losey (1974). As a matter of record it should be noted that Quoy and Gaimard (1834: 719) were the first authors to remark on the similarity of the blenny to the wrasse. In their original description of *Aspidontus taeniatus* they stated "Ce Poisson a quelque ressemblance avec le Labre large-raie."

All species of *Plagiotremus* are obligatory, scale-eating predators (see general section on biology). It is likely that all mimetic interactions involving species of *Plagiotremus* are primarily of the aggressive type. When the model is a species of *Meiacanthus,* Batesian or Müllerian mimicry possibly is of secondary importance.

SYSTEMATIC ACCOUNTS

Tribe NEMOPHINI Kaup, 1858

Type-genus. — *Xiphasia* Swainson

Diagnosis. — Fishes of the family Blenniidae with (1) cirri associated with the symphyseal mandibular sensory canal opening in at least mature males (refer to discussion under general account of sexual dimorphism); (2) supracleithrum articulating with epiotic and posttemporal; (3) basisphenoid absent; (4) intercalar absent (except subgenus *Dasson*); (5) cranial bones of adults ornamented with tiny shallow depressions to deep pits (except *Xiphasia*); (6) pterosphenoid well developed and visible on external surface of skull; (7) interopercle without posteroventral spur; (8) pelvice fin absent or I, 3; (9) two well-developed postcleithra; (10) premaxillae and dentaries not excavated; (11) all premaxillary incisors firmly ankylosed to premaxillae; (12) dentaries joined by suturing joint; (13) caudal-fin rays all simple.

The importance of the above characters in distinguishing the Nemophini from the other tribes of the Blenniidae is summarized below.

character	*tribes excluded*
1, 2, 3	all others
4, 5 (excluding exceptions)	all others
6	Omobranchini and Phenablenniini
7, 8	Omobranchini
9	Phenablenniini
10, 11	Salariini
12	Salariini and Blenniini (except *Blennius ocellaris* and *B. normani*)
13	Blenniini and most Salariini

The occurrence of these and other characters useful in defining the blenniid tribes is summarized in Table 28. One possible interpretation of the phylogenetic position of the Nemophini is shown in Figure 88.

Etymology. — The tribal name is derived from *Nemophis* (= *Xiphasia*) upon which Kaup founded the Nemophidae, and ultimately from the Greek *nema* (thread) and *ophis* (snake).

Description. — Diagnostic characters are not repeated. Dorsal-fin spines III-XIV; segmented dorsal-fin rays 14-119; total dorsal-fin elements 25-133; anal fin II, 14-119; pectoral fin 12-16; segmented caudal-fin rays (dorsal + ventral) 5-7 + 5-6 = 10-13; procurrent caudal-fin rays 1-9 + 1-9 = 2-18; all fin elements unbranched; last dorsal-fin spine not reduced in size; dorsal fin without notch between spinous and rayed portions; last anal-fin pterygiophore supports a single ray; 1-5 proximal dorsal-fin pterygiophores dorsal to supraoccipital; branchiostegal rays 6; precaudal vertebrae 11-16; caudal vertebrae 19-119; total vertebrae 30-135; pleural ribs on vertebrae 3 to 10-16 (rarely 17); epipleural ribs on vertebrae 1 to 8-36; neural arch of first vertebrae complete or open dorsally; hypural 5 absent; one epural (rarely 2); infraorbital bones 3-4; median ethmoid present; rostral cartilage and kinethmoid absent; nasal bones separate, and trabeculate (at least dorsolaterally); posttemporal forked or unforked; supracleithrum without tube for sensory canal; ascending process of premaxillary arising broadly and firmly from body of premaxillary; replacement incisor teeth developing internally in hollow of premaxillae and dentaries (encapsulated by bone of jaws); dentary incisors firmly ankylosed to dentaries (except *Plagiotremus* which has functional incisors ligmentously attached to dentaries, Fig. 73); number of incisors in both jaws increase with increase in standard length; total dentary incisors in specimens 30 mm SL and larger 10-87; total premaxillary incisors in specimens 30 mm SL and larger 12-45; incisor teeth ratio (lower/upper) 0.7-2.4; dentary with greatly enlarged canine posteriorly; premaxillary with relatively small canine posteriorly (except *Plagiotremus* which lacks premaxillary canines); vomerine teeth absent; cephalic sensory pores simple (no horizontal pairs or multiples); circumorbial pores (infraorbital + supraorbital) 7-11; interorbital pores 2 or 4; mandibular pores 2 or 3; preopercular pores 5 or 6; lateral line present or absent, when present consisting of 4-10 simple, bipored tubes restricted to the body anteriorly; anterior and posterior nostrils present; orbital cirri absent (except some species of *Petroscirtes*); no raised, bladelike, fleshy crest on top of head; lower jaw without labial flap or well-developed free fold; gill opening varying from restricted entirely above level of dorsalmost pectoral-fin ray to extending ventrally to level of ventralmost pectoral-fin ray; last ray of dorsal and anal fins broadly bound by membrane to caudal peduncle or caudal fin; swim bladder present (except in *Xiphasia*).

Major characters useful in distinguishing the genera of Nemophini are summarized in Tables 1, 3, 4 and in the following key.

LIST OF NOMINAL GENERA AND SUBGENERA WITH PRESENT ALLOCATIONS

Genus or (subgenus)	present allocation
(*Allomeiacanthus*) Smith-Vaniz	subgenus of *Meiacanthus*
Aspidontus Cuvier, 1834	*Aspidontus*
Atopoclinus Vaillant, 1894	*Plagiotremus*
Blennechis Valenciennes, 1836	*Aspidontus*
Dasson Jordan and Hubbs, 1925	subgenus of *Petroscirtes*
(*Escadotus*) Smith, 1959	*Aspidontus*
(*Holomeiacanthus*) Smith-Vaniz	subgenus of *Meiacanthus*
Lembeichthys Herre, 1936	*Plagiotremus*
Macrurrhynchus Ogilby, 1896	*Plagiotremus*
Meiacanthus Norman, 1943	*Meiacanthus*
(*Musgravius*) Whitley, 1961	subgenus of *Plagiotremus*
Nemophis Kaup, 1858	*Xiphasia*
Oncesthes Jordan and Hubbs, 1925	*Aspidontus*
(*Ostreoblennius*) Whitley, 1930	*Petroscirtes*
Petroscirtes Rüppell, 1830	*Petroscirtes*
Plagiotremus Gill, 1865	*Plagiotremus*
Runula Jordan and Bollman, 1865	subgenus of *Plagiotremus*
Xiphasia Swainson, 1839	*Xiphasia*
Xiphogadus Günther, 1862	*Xiphasia*

TABLE 3. — Summary of selected characters used to define genera of Nemophini.

Character	Petroscirtes	Aspidontus	Xiphasia	Meiacanthus	Plagiotremus
1. Dorsal-fin spines	10-12	9-12	11-14	3-10	6-12
2. Segmented dorsal-fin rays	14-21	26-34	99-119	20-28	25-61
3. Total dorsal-fin elements	25-33	37-44	110-133	28-32	32-71
4. Segmented anal-fin rays	14-21	25-30	97-119	14-19	19-58
5. Pectoral-fin rays	13-16	13-15	10-14	12-16	11-13
6. Segmented caudal-fin rays	11	11	10	11-13	11
7. Procurrent caudal-fin rays	5-8	6-8	2-3	9-15	10-18
8. Precaudal vertebrae	11-13	13-15	13-16	11-14	13-16
9. Caudal vertebrae	19-25	27-34	100-119	19-25	24-60
10. Total vertebrae	30-37	40-48	113-135	32-38	38-76
11. Pleural ribs	9-11	12-14	11-15	8-12	11-15
12. Epipleural ribs	12-15	15-36	8-13	11-16	14-20
13. Total premaxillary incisors[1]	20-45	16-39	12-30	15-26	18-44
14. Total dentary incisors[1]	20-48	11-36	10-34	14-28	20-87
15. Premaxillary canines	yes	yes (frequently lost in adult *taeniatus*)	yes	yes	no
16. Dentary incisors with broad base, firmly ankylosed in dentaries	yes	yes	yes	yes	no (Figs. 72-73)
17. Dentary canines with prominent groove and associated gland	no	no	no	yes (Figs. 43-44)	no
18. Dentary canines of adults and prejuveniles essentially identical	yes	no (Fig. 29)	yes	yes	yes
19. Prootic and ascending wing of parasphenoid form continuous external ridge	no	no	no	yes	no
20. Nasals, median ethmoid and lateral ethmoids fused	no	no	no	no	yes
21. Ornamentation of cranial bones	weak to moderate	well developed	weak	well developed	well developed
22. Frontals fused	no	no	no	no	yes
23. Supraoccipital depressed, forming deep trough	no	yes, *taeniatus* slight, *dussumieri*	yes	no	no
24. Pterygiophores of dorsal-fin spines dorsal to skull (Fig. 94)	1-3	2-3	4-5	1-3	1-3
25. Parietal crest (in large adults)	yes	no	no	no	no
26. Synostosis of lateral extra-scapulars with parietals[2]	no	no, *taeniatus* yes, *dussumieri*	no	no	yes

TABLE 3. — Continued.

27. Intercalar bone	yes, Subg. *Dasson* no, Subg. *Petroscirtes*	no	no	no	no
28. Posttemporal forked (ventral arm present)	no	yes, *taeniatus* no, *dussumieri*	yes	yes (except Subg. *Allomeiacanthus*)	no
29. Infraorbital bones (Fig. 93)	4	4	4	4 or 3 (Subg. *Meiacanthus* only)	4
30. Neural arch of first vertebra	incomplete (open dorsally)	incomplete (open dorsally)	incomplete (touching dorsally)	complete (closed dorsally)	incomplete (open dorsally)
31. "Neural arch processes" present in adults[3]	yes	yes	no	yes	yes
32. Ventral hypural plate fused to urostylar centrum	no	no	yes (Fig. 95c)	no	no
33. Neural and haemal spines of last few vertebrae broad and bladelike	no	no	no	no	yes (Figs. 95f-g)
34. Posterior extensions of proximal pterygiophores of dorsal-fin spines narrow and elongate	no	no	no	no	yes (Fig. 97b)
35. Supratemporal canal of adults continuous anteriorly	no	no	yes (Fig. 90)	no	no
36. Lateral line present	yes	yes	yes	yes (except Subg. *Allomeiacanthus*)	no
37. Restriction of gill opening ventrally[4]	entirely above pectoral-fin base	opposite 3rd to 14th pectoral-fin ray	opposite 1st to 7th pectoral-fin ray	entirely above pectoral-fin base	opposite 1st to 6th pectoral-fin ray
38. Fleshy flap on gill opening	lateral only	lateral only	mesial and lateral	lateral only	lateral only
39. Membranous attachment of last dorsal- and anal-fin rays	caudal peduncle	caudal peduncle	caudal fin (Fig. 36)	caudal peduncle	caudal peduncle
40. Orbital cirrus	yes or no (see Table 5)	no	no	no	no
41. Posttemporal cirrus	yes (except *P. marginatus*)	no	no	no	no
42. Peritoneal pigment (in alcohol)	solid black or brown	speckled, *taeniatus* black or brown, *dussumieri*	speckled	solid black or brown	white
43. Swimbladder (Fig. 98)	yes	yes	no	yes	yes
44. Preanal length (as percent SL)	47.5-61.2	43.6-55.6	11.7-19.3	41.5-59.4	22.8-51.3

1 Number of incisors increases with increase in SL; counts record minimum and maximum values obtained for specimens 30 mm SL and larger.
2 Does not apply to large adults in which synostosis of extrascapulars may occur secondarily.
3 Refer to Methods and Materials Section for discussion.
4 Pectoral-fin rays are counted dorsally to ventrally.

TABLE 4. — Frequency distributions of relative number of incisors in each jaw and incisor teeth ratio (lower/upper) in selected species of *Petroscirtes* and *Meiacanthus* and all species of *Aspidontus, Xiphasia* and *Plagiotremus*.[1]

	relative number of incisors in each jaw			incisor teeth ratio (lower/upper)																			
	Lower < upper	Equal number	Lower > upper	0.7	0.8	0.9	1.0	1.1	1.2	1.3	1.4	1.5	1.6	1.7	1.8	1.9	2.0	2.1	2.2	2.3	2.4	N	x̄
Petroscirtes																							
mitratus	1	3	89				18	67	8													93	1.09
xestus	—	3	17				8	11	1													20	1.07
breviceps	3	10	59			1	19	46	5	1												72	1.08
variabilis	—	3	83				9	49	26	1												86	1.12
thepassi	—	2	22				3	16	5													24	1.11
lupus	—	5	25				14	13	3													30	1.07
Aspidontus																							
taeniatus	139	30	23	4	61	52	63	10	2													192	0.91
dussumieri	64	7	5	2	11	43	18	1	1													76	0.91
Xiphasia																							
setifer	6	18	35		1	4	27	22	4	1												59	1.05
matsubarai	3	2	4			2	2	4														9	1.02
Meiacanthus																							
anema	6	27	51			2	31	45	6													84	1.05
ditrema	5	11	7				16	6	1													23	1.01
grammistes	—	7	62				17	40	12													69	1.09
atrodorsalis	1	6	37				8	20	13	3												44	1.12
Plagiotremus																							
spilistius	—	—	2					1	1													2	1.14
celebesensis	1	—	1				2															2	1.00
iosodon	—	—	2					2														2	1.10
laudandus	—	—	47										1	5	13	14	13	1				47	1.87
phenax	—	—	31									5	11	12	1	2						31	1.64
townsendi	—	—	56								1	11	16	19	6	3						56	1.64
rhinorhynchos	—	—	161							6	45	53	38	14	5							161	1.51
ewaensis	—	—	27							3	8	8	7	1								27	1.48
tapeinisoma	—	—	206											1	3	45	57	50	38	9	3	206	2.06
goslinei	—	—	36											2	3	12	10	3	5	—	1	36	1.98
azaleus	—	—	108										2	5	40	31	21	7	1			108	1.89

[1] Dashed lines separate species groups of *Plagiotremus*.

KEY TO GENERA OF NEMOPHINI

1. Origin of dorsal fin over orbits in adults (Figs. 33-34); last ray of dorsal and anal fins broadly bound by membrane to caudal fin (Fig. 36); segmented anal-fin rays 97-119; segmented caudal-fin rays 10 (one of 61 with 11) .... *Xiphasia*

Origin of dorsal fin distinctly behind or over posterior rim of orbits in adults; last ray of dorsal and anal fins broadly bound by membrane to caudal peduncle; segmented anal-fin rays 14-58; segmented caudal-fin rays 11-13 .... **2**

2. Dentary canines with deep groove along anterior surface (Fig. 43); prominent gland on each dentary extending into base of canine (Fig. 44) .... *Meiacanthus*

Dentary canines without groove along anterior surface; no glands associated with dentary canines **3**

3. Pectoral-fin rays 11-13 (typically 12); total procurrent caudal-fin rays 10-18; lateral line absent; adults with frontal bones fused (no external joint); premaxillae completely shielded dorsally by median ethmoid and wide nasals (Fig. 92) .... *Plagiotremus*

Pectoral-fin rays 13-16; total procurrent caudal-fin rays 5-8; lateral line extending posteriorly beneath dorsal fin at least to base of last spine; adults with frontal bones separate; premaxillae not completely shielded dorsally by median ethmoid and nasals **4**

4. Lower margin of gill opening entirely above pectoral-fin base; segmented anal- or dorsal-fin rays 14-21; orbital cirrus present or absent; posttemporal cirrus present (except *P. marginatus*) .... *Petroscirtes*

Lower margin of gill opening extending ventrally at least to point opposite base of third pectoral-fin ray (counting ventrally); segmented anal- and dorsal-fin rays 25-30 and 26-34, respectively; orbital cirrus absent; posttemporal cirrus absent .... *Aspidontus*

PETROSCIRTES Rüppell

Petroscirtes Rüppell, 1830: 110 (type-species: *P. mitratus* Rüppell, 1830, by monotypy).

Dasson Jordan and Hubbs, 1925: 318 (type-species: *Aspidontus trossulus* Jordan and Snyder, 1902 [= *Blennechis breviceps* Valenciennes, 1836], by original designation).

Ostreoblennius Whitley, 1930: 20 (as a subgenus of *Petroscirtes;* type-species: *Petroscirtes* (*Ostreoblennius*) *steadi* Whitley, 1930 [=*Salarius lupus* De Vis, 1886], by monotypy).

Diagnosis. — Major characters useful in distinguishing the genera of Nemophini are summarized in Tables 1, 3 and 4. *Petroscirtes* is the most generalized genus of Nemophini and as such has few unique characters. The dorsal-fin spines are sexually dimorphic in length in several species (including both subgenera), a character not observed in the other genera. Only *Petroscirtes* and *Meiacanthus* have the gill opening entirely above the pectoral fin. *Petroscirtes* lacks the grooved canines and dentary glands of *Meiacanthus* and has more dorsal-fin spines, 10-12 versus 3-10 (rarely 10). The possession of either posttemporal or orbital cirri (see Table 5) will distinguish all species of *Petroscirtes* (except *marginatus*) from the other genera of Nemophini. Large males (only) of *Meiacanthus anema* also occasionally have cirri (Fig. 51), but orbital cirri are never present. The general physiognomy of the heads of representative species of *Petroscirtes* is shown in Figures 10-13.

Nomenclature. — Jordan and Hubbs (1925) distinguished *Dasson* from *Petroscirtes* solely on the

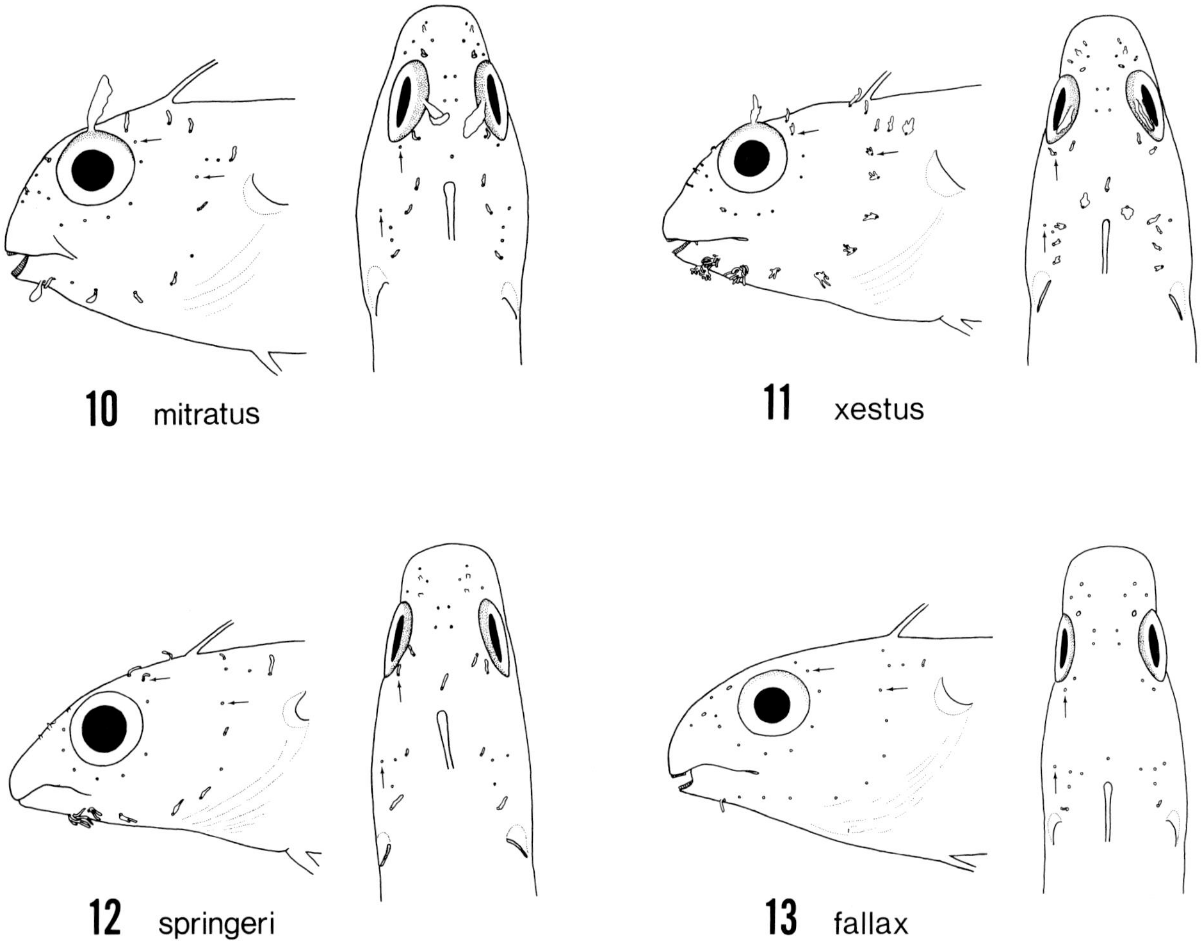

Figures 10-13. — General physiognomy and cephalic sensory pore distributions in selected species of *Petroscirtes:* Fig. 10, *mitratus,* USNM 199433, male, 52.7 mm, Madagascar, Tulear; Fig. 11, *xestus,* BPBM 12462, male, 53.3 mm, Marshall Islands, Eniwetok; Fig. 12, *springeri,* USNM 203279, holotype, male, 49.7 mm, Taiwan; Fig. 13, *fallax,* USNM 201365, holotype, male, 68.1 mm, Australia, Queensland. Small arrows identify ventralmost supraorbital and dorsalmost preopercular pores in both lateral and dorsal views.

basis of its nonelevated dorsal fin. The type-species of *Petroscirtes* also differs in other characters (refer to discussion of intrageneric relationships). I recognize *Dasson* as a subgenus of *Petroscirtes* because I believe that the characters shared by the two taxa should be emphasized. The morphological gap dividing *Petroscirtes* and *Dasson* is not nearly as pronounced as that separating the other genera of Nemophini, and the life histories and behavior of the two taxa are essentially identical.

Whitley proposed *Ostreoblennius* as a new subgenus for reception of his new species *Petroscirtes steadi.* He gave no basis for his action other than stating that his new species was sufficiently different from *P. variabilis* to merit subgeneric distinction. There are no important differences between *Ostreoblennius* and the type-species of *Dasson* (or, for that matter, *variabilis*), thus Whitley's action was unjustified.

Etymology. — The generic name is a combination of the Greek *petro* (stone) and *scirtes* (to leap); gender masculine.

Biology. — Most species of *Petroscirtes* are usually found near cover, often in the immediate vicinity

TABLE 5. — Summary of cirri development in species of *Petroscirtes*.[1]

Cirri	mitratus	xestus	springeri	breviceps	fallax
symphyseal mandibular	simple	multifid (8-26 cirri)	multifid (3-4 cirri)	simple	simple (minute in females)
preopercular	present	present	present	absent	absent
orbital	present	present	absent	Present or absent (usually present but often minute in adults)	absent
postorbital	present	present	present	Present or absent (frequently absent)	absent
predorsal	absent	present (minute)	present (minute)	absent	absent
nuchal	present	present	present	absent	absent
posttemporal	present	present	present	present	present (minute)

Cirri	marginatus	variabilis	thepassi	ancylodon	lupus
symphyseal mandibular	absent	simple	simple	simple	simple
preopercular	absent	absent	absent	absent	absent
orbital	absent	Present or absent (typically present)	Present or absent (typically present)	Present or absent (typically present)	Present or absent (usually present but often minute in adults)
postorbital	absent	Present or absent (typically present)	Present or absent (typically present)	Present or absent (typically present)	Present or absent (usually present)
predorsal	absent	absent	absent	absent	absent
nuchal	absent	Present or absent (usually present)	Present or absent (frequently absent)	Present or absent (typically present)	Present or absent (frequently absent)
posttemporal	absent	present	present	present	present

[1] Refer to methods section for discussion and definitions.

of grass beds (*Thalassia* or *Zostera*) or flotsam such as *Sargassum*. *P. mitratus* and, especially, *xestus* seem to prefer sandy bottom habitats and frequently occupy empty pelecypod shells. The deepest collection of a species of *Petroscirtes* was 18 meters but most collections have come from depths shallower than five meters. Most species are cryptically colored and seldom venture far from cover. *P. fallax*, a boldly pigmented species, probably avoids predation by mimicking *Meiacanthus lineatus*. The dentary canines are especially well developed in *Petroscirtes;* as noted by Herre (1939: 326) and Smith (1959: 233), when handled *Petroscirtes* does not hestitate to use them for protection.

The following observations provided by Dr. Alan R. Emery (*in litt.*) dramatically illustrate the general behavior of *P. thepassi:*

"I noticed the blenny swimming free in the water — head up, vertically oriented. When I glanced in his direction he darted — still head up — onto a frond of *Thalassia*. I saw where he went and shot blindly into the frond behind which he had hidden. I then broke off the *Thalassia*. When I attempted to put him in the bottle, however, he would not allow my fingers close without snapping at them. . . . Later, I observed several additional blennies resting on *Thalassia* blades. Occasionally they would bounce free, hang vertically while obviously watching something then pop back onto the blade."

The above observations were made while collecting fish in the Palau Islands using a small multi-prong spear. The depth of the water was approximately one meter where the *thepassi* were observed.

My casual survey of the gut contents of *Petroscirtes* indicates that crustaceans, especially copepods, and diatoms are frequently ingested. The black

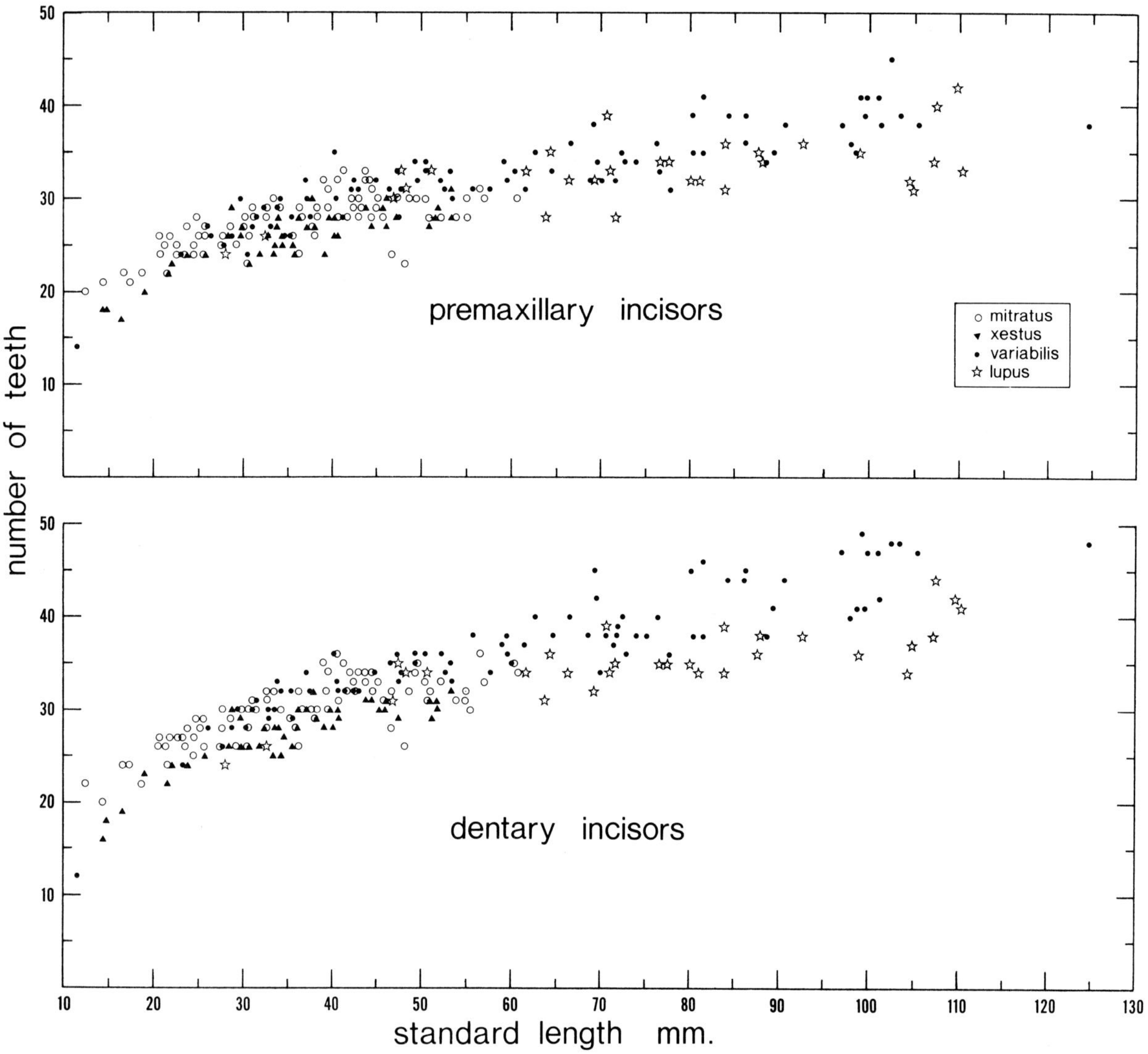

Figure 14. — Relationship between number of incisor teeth versus standard length in selected species of *Petroscirtes.*

peritoneum and relatively elongate gut (Figs. 99a-b) suggest that diatoms or other plant material may be a major item in the diet (refer to section on evolutionary trends for discussion). At least on the basis of relative number of incisors (Fig. 14), there appears to be no fundamental difference in the feeding habits of the various species of *Petroscirtes.*

Munro (1954) described and illustrated the eggs and larvae of *Petroscirtes lupus.* He noted that in all instances the eggs were attached to the inner surfaces of shells of dead bivalve mollusks and guarded by the male. Males were observed utilizing the valves of the Sydney cockle, *Anadara trapezia* (Deshayes) and the razor shell, *Pinna menkei* (Reeve). He noted that nests contained eggs in several stages of development, suggesting the possibility of deposition or fertilization on different days. One male that was observed in the laboratory fanned the eggs continuously with its tail and was not observed to leave the nest or feed for 17 days. Immediately after hatching the larvae congregated near the surface of the aquarium; after about one week the yolk-sac was completely absorbed and they began to feed on the bottom.

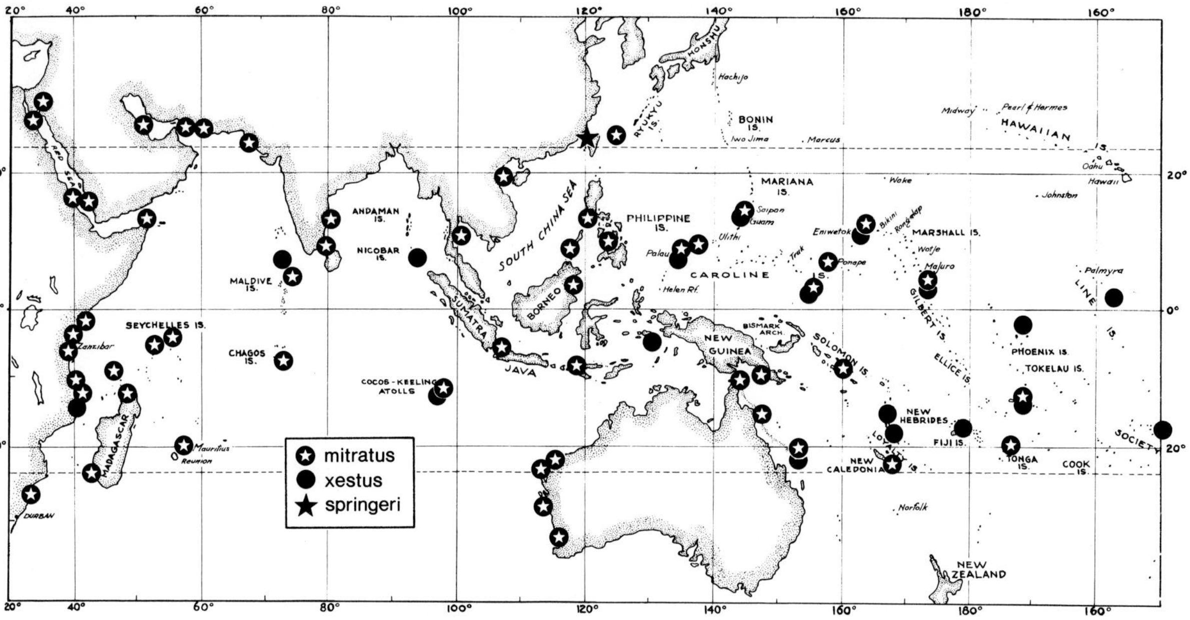

Figure 15. — Distributional records for selected species of *Petroscirtes*.

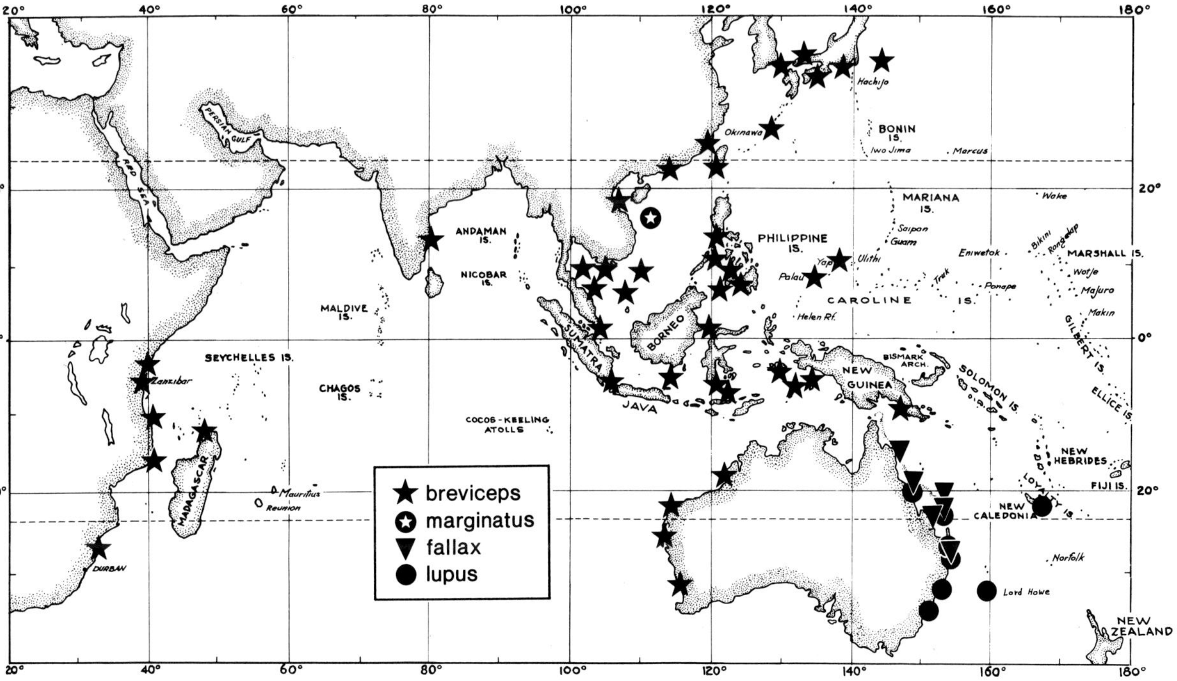

Figure 16. — Distributional records for selected species of *Petroscirtes*.

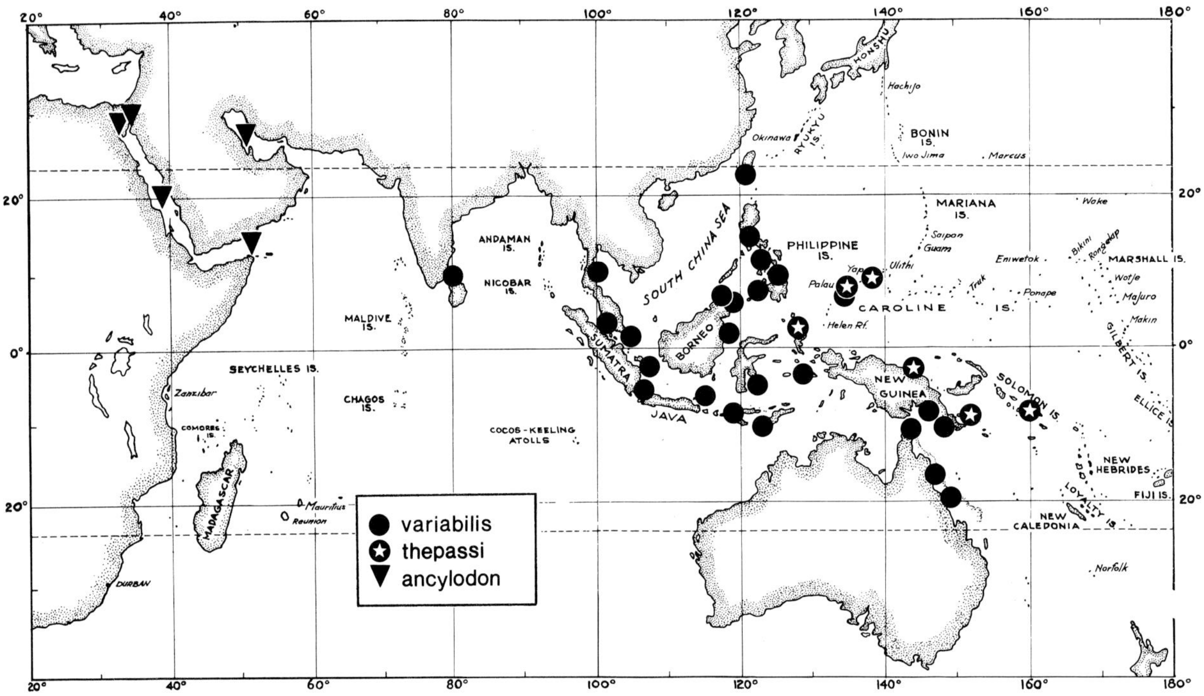

Figure 17. — Distributional records for three closely related species of *Petroscirtes.*

Zoogeography. — The distribution of *Petroscirtes* (Figs. 15-17) extends from the northern Red Sea and southern Africa in the western Indian Ocean to the Society and Line Islands in the central Pacific Ocean. The juveniles and larvae of several species, including *breviceps,* have been collected beneath drifting *Sargassum* which might serve as a vehicle for dispersal. However, it is important to note that only *mitratus* and *xestus* have broad distributions in Oceania and have become established on atolls and volcanic islands in the Indian Ocean. The record of *breviceps* from southern New Guinea is most probably the result of an introduction (see material examined section). Springer and Gomon (1975: 59) give a thorough discussion of the artifical range extension of the blenniid *Omobranchus punctatus* (Valenciennes) via unintentional ship transport. Two species, *P. marginatus* and *springeri,* are known from single collections in the South China Sea. The complex color patterns of *ancylodon* and *lupus* are so similar (at least in preservation) and their respective distributions so isolated, that the possibility that they evolved from a broadly distributed common ancestor immediately comes to mind.

Intrageneric relationships. — *Petroscirtes mitratus* is accorded separate subgeneric distinction because it is the only member of the genus known to lack an intercalar bone[1] (Fig. 18b) and it alone has the first dorsal-fin spine longer than the second (second spine longer than first in the other species). It has the narrowest interorbital, best development of parietal crests (apparent only in large specimens), and is also unique among *Petroscirtes* in having the anterolateral profile of the head almost vertical in large males.

---

[1] The importance of the loss of the intercalar bone, an obvious specialization, should not be overemphasized. Although the left intercalar bone was readily apparent in an illustrated specimen of *P. variabilis* (Fig. 18b), I could detect no trace of a right intercalar.

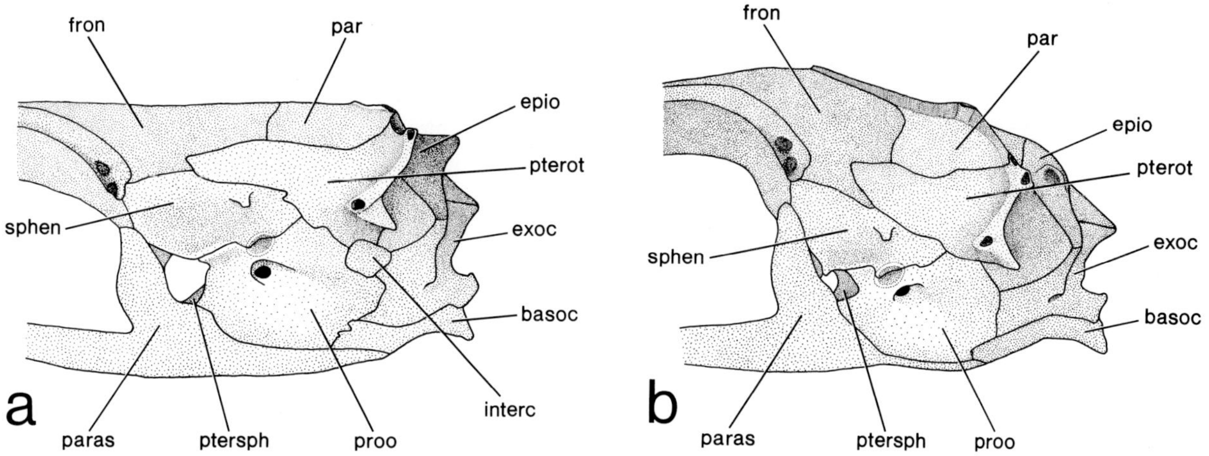

Figure 18.—Lateral view of skull in two species of *Petroscirtes:* a, *P.* (*Dasson*) *variabilis,* USNM 138250, 49.9 mm; b, *P.* (*Petroscirtes*) *mitratus,* ANSP 128560, 39.8 mm. Abbreviations: basoc = basioccipital; epio = epiotic; exoc = exoccipital; fron = frontal; interc = intercalar; par = parietal; paras = parasphenoid; proo = prootic; pterot = pterotic; ptersph = pterosphenoid; sphen = sphenotic.

Dasson and *Petroscirtes* are not recognized as separate genera primarily because both taxa are unique among the Nemophini in having posttemporal cirri (secondarily lost in *marginatus*) and, in most species, supratemporal and orbital cirri and sexual dimorphism of the dorsal-fin spines. They also differ from the other genera of Nemophini in having the symphyseal mandibular cirri present in juveniles as well as the adults of both sexes (presumably a plesiomorphic condition).

The species of *Petroscirtes* have few characters that permit identification of synapomorphies. *P. fallax* and *marginatus* are believed to be closely related because cirri are completely lacking in *marginatus* and those that are present in *fallax* (see Table 5) are very minute. *P. ancylodon, variabilis* and *thepassi* comprise a trio of allopatric species that have essentially lanceolate caudal fins, mottled color patterns and frequently or typically two pores on one or both sides of the mid-dorsal supratemporal pore (see Table 7). Other possible relationships, based largely on morphological similarity, are suggested in the species accounts.

KEY TO SPECIES OF *PETROSCIRTES*

1. First dorsal-fin spine distinctly longer than fourth giving fin notched appearance in all but smallest juveniles; first dorsal-fin spine always longer than second (occasionally subequal); outer ray of pelvic fin with dark spot (indistinct in juveniles) near base of fin; (widespread Indo-Pacific) .... *mitratus*

First dorsal-fin spine subequal or shorter than fourth; first dorsal-fin spine always shorter than second; outer ray of pelvic fin without a dark basal spot **2**

2. Symphyseal mandibular cirrus multifid; preopercular pores with cirri, 3-6 pores each side (Figs. 11-12) **3**
Symphyseal mandibular cirrus simple, if present; preopercular pores without cirri **4**

3. Orbital cirrus present; total dorsal-fin rays 25-27; segmented anal-fin rays 14-16; (Indian and Pacific oceans) .... *xestus*
Orbital cirrus absent; total dorsal-fin rays 33; segmented anal-fin rays 21; (Taiwan) .... *springeri*

4. Body of adults with two or more dark stripes against pale background **5**
Pigmentation not as above, at most a single stripe present **7**

5. Broad dark stripe along entire base of dorsal fin; mid-lateral stripe distinctly better developed than ventral stripe, especially anteriorly; orbital cirri usually present; (Indian and western Pacific oceans, excluding eastern Australia) .... *breviceps*
Pigmentation of dorsal fin not as above; mid-lateral stripe not distinctly better developed than ventral stripe; orbital cirri absent **6**

6. Distal third of dorsal fin mostly pale; segmented anal-fin rays 18-20; (eastern Australia) .... *fallax*
Distal third of dorsal fin heavily and uniformly pigmented; segmented anal-fin rays 17; ("China") .... *marginatus*

7. Broad dark stripe along entire base of dorsal fin; typically 3 supratemporal pores; (Indian and western Pacific oceans, excluding eastern Australia) .... *breviceps*
Pigmentation of dorsal fin not as above; 3-5 supratemporal pores **8**

8. Snout elongate, bulbous (Figs. 24d-f); second dorsal-fin spine 6.6-8.6 percent SL (Fig. 25); depth at anal-fin origin 12.3-16.9 percent SL (Fig. 26); (eastern periphery of Indo-Australian Archipelago) .... *thepassi*
Snout not as above; second dorsal-fin spine 8.1-31.9 percent SL; depth at anal-fin origin 14.2-22.3 percent SL **9**

9. Posterior nostril usually with a short flap on anterior rim; dorsal- and anal-fin rays barred or spotted, frequently forming diagonal rows posteriorly; segmented anal-fin rays 16-19, usually 17; (eastern Indian and Pacific oceans) .... *variabilis*
Posterior nostril a simple hole, without flap on anterior rim; pigmentation of dorsal and anal fins not as above; segmented anal-fin rays 17-20 (rarely 17) **10**

10. Typically 5 supratemporal pores; segmented anal-fin rays 17-19, usually 18; second dorsal-fin spine of males 13.3-17.7 percent SL; (Red Sea-Persian Gulf) .... *ancylodon*
Typically 3 supratemporal pores; segmented anal-fin rays 18-21 (rarely 18); second dorsal-fin spine of males 10.9-31.9 percent SL (Fig. 22); (eastern Australia, New Caledonia and Lord Howe Is.) .... *lupus*

TABLE 6. — Frequency distributions of dorsal-fin rays and vertebrae in species of *Petroscirtes*.[1]

Species	Dorsal-fin spines					Segmented dorsal-fin rays										Total dorsal-fin elements										
	10	11	12	N	x̄	14	15	16	17	18	19	20	21	N	x̄	25	26	27	28	29	30	31	32	33	N	x̄
mitratus	12	284	1	297	11.0	24	214	52	7					297	15.1	29	215	47	6						297	26.1
xestus	1	60		61	11.0	5	53	3						61	15.0	5	54	2							61	26.0
springeri			1	1	12.0								1	1	21.0									1	1	33.0
breviceps	7	227	3	237	11.0				4	31	138	52	12	237	19.2				4	29	144	49	11		237	30.1
fallax		52		52	11.0					5	40*	7		52	19.0					5	40*	7			52	30.0
marginatus		1		1	11.0						1			1	19.0						1				1	30.0
variabilis	7	207		214	11.0			8	138	65	3			214	17.3			10	141	60	3				214	28.3
thepassi	1	42		43	11.0			11	29	3				43	16.8			12	28	3					43	27.8
ancylodon	1	84	1	86	11.0				1	42	43			86	18.5				1	43	41	1			86	29.5
lupus	1	107		108	11.0						50	57	1	108	19.5						51	56	1		108	30.5

Species	Precaudal vertebrae					Caudal vertebrae									Total vertebrae									
	11	12	13	N	x̄	19	20	21	22	23	24	25	N	x̄	30	31	32	33	34	35	36	37	N	x̄
mitratus	271			271	11.0	22	189	57	3				271	20.2	22	189	57	3					271	31.2
xestus	55	4		59	11.1	7	49	3					59	19.9	3	53	3						59	31.0
springeri		1		1	12.0							1	1	25.0								1	1	37.0
breviceps	15	197		212	11.9				17	145	47	3	212	23.2					19	154	38	1	212	35.1
fallax	4	39*		43	11.9			1	5*	32	5		43	23.0				1	8*	30	4		43	34.9
marginatus			1	1	13.0				1				1	22.0						1			1	35.0
variabilis	43	161		204	11.8		8	115	77	5			205	21.4			11	148	44	2			205	33.2
thepassi	4	38		42	11.9		9	24	9				42	21.0			10	26	6				42	32.9
ancylodon	3	72		75	12.0				15	55	5		75	22.9					16	56	3		75	34.8
lupus	4	93		97	12.0					47	49	1	97	23.5						50	47		97	35.5

[1] When counts vary, those for holotype of *P. fallax* are indicated by an asterisk.

TABLE 7. — Frequency distributions of certain meristic characters in species of *Petroscirtes*.[1]

	Segmented anal-fin rays										Total pectoral-fin rays								
Species	14	15	16	17	18	19	20	21	N	x̄	26	27	28	29	30	31	32	N	x̄
mitratus	37	217	40	2	1				297	15.0		1	100	39	130	8	4	282	29.2
xestus	3	51	6						60	15.1	2	3	47	5				57	28.0
springeri								1	1	21.0			1					1	28.0
breviceps				2	53	137	38	6	236	19.0	5	9	123	19	77	3		236	28.7
fallax					14	33*	3		50	18.8	2	2	39*	5	2			50	28.1
marginatus				1					1	17.0	1							1	26.0
variabilis			20	153	39	2			214	17.1	6	11	178	7	4			206	28.0
thepassi			10	30	3				43	16.8	2	1	35	3				41	28.0
ancylodon				3	53	29			85	18.3		1	74	8	1			84	28.1
lupus					2	59	47		108	19.4	1	2	93	7	5			108	28.1

	Procurrent caudal-fin rays												Supratemporal pores			
	Dorsal		Ventral			Total							1-1-1	1-2-1	1-1-2	2-1-2
Species	3	4	2	3	4	5	6	7	8	N	x̄		(3)	(4)	(4)	(5)
mitratus	206	29	2	214	19	2	194	30	9	235	6.2		8	—	20	173
xestus	52	2		54			52	2		54	6.0		—	—	2	46
springeri	1			1			1			1	6.0		1	—	—	—
breviceps	159	12	2	162	7	2	155	9	5	171	6.1		174	1	10	—
fallax	36*	2		31*	7		31*	5	2	38	6.2		42	—	—	—
marginatus	1			1			1			1	6.0		—	?	—	—
variabilis	173	10		170	13		166	11	6	183	6.1		4	—	25	176
thepassi	39		1	38		1	38			39	6.0		15	2	8	10
ancylodon	59			58	1		58	1		59	6.0		4	—	11	54
lupus	82	12		89	5		80	11	3	94	6.2		94	—	2	5

[1] When counts vary, those for holotype of *P. fallax* are indicated by an asterisk.

Petroscirtes (*Petroscirtes*) *mitratus* Rüppell
Specimen Figures 10, 100-102; Tables 4-8.

Petroscirtes mitratus Rüppell, 1830: 111, pl. 28, Fig. 1 (type locality: Red Sea; holotype NFIS 1853).

Petroskirtes hypselopterus Bleeker, 1855: 423 (type locality: Amboina; syntypes lost ?).

Petroskirtes barbatus Peters, 1855a: 248; 1855b: 440 (type locality: Mozambique; holotype lost ?).

Petroscirtes longifilis Kner and Steindachner, 1866: 367, Fig. 5 (type locality: Samoa; holotype NMV 71777).

Petroskirtes marmoratus Bleeker, 1875: 74 (based solely on description of "Blennechis marbre" *in* Lienard, 1839: 39-40, Mauritius).

Petroscirtes lienardi Day, 1878: 327, pl. 39, Fig. 8 (type locality: illustrated specimen from Karachi; holotype lost ?).

Petroscirtes elatus Jordan and Snyder, 1902: 452, Fig. 5 (type locality: Yaeyama, Ishigaki Island, Ryukyu Islands; holotype SU 7071).

Description. — Intercalar absent. Dorsal fin X-XII, 14-16; total elements 25-28. Anal fin II, 14-16. Pectoral fin 13-16. Vertebrae: precaudal 11; caudal 19-22; total 30-33. Pleural ribs on vertebrae 3-11 or 12; epipleural ribs 12-15. Incisor teeth (Fig. 14) in specimens 12-61 mm, n = 93: lower jaw 20-36; upper jaw 20-33.

Symphyseal mandibular cirrus simple and broad; preopercular (3-5 each side), orbital, postorbital, nuchal and posttemporal cirri typically present. Posterior nostril typically with a well-developed flap on anterior rim. Supratemporal pores 3-5, typically 5. Depth at anal-fin origin 20.5-27.9% SL. First dorsal-fin spine elongate, subequal or longer than second, and longer than fourth, giving fin a notched appearance in all but smallest juveniles.

Sexual dimorphism well developed in first and second dorsal-fin spines (Figs. 19-20), moderate in

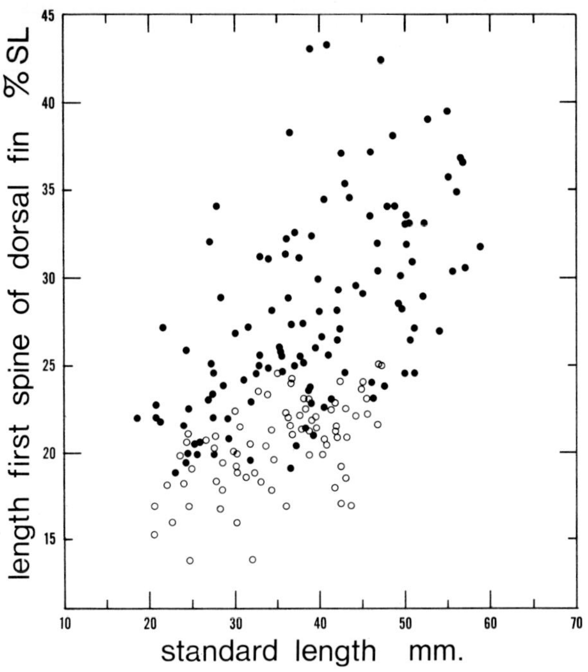

Figure 19. — Relationship between length of first spine of dorsal fin versus standard length in *Petroscirtes mitratus.* Closed symbols = males; open symbols = females.

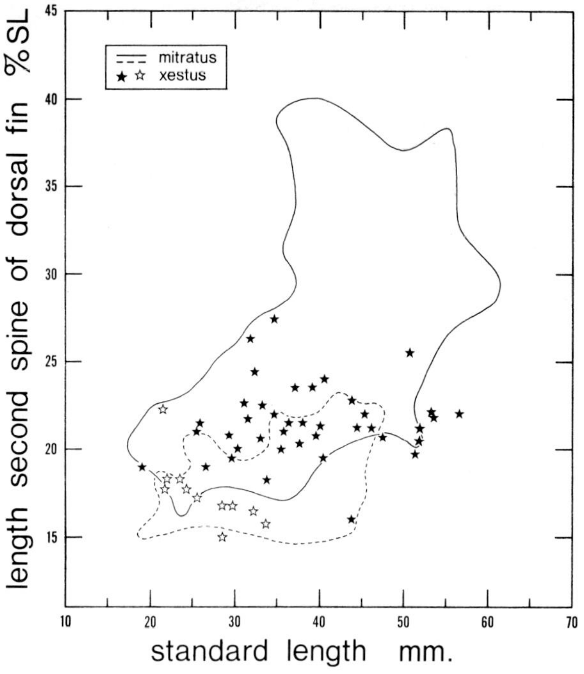

Figure 20. — Relationship between length of second spine of dorsal fin versus standard length in *Petroscirtes mitratus* and *xestus.* Closed symbols and solid lines = males of *xestus* and *mitratus* respectively; open symbols and dashed line = females.

pelvic fins (Fig. 21), and slight in caudal fin in specimens > 30 mm SL. First dorsal-fin spine: males 19.2-43.2% SL, females 15.9-25.2% SL; second dorsal-fin spine: males 17.5-39.8% SL, females 14.8-24.2% SL; pelvic fin: males 17.8-35.4% SL, females 16.0-26.0% SL; caudal fin: males 22.4-34.8% SL, females 22.8-28.2% SL. Outer lobes of caudal fin not elongate. Anterior margin of first dorsal-fin spine with slight, fleshy flap in largest males only.

Color pattern. — Brownish, speckled and mottled, and with 5-6 indistinct dark blotches or bands on sides, usually with dark edged ocelli dorsally. Dorsal, anal and caudal fins spotted and banded, the interradial membranes of the first several dorsal-fin spines more heavily pigmented than remainder of fin. Outer ray of pelvic fins with a dark spot (indistinct in juveniles, conspicuous in large adults) at fin base. A good illustration of an adult male accompanied the description of *Petroscirtes elatus.*

The life colors consist of various shades of brown, yellow, green, black and white; the predominant color is brown, with yellowish-white spots and speckles. Color illustrations of the species are given in Smith (1949: pl. 74), Burgess and Axelrod (1973b: 639) and Masuda *et al* (1975: 85).

Geographic variation. — Ocelli appear to be best developed on the sides of specimens from the Red Sea and western Indian Ocean (including Sri Lanka). Populational variation in frequencies of selected meristic characters is shown in Table 8.

Species comparisons. — *P. mitratus* is the only species of *Petroscirtes* in which the first dorsal-fin spine is longer than the fourth spine. The development and distribution of cirri on the head (see Table 5) will also distinguish *mitratus* from its congeners.

Nomenclature. — The date of publication for *P. mitratus* is usually given as 1828, but see Smith-Vaniz and Springer (1971: 19). The descriptions and figures of the various synonyms of *mitratus* are all unmistakably based on the same species. Nominal species synonymous with *mitratus* either were

TABLE 8. — Geographic variation in frequencies of certain meristic characters in *Petroscirtes mitratus* as indicated by specimens from selected localities.

Localities	Total vertebrae						Total dorsal-fin elements					
	30	31	32	33	N	x̄	25	26	27	28	N	x̄
Red Sea	2	43	7		52	31.1	4	42	6		52	26.0
Persian Gulf	2	9	3		14	30.9	2	9	4		15	26.1
Africa		45	6		51	31.1	2	47	5		54	26.1
Madagascar		8	2		10	31.2		12	5		17	26.3
Mauritius			3		3	32.0			3		3	27.0
Chagos Is.	2	4	1		7	30.9	1	6			7	25.9
India		3			3	31.0		4			4	26.0
W. Australia		2	6	3	11	32.1		1	8	6	15	27.3
Indonesia	2	1			3	30.3	2	3	1		6	25.8
Gulf of Thailand	2	13			15	30.9	3	12	1		16	25.9
Borneo		3			3	31.0		3			3	26.0
Philippines	2	2			4	30.5	2	3			5	25.6
Palau Is.	6	22	1		29	30.8	5	26	1		32	26.7
Yap Is.		6			6	31.0	2	4			6	25.7
Marianas Is.	1	7			8	30.9	3	5			8	25.6
Kapingamarangi	2	6	1		9	30.9		8			8	26.0
Samoa		2	1		3	31.3		3			3	26.0
Tonga		2	7		9	31.8		6	3		9	26.3
New Guinea	1	4			5	30.8		5			5	26.0
E. Australia		5	18		23	31.8		13	10		23	26.4

Localities	Segmented anal-fin rays							Total pectoral-fin rays							
	14	15	16	17	18	N	x̄	27	28	29	30	31	32	N	x̄
Red Sea	6	43	3			52	14.9	1	35	3	2			41	28.1
Persian Gulf	3	10	2			15	14.9		8	3	4			15	28.7
Africa	2	47	5			54	15.1		5	8	37	2	1	53	29.7
Madagascar		16	1			17	15.1			5	11			16	29.7
Mauritius			3			3	16.0				2	1		3	30.3
Chagos Is.	2	5				7	14.7		2	–	5			7	29.4
India		4				4	15.0				4			4	30.0
W. Australia		2	10	2	1	15	16.1		2	–	9	3	1	15	30.1
Indonesia	1	5				6	14.8		3	2	1			6	28.7
Gulf of Thailand	5	11				16	14.7		6	5	5			16	28.9
Borneo	1	2				3	14.7		3					3	28.0
Philippines	2	3				5	14.6		3	1				4	28.2
Palau Is.	8	24				32	14.7		21	4	7			32	28.6
Yap Is.	2	4				6	14.7		3	–	3			6	29.0
Marianas Is.	3	4	1			8	14.8			3	5			8	29.6
Kapingamarangi	1	8				9	14.9		4	1	4			9	29.0
Samoa		3				3	15.0		1	–	2			3	29.3
Tonga		6	3			9	15.3				9			9	30.0
New Guinea	1	4				5	14.8			2	2			4	29.5
E. Australia		11	12			23	15.5			1	17	2	2	22	30.2

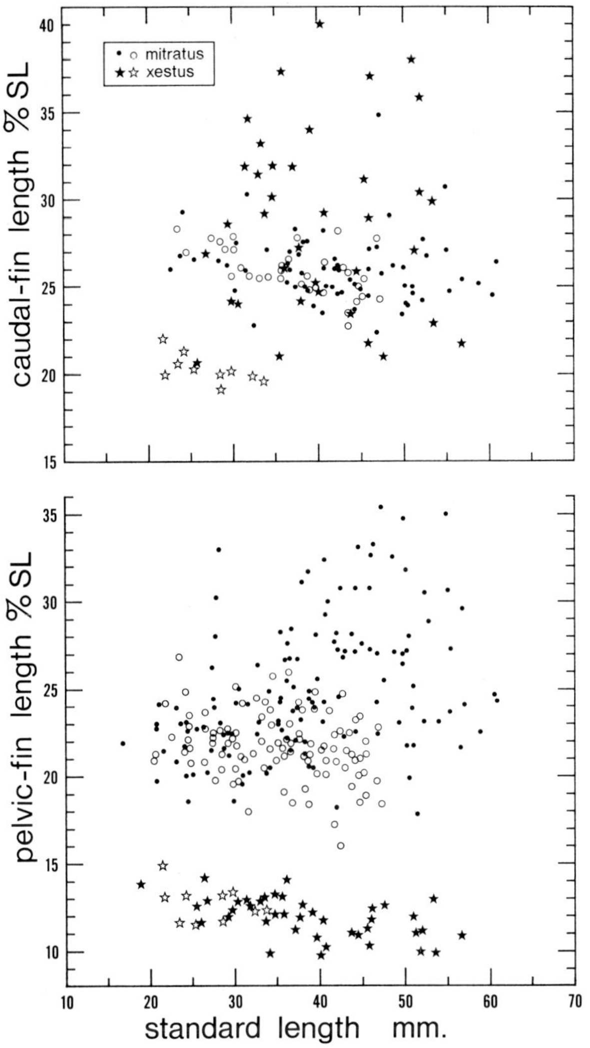

Figure 21. — Relationship between pelvic-fin and caudal-fin lengths versus standard length in *Petroscirtes mitratus* and *xestus*. Closed symbols = males; open symbols = females.

not compared with *mitratus* in their original descriptions, or the supposed differences are invalid because the authors failed to appreciate the changes in morphometry due to sexual dimorphism.

Etymology. — The Latin specific name, meaning to wear a miter or other head-dress, undoubtedly refers to the well-developed orbital cirrus.

Distribution (*Fig. 15*). — Widely distributed in the Indo-Pacific region. *P. mitratus* and *xestus* are the only species of *Petroscirtes* that commonly occur in insular habitats and have become established east of the Andesite Line (refer to general section on zoogeography and speciation). The following acceptable distributional records for *mitratus,* for which no specimens were available, are plotted on the distributional map: Karachi (Day, 1878: 327) and Port Okha, Gulf of Kutch (Mohan, 1968: 122).

Material examined. — 376 specimens (7.2-60.9 mm SL) from 133 collections.

PERSIAN GULF: USNM 147979 (1:55), Saudi Arabia, Tarut Bay.

GULF OF OMAN: BMNH 1899.5.8.96-99 (4:21-36), Jask; BMNH 1901.1.30.72-75 (4:27-45), Mekran Coast, "Charbar."

GULF OF AQABA: HUJ E56/159 (1:22), HUJ E56/293 (1:55), HUJ E57/281 (1:19), HUJ E57/293.1 (1:24), HUJ E62/638 (1:38), HUJ E62/1219 (12:7-9), HUJ E62/3617 (2:24-27), HUJ E62/3660 (1:24), HUJ F4358 (1:36), HUJ F4802 (1:37), UTAI 2111 (1:32).

SINAI PENINSULA: HUJ F4356 (1:37), HUJ F4372 (16 of 47:23-36), USNM uncat. (9:18-28), sta. VGS 69-31, Ras Muhammad; HUJ uncat. (2:35-38), HUJ SLR2126 (1:31), USNM uncat. (1:12), sta. VGS 69-29, Gulf of Suez, El Tur.

STRAITS OF JUBAL: USNM 200567 (4:17-47), USNM uncat. (1:41), sta. HA-34, USNM 167026 (6:37-43), Hurghada.

SOUTHERN RED SEA: NFIS 1853 (48.1, holotype of *Petroscirtes mitratus*), "Red Sea"; BMNH 1888.6.12.14 (1:47), Massawa; HUJ F4623 (11:14-30), Zubair Is., 15°0′N, 42°10′E.

GULF OF ADEN: BMNH 1939.5.24.1433-35 (3:26-35), 12°13′42″N, 50°54′18″ E.

SEYCHELLES: ANSP 114823 (2:31-33), Mahe I.; USNM uncat. (1:47), Amirante Is., sta. HA-19.

ALDABRA IS.: USNM 200599 (4:39-55), USNM uncat. (1:38), sta. HA 67-55, USNM uncat. (2:50), sta. HA 67-58, USNM uncat. (13:27-43), sta. HA 67-66, USNM 214367 (7:38-46).

E. AFRICA: RUSI M.1091 (1:36), Malinda; RUSI uncat. (1:50), Mombasa; RUSI S.67 (1:39), Shimoni; RUSI uncat. (1:36), RUSI P.75 (1:47), Pemba I.; BMNH 1867.3.4.613-614 (5:36-48), RUSI Z.111 (1:39), RUSI Z.750 (1:37), RUSI Z.1179 (1:37), Zanzibar; RUSI uncat. (3:24-61), Palma; RUSI uncat. (1:57), Ibo; RUSI B.P. 580 (1:52), RUSI B.P. 693 (1:38), RUSI B.P. 791 (1:44), RUSI B.P. 1293 (1:54), RUSI B.P. 1298 (1:51), RUSI uncat. (2:45-50), Pinda; ISZZ 1941 (3:33-48), "Mozambique"; RUSI N.816 (1:41), Delagoa Bay; RUSI uncat. (2:29-43), Inhaca.

MAURITIUS: BPBM 15948 (3:21-31).

ST. BRANDON SHOALS (CARGADOS CARAJOS): USNM 215440 (10:25-44).

CHAGOS IS.: USNM 214368 (7:17-42), Diego Garcia Atoll.

MALDIVE IS.: NFIS 9764 (1:11), Fadiffulu Atoll.

INDIA: FMNH 71346 (1:36), FMNH uncat. (1:19), sta. LW-3, FMNH uncat. (1:23), sta. LW-21, Gulf of Mannar, Hare I. (Musal tivu) across channel from Mandapam fish camp; USNM uncat. (3:42-51), Ennore area, Madras vicinity.

COCOS-KEELING IS.: BMNH 1949.11.29.637-638 (2:18-22), ANSP 124777 (1:25).

INDONESIA: RMNH 16774 (2:21-25), Batavia; ZMA 109.347 (4:10-25), Sumbawa I., Bima; ZMA 109.341 (3:13-26), Pater-Noster Is., ca. 7°38′S, 117°28′E; USNM uncat. (4:26-38), Te Vega Cr. 6, sta. 218, Borneo, Darvel Bay; USNM 137848 (1:59), Pandanan I., 8°17′N, 117°13′E.

W. AUSTRALIA: WAM P.4844 (2:21-23), Rottnest I., 23°1′S, 115°28′E; WAM P.4843 (1:21), WAM P.4849 (5:24-34), WAM P.10600 (1:28), WAM P. 11172 (2:42-44), Houtman Abrolhos Is., Easter Group; WAM P.10712-15 (4:28-31), Point Cloates, 22°41′S, 113°41′E; WAM P.11174 (1:24), Locker I., near Onslow.

GULF OF THAILAND: CAS 28987 (1:17), CAS 28989 (1:32), CAS 28990 (2:39-41), CAS 28992 (1:37), CAS 29007 (7:24-37), CAS 29008 (2:25-39), CAS 29009 (1:52), CAS 29010 (3:40-51), USNM 179819 (1:44).

PHILIPPINE IS.: CAS 29011 (4:14-34), Negros, Apo I.; USNM 137843 (3:30-57), USNM 137846 (1:56), USNM 138301 (1:42), USNM 139685 (1:32), Cebu Mkt.; USNM 137849 (1:23), Mactan I., 10°16′N, 123°58′E; USNM 137847 (1:21), Tara I., 12°20′N, 120°16′E.

GULF OF TONGKIN: USNM 197981 (1:12).

RYUKYU IS.: HA 7302 (1:36), "Ryukyu Is."; SU 7071 (56.0, holotype of *Petroscirtes elatus*), Ishigaki I., Yayeyama.

MARIANA IS.: UG 4518 (1:24), CAS 28995 (7:29-43), Guam I.

YAP IS.: CAS 28999 (3:23-35), CAS 29002 (2:26-39), CAS 29003 (1:30).

PALAU IS.: ANSP 126028 (7:24-50), CAS 28978 (1:34), CAS 28979 (1:41), CAS 28982 (2:21-41), CAS 28984 (17:17-40), CAS 28996 (3:36-39), CAS 28998 (3:29-41).

KAPINGAMARANGI ATOLL: CAS 28969 (5:21-28), CAS 28971 (3:23-31), CAS 28972 (1:25).

PONAPE IS.: CAS 28974 (1:37).

MARSHALL IS.: USNM 152924 (1:27), Eniwetok Atoll.

GILBERT IS.: AMS uncat. (1:57), sta. DFH 73-25, Apaiang Atoll; AMS uncat. (3:27-37), sta. DFH 73-53, Tarawa Atoll.

SAMOA IS.: NMV 71777 (35.4, holotype of *Petroscirtes longifilis*), "Samoa Is."; BPBM 13230 (2:28-44), Tutuila.

TONGA IS.: USNM 214369 (9:24-57), Vavau Harbor.

SOLOMON IS.: USNM 123462 (2:11-14), Tulagi I. 9°6′S, 160°8′E.

NEW GUINEA: DASF FO567 (1:31), DASF FO856 (1:45), DASF FO1918 (1:25), Port Moresby; USNM 205675 (2:41-51), Kapa-Kapa, ca. 80 km SE Port Morseby.

E. AUSTRALIA: AMS I. 11795 (1:27), Torres Straits, Murray I.; ANSP 124769 (17 of 51:33-50), ANSP 128560 (3:33-42), Queensland, Little Hope I.; AMS IA.4445 (1:43), Snapper I., near Low Is.; AMS I.15414-1 (1:42), Batt Reef, near Low Is.; BPBM 14333 (2:33-40), USNM uncat. (1:46), sta. VGS 66-3, USNM uncat. (1:49), sta. VGS 66-4, Great Barrier Reef, Capricorn Group, One Tree I.

NEW CALEDONIA: SU 47021 (1:46), Noumea.

Petroscirtes (*Dasson*) *xestus* Jordan and Seale
Specimen Figures 11, 103; Tables 4-7

Petroscirtes xestus Jordan and Seale, 1906: 433, Fig. 110 (type locality: Pago Pago, Samoa; holotype USNM 151797).

Petroscirtes pindae Smith, 1959: 231, pl. 17, Fig. 4 (type locality: Pinda, Mozambique; holotype RUSI B.P. 2146).

Description. — Intercalar present. Dorsal fin X-XI, 14-16; total elements 25-27. Anal fin II, 14-16. Pectoral fin 13-15. Vertebrae: precaudal 11-12; caudal 19-21; total 30-32. Pleural ribs on vertebrae 3-11 or 12; epipleural ribs 13-14. Incisor teeth (Fig. 14) in specimens 16.5-53 mm, n = 20; lower jaw 19-32; upper jaw 17-30.

Symphyseal mandibular cirrus multifid (8-26 cirri, the number of cirri increasing with SL); preopercular (3-6 each side), orbital, postorbital, predorsal (minute), nuchal and posttemporal cirri present. Posterior nostril usually with a short flap on anterior rim. Supratemporal pores typically 5. Depth at anal-fin origin 15.1-19.7% SL. First dorsal-fin spine moderately elongate (adult males) but shorter than second (which may be noticeably elongate in adult males) and fourth spines.

Sexual dimorphism well developed in first and second dorsal-fin spines (Fig. 20) and caudal fin (Fig. 21), absent in pelvic fin in specimens > 25 mm SL. First dorsal-fin spine: males 14.7-23.0% SL, females 13.0-15.6% SL; second dorsal-fin spine: males 16.0-27.3% SL, females 15.0-17.2% SL; pelvic fin: males 9.7-14.2% SL, females 11.5-13.5% SL; caudal fin: males 21.1-40.0% SL, females 19.2-20.3% SL. Outer lobes of caudal fin elongate in adult males only. Anterior margin of first dorsal-fin spine with well-developed fleshy flap in adult males only.

Color pattern. — The typical color pattern of adults is shown in Figure 103; the most diagnostic feature is a series of submarginal black spots in the dorsal fin. Except for the spot between the 9th and 11th spines (the largest usually slightly exceeding the pupil diameter), the spots are largely confined to the ray elements and occur on alternate or every third or fourth ray. A black spot is also present distally between the first two spines. A very narrow, irregular, dark stripe extends from the eye to the caudal fin where it may terminate as an indistinct basicaudal spot. The remainder of the body and fins are lightly spotted and speckled. A good illustration of the juvenile holotype accompanied the original description, and Jones and Kumaran (1964: 117) present an illustration of an adult from Minicoy in the Laccadive Islands.

Smith (1959: pl. 17) gives a color illustration of the holotype of *P. pindae* in which the body and fins have greenish blotches and speckles, the largest black spot in the dorsal fin has a blue border and the distal fourth of the dorsal fin is orange. Adult males of *xestus* collected from the Cocos-Keeling Islands during March 1974 were light tan, white ventrally, with brown markings; the largest black spot in the dorsal fin had a white border and distally the fin was pale yellowish instead of orange.

Species comparisons. — The pigmentation of the dorsal fin easily distinguishes *xestus* from the other species of *Petroscirtes.* Only *xestus* and *springeri* have multifid, symphyseal mandibular cirri, but the two species are easily distinguished by a large number of characters.

Nomenclature. — In the original description, Smith stated that *pindae* was related to *xestus,* but the latter differed in having a deeper body and a wider interorbital. Both of these differences merely reflect the juvenile status of the two type specimens of *xestus.*

Etymology. — The specific name is from the Greek *xestos,* meaning polished by scraping or filing. Presumably the name alludes to the coloration of the dorsal surface of the head of the juvenile type, which was described as "bright silvery, as if coated with mercury."

Distribution (Figure 15). — Widely distributed in the Indo-Pacific, but unreported from western Australia, South China Sea, Japan or the Red Sea, probably as a result of inadequate collecting. *P. xestus* is usually collected in sandy, flat-bottom habitats, often in association with empty mollusk shells which provide shelter and nesting sites.

Material examined. — 134 specimens (9.3-56.8 mm SL) from 38 collections.

E. AFRICA: RUSI BP.2146 (34.3, holotype of *Petroscirtes pindae*), Pinda; RUSI uncat. (1:17), taken with holotype.
MALDIVE IS.: BPBM 1901.12.31.76 (1:23).
NICOBAR IS.: NFIS 9765 (1:57).
COCOS-KEELING IS.: ANSP 128045 (2:44-52), ANSP 128046 (1:51), ANSP 128049 (1:36), ANSP 128416 (1:14).
INDONESIA: USNM 195698 (3:26-34), "Dutch East Indies-Longley", presumably from Banda Sea.
MARIANA IS.: ANSP 85879 (1:48), ANSP 124874 (1:46), BPBM 4913 (1:54), BPBM 7412 (1:18), CAS 29030 (2:24-26), CAS 29031 (4:26-38), CAS 29032 (1:26), all from Guam I.
PALAU IS.: CAS 29028 (2:29-41), Chol I.
KAPINGAMARINGI ATOLL: CAS 29027 (1:22), CAS GVF410 (1:29).
MARSHALL IS.: ANSP 124844 (1:52), BPBM 12461 (1:46), BPBM 12462 (2:51-53), Parry I.; BPBM 12460 (1:37), Japtan I.; USNM 142229 (1:11), Teiteiripucchi I., all from Eniwetok Atoll.
GILBERT IS.: AMS uncat. (8:29-45), sta. DFH 73-24, AMS uncat. (1:40), sta. DFH 73-25, AMS uncat. (2:15-25), sta. DFH 73-41, Apaiang Atoll; AMS uncat. (7:19-40), sta. DFH 73-53, Tarawa Atoll.
PHOENIX IS.: USNM 115531 (3:44-46), Canton I.
LINE IS.: ANSP 126213 (58:9-10), NMFS 2757 (5:10), Christmas I.
SOCIETY IS.: BPBM 11611 (1:22), Bora Bora I.; BPBM 12463 (1:16), Moorea I.
SAMOA IS.: USNM 151797 (12.4, holotype of *Petroscirtes xestus*), Pago Pago; BPBM 6134 (5:23-41), Apia.
FIJI IS.: SU 24829 (1:16), Suva.
NEW HEBRIDES IS.: CAS 29029 (5:22-33), USNM 195788 (1:33), Espiritu Santo I.; USNM 198963 (1:24), Efate I.
E. AUSTRALIA: BPBM 14332 (1:40), Great Barrier Reef, Capricon Group, One Tree I.

Petroscirtes (Dasson) springeri, new species
Specimen Figures 12, 104-105; Tables 5-7

Description. — Intercalar presumably present. Dorsal fin XII, 21. Anal fin II, 21. Pectoral fin 14. Vertebrae: precaudal 12; caudal 25. Pleural ribs on vertebrae 3-13; epipleural ribs 15. Incisor teeth: lower jaw 30; upper jaw 24.

Symphyseal mandibular cirrus multifid (3-4 cirri);

preopercular (3 left; 4 right), interorbital (only anterior pair of pores), postorbital, predorsal (minute), nuchal and posttemporal cirri present; orbital cirri absent. Posterior nostril with a short flap on anterior rim. Supratemporal pores 3. Depth at anal-fin origin 16.0% SL. First dorsal-fin spine not elongate, shorter than second and fourth spines.

Sexual dimorphism (male holotype only known specimen) probably absent in length of second dorsal-fin spine and pelvic fin, and well developed in caudal fin. Second dorsal-fin spine 9.5% SL; pelvic fin 12.1% SL; caudal fin 25.9% SL. Outer lobes of caudal fin elongate. Anterior margin of first dorsal-fin spine without fleshy flap.

Color pattern. — The most conspicuous feature of the color pattern is a dark spot, slightly smaller than eye diameter, on the operculum. A diffuse stripe extends from snout through eye and opercular spot and ends as a dark basicaudal spot. Except for a minute dark spot on the distal margin of first dorsal-fin spine, the fins and remainder of the body are rather uniformly pigmented.

The coloration of the holotype, based on field notes taken by V. G. Springer soon after capture, is as follows: Opercular and basicaudal spots dark blue; stripe pale blue. Body pale pinkish brown with faint yellowish cast above stripe. Dorsal fin pale olive-yellow with pale pinkish brown distal margin. Anal fin pale olive with dusky cast. Caudal fin with slight pinkish cast, pectoral and pelvic fins immaculate.

Species comparisons. — *P. springeri* and *xestus* are the only species of *Petroscirtes* that have multifid, symphyseal mandibular cirri and a minute predorsal cirrus; they are easily distinguished by differences in color pattern and meristic characters.

Etymology. — Named in honor of Dr. Victor G. Springer, National Museum of Natural History, who first introduced me to blenniid fishes and suggested the present study.

Distribution (*Fig. 15*). — Known only from Taiwan.

Material examined. — Holotype, only known specimen.

Holotype. — USNM 203279, male, 49.7 mm, Taiwan, cove on point just SW of Yeh-Lin, 25°12'N, 121°41'E; rocky bottom; depth 7.5-9 m; 19 May 1968; V. G. Springer; original no. VGS 68-30.

Petroscirtes (*Dasson*) *breviceps* (Valenciennes)
Specimen Figures 106-108, 110-111; Tables 4-7, 9.

Blennechis breviceps Valenciennes *in* Cuvier and Valenciennes, 1836: 283 (type locality: Gulf of Bengal; holotype MNHN 999).

Blennechis cyprinoides Valenciennes *in* Cuvier and Valenciennes, 1836: 286 (type locality: not stated, "Voyage de Peron"; syntypes MNHN 717).

Petroscirtes bankieri Richardson, 1845: 136, pl. 64, Figs. 8-10 (type locality: Hong Kong; holotype lost ?).

Blennechis polyodon Bleeker, 1851a: 254 (type locality: Batavia; holotype RMNH 26471).

Petroscirtes cynodon Peters, 1855a: 247; 1855b: 439 (type locality: Mossambique; holotype ISZZ 1939).

Aspidontus trossulus Jordan and Snyder, 1902: 455, Fig. 7 (type locality: Misaki, Japan; holotype SU 7114).

Petroscirtes annamensis Chabanaud, 1924: 363 (type locality: coast of Annam [= North Vietnam]; holotype MNHN 1924.161).

Dasson duperreyi Whitley, 1945: 31 (type locality: Sharks Bay, Western Australia; holotype WAM P.2152).

Description. — Intercalar present. Dorsal fin X-XII, 17-21; total elements 28-32. Anal fin II, 17-21 (rarely 17). Pectoral fin 13-16. Vertebrae: precaudal 11-12; caudal 22-25; total 34-37. Pleural ribs on vertebrae 3-12 or 13; epipleural ribs 13-15. Incisor teeth in specimens 18-95 mm, n = 72: lower jaw 17-44; upper jaw 17-41.

Symphyseal mandibular cirrus simple; other cirri absent except as follows: orbital cirri present or absent, usually present but often minute in adults; postorbital cirri present or absent, frequently absent; posttemporal cirrus present. Posterior nostril a simple hole, without flap on anterior rim. Supratemporal pores 3-4, typically 3. Depth at anal-fin origin 16.4-21.0% SL. First dorsal-fin spine not elongate, shorter than second and fourth spines.

Sexual dimorphism slight in second dorsal-fin spine (Fig. 22) and pelvic fin, moderate in caudal fin (Fig. 23) in specimens > 30 mm SL. Second dorsal-fin spine: males 7.8-17.3% SL, females 7.1-13.6% SL; pelvic fin: males 11.9-24.9% SL, females 12.8-17.8% SL; caudal fin: males 19.2-42.8% SL, females 18.6-25.5% SL. Outer lobes of caudal fin elongate in adults. Anterior margin of first dorsal-fin spine with slight, fleshy flap (distally) in large males only.

Color pattern. — Broad dark stripe extends from snout (where often indistinct) through eye, envelops the gill-opening and ends at base or in the proximal

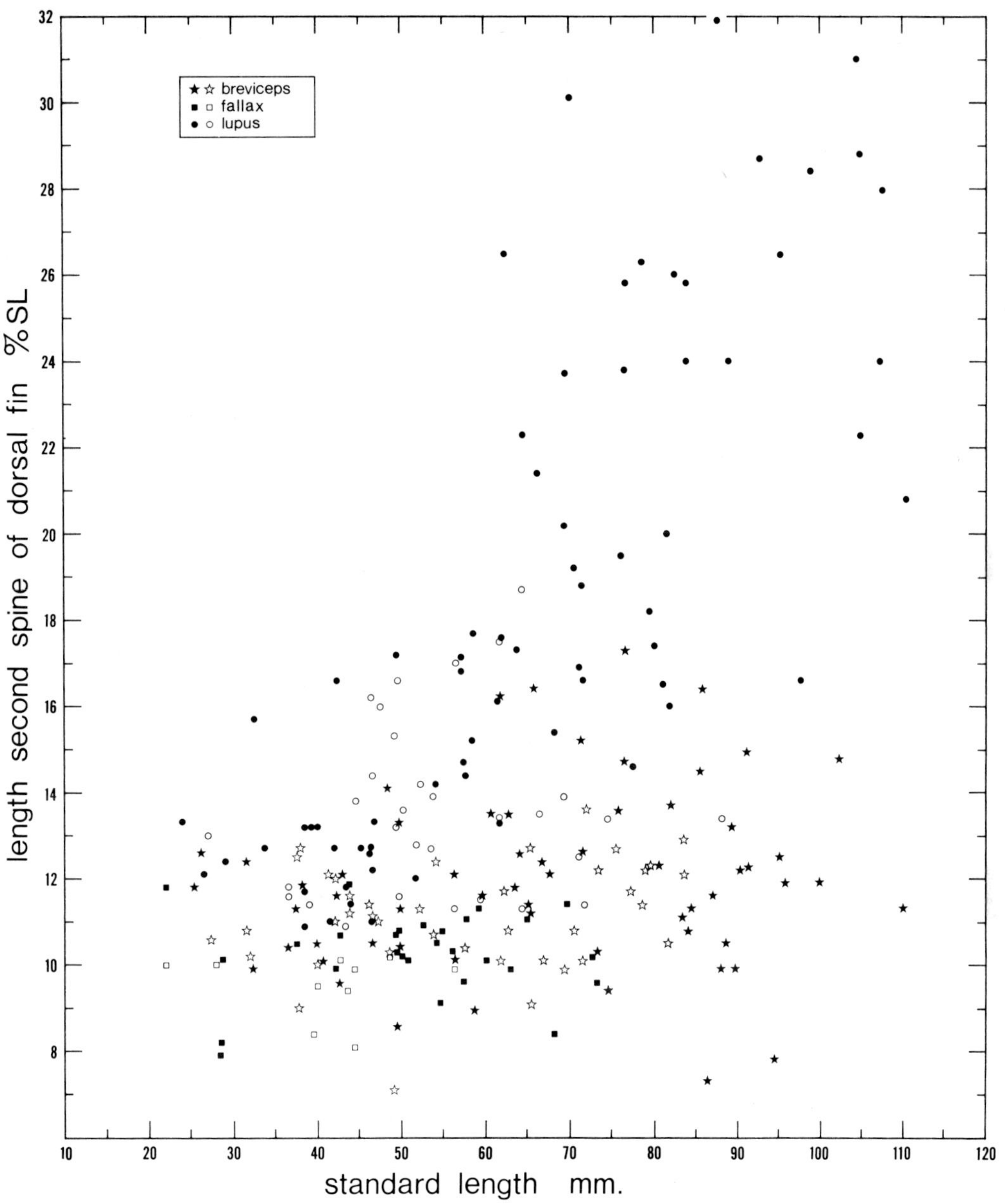

Figure 22. — Relationship between length of second spine of dorsal fin versus standard length in *Petroscirtes breviceps, fallax* and *lupus*. Closed symbols = males; open symbols = females.

region of caudal fin; in occasional specimens the lateral stripe may be broken into a series of bands. Specimens from the central Pacific Ocean often have a paler ventral stripe extending from the lower jaw through lower half of pectoral-fin base to end of anal fin. On the head and pectoral-fin base the ventral stripe occasionally consists of a series of dark spots, and the ventral margin of the stripe is poorly defined. A narrow, dark stripe follows the dorsal contour of the body and extends well onto the base of the dorsal fin for its entire length; distally the dorsal fin is spotted or heavily reticulated. The anal fin may

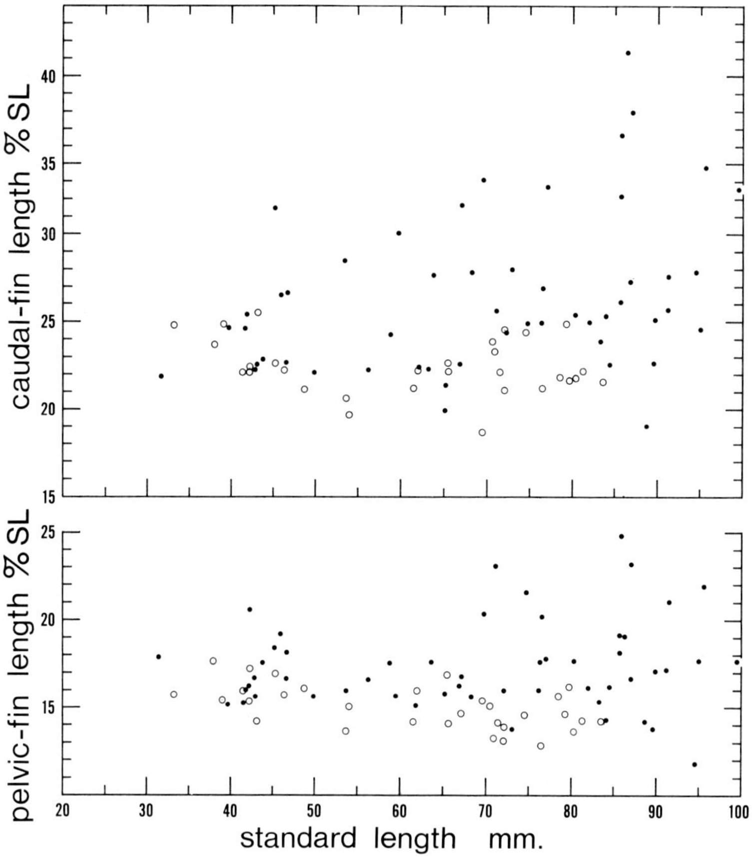

Figure 23. — Relationship between pelvic-fin and caudal-fin lengths versus standard length in *Petroscirtes breviceps*. Closed symbols = males; open symbols = females.

be spotted, reticulated or nearly uniformly dusky. In adults, the distal half of the caudal fin is rather uniformly pigmented (the interradial membranes may be dusky) and the rays are not spotted.

In juveniles, the lateral stripe is typically present but there also may be a series of well-defined bands on the sides. The caudal-fin rays are also frequently barred in juveniles.

According to Smith (1959: 233), the life colors of Indian Ocean specimens are "highly variable, greenish yellow . . ., from eye to Cd a series of connected darker blotches, D basally dark, distally speckled red and yellow." The holotype of *Aspidontus trossulus* was reported to be grayish with two stripes of violet black; the interspace between the stripes yellowish-white and the stripe at the dorsal-fin base blackish. See also discussion of mimicry below.

Smith (1959: pl. 19) gives a photograph of an Indian Ocean specimen (identified as *Dasson variabilis*), and a photograph of a specimen from Palau is given in Springer and Smith-Vaniz (1972a: 34). Excellent drawings of specimens from Japan (juvenile and adults of each sex) showing a well-developed ventral stripe are given in Tomiyama (1951). Color illustrations of *breviceps* appear in Smith (1949: pl. 82) and Burgess and Axelrod (1971: 289; 1974: 1333). Masuda *et al.* (1975: 86) give a color photograph of a specimen of *breviceps* (identified as *Dasson trossulus*) from southern Japan.

Geographic variation.—Specimens of *P. breviceps* from the Indian Ocean (including western Australia)

TABLE 9. — Geographic variation in frequencies of certain meristic characters in *Petroscirtes breviceps* as indicated by specimens from selected localities.

	Total vertebrae						Total dorsal-fin elements						
Localities	34	35	36	37	N	x̄	28	29	30	31	32	N	x̄
Africa	3	38	6		47	35.1		3	43	3		49	30.0
India		3			3	35.0			3	1		4	30.2
Java Sea	1	7	1		9	35.0			6	3		9	30.3
Celebes		13			13	35.0		1	11	1		13	30.0
Banda Sea		4			4	35.0			4	1		5	30.2
W. Australia		12	11		23	35.5			11	9	8	28	30.9
Singapore	1	2	1		4	35.0			3	1		4	30.2
Gulf of Thailand		11	3		14	35.2			11	3		14	30.2
Philippines	4	32	2		38	34.9		2	31	5		38	30.1
Hong Kong		18	6		24	35.2		2	7	14	1	24	30.9
Japan		2	6	1	9	35.9			1	6	2	9	31.1
Palau-Yap	10	3			13	34.2	4	20	4			28	29.0

	Segmented anal-fin rays							Total pectoral-fin rays							
Localities	17	18	19	20	21	N	x̄	26	27	28	29	30	31	N	x̄
Africa		8	35	5		48	18.9	1	–	2	4	41	1	49	29.8
India			4			4	19.0					4		4	30.0
Java Sea		1	7	1		9	19.0			5	1	3		9	28.4
Celebes		4	9			13	18.7			8	3	2		13	28.5
Banda Sea			5			5	19.0		1	3				4	27.8
W. Australia		2	11	12	3	28	19.6	1	1	25	–	1		28	28.0
Singapore		1	3			4	18.7			1	–	3		4	29.5
Gulf of Thailand		2	8	3	1	14	19.2			2	1	10	1	14	29.7
Philippines		8	29	1		38	18.8		5	25	2	2		34	28.0
Hong Kong			13	11		24	19.5	1	1	13	4	4	1	24	28.5
Japan		1	3	3	2	9	19.7	1	–	7	–	1		9	28.0
Palau-Yap	2	25	1			28	18.0	1	1	24	1	1		28	28.0

lack the ventral stripe (see above) usually present in specimens from Indonesia, New Guinea, Philippines and the Palau Islands. The ventral stripe is variously developed, however, in specimens from the Gulf of Thailand and populations in the South China Sea. The specimen from Hong Kong illustrated in Figure 108 exhibits an intermediate condition with regard to the development of the stripe.

Populational variation in frequencies of selected meristic characters is shown in Table 9.

Species comparisons. — The presence of a broad, dark stripe along the entire dorsal-fin base distinguishes *breviceps* from all other species of *Petroscirtes*. See also discussion under account of *P. fallax*.

Nomenclature. — The holotype of *P. bankieri* is apparently lost but the figure and description clearly apply to *breviceps*. I have examined the primary type material upon which all of the other nominal species were based. *Blennechis cyprinoides* was based on the banded juveniles of *breviceps*, and *P. cynodon* and *Dasson duperreyi* were not compared with *breviceps* in the original description. The other names apply to specimens of *breviceps* with a moderately well-developed ventral stripe (refer to discussion of geographic variation).

Etymology. — The specific name is a combination of the Latin *brevis* (short) and *-ceps* (head).

Mimicry. — Refer to comments under account of *Meiacanthus grammistes*.

Distribution (Fig. 16). — Widely distributed in the Indian and west Pacific Oceans, but unknown from the Red Sea, Persian Gulf or eastern Australia.

Material examined. — 321 specimens (9.7-110 mm SL) from 122 collections.

LOCALITY UNKNOWN: MNHN 717 (5:29.0-31.8, syntypes of *Blennechis cyprinoides*), "Voyage de Peron."

E. AFRICA: RUSI uncat. (1:40), Mombasa; NMV 71799 (1:54), RUSI Z.1130 (1:63), Zanzibar; RUSI uncat. (1:50), Palma, 10°48′S, 40°29′E; RUSI uncat. (1:87), Pinda; ISZZ 1939 (94.5, holotype of *Petroscirtes cynodon*), (3:72-80, in same jar with holotype), ISZZ 1938 (4:72-79), BMNH 1861.5.2.78-79 (2:86-93), Mozambique, Peters Collection; SAM 12416 (3:76-81), Mossuril, 14°55′S, 40°41′E; ANSP 122778 (1:84), RUSI uncat. (1:90), Mozambique; ANSP 97449 (2:65-77), ANSP 122785 (1:80), HUJ F494 (1:80), RUSI N.816 (1:83), Delagoa Bay; HUJ F493 (1:67), SAM 25618 (7:38-56), SAM 26412 (1:74), RUSI uncat. (2:47-89), Inhaca, 26°1′S, 32°58′E.

MADAGASCAR: USNM uncat. (12:14-30), sta. JR-57, Nossi-Be'; ZSZM 11424 (2:21-31), "Bay of Madagascar."

INDIA: MNHN 999 (48.2, holotype of *Blennechis breviceps*), Bay of Bengal; BMNH 1888.11.6.30-33 (2:84-91), Madras, E. coast; USNM 202741 (1:82), Ennore area, Madras vicinity.

INDONESIA: RMNH 26471 (72.0, holotype of *Blennechis polyodon*), Batavia; RMNH 4773 (2:55-76), originally in same jar with holotype); RMNH 16438 (3:48-67), Java Sea, probably Batavia; USNM 211998 (1:31), Seribu Is., 5°51′S, 106°34′E; USNM 211973 (1:40), Bawean I., 5°52′S, 112°37′E; ZMA 109.334 (2:38-41), Salembu Besear, 5°30′S, 114°30′E; NMV 71761 (1:75), Borneo, "Salanga"; NMV 71801 (1:59), "Celebes"; RMNH 20650 (21:22-41), RMNH 20820 (4:10-19), Celebes, Makasar; RMNH 21274 (1:25), Makassar Strait 1°16′N, 119°57′E; ZMA 109.364 (1:11), Salajar, 6°7′S, 120°28′E; USNM 211964 (12:16-34), Buton I., 5°27′20″S, 122°37′5″E; RMNH 20316 (2:21-31), Flores Sea, 7°55′S, 122°12.5′E; NMV 71800 (1:71), Ambon I., (Amboina); USNM 205287 (3:70-85), "Dutch East Indies — Longley", presumably Banda Sea; RMNH 20120 (1:33), Banda Sea, 6°22′S, 131°52′E; RMNH 20600 (1:14), Aru (Aroe) Is., 6°10′S, 134°20′E.

W. AUSTRALIA: AMS IA.4661 (1:32), Gantheaume Point, Broome; CSIRO A.3380 (1:80), CSIRO A.3381 (1:97), CSIRO A.3428 (1:68), CSIRO A.3431 (1:41), WAM P.11175-208 (16 of 33:22-49), WAM P.11233 (1:20), Exmouth Gulf; WAM P.2152 (89.1, holotype of *Dasson duperreyi*), WAM P.4663 (3:42-90), WAM P.5381 (3:73-110), WAM P.12074 (1:17), AMS IB.323-5 (3:37-66), AMS IB.347 (1:62), AMS IB.3046 (1:71), AMS I.13172 (1:100), Shark Bay; WAM uncat. (1:44), Lancelin I., 31°33′S, 115°19′E; WAM P.11213-6 (4:19-23), Rottnest I., 32°1′S, 115°28′E; WAM P.5516 (1:96), Woodmans Point, just S of Perth.

SINGAPORE: SU 67354 (4:43-71).

GULF OF THAILAND: CAS 28948 (1:26), CAS 28951 (1:54), CAS 28952 (2:25-37), CAS 28593 (3:38-49), CAS 28954 (1:28), CAS 28956 (1:30), NMV 71796 (1:44), SIO 65-39 (1:24), SIO 65-40 (1:24).

SOUTH CHINA SEA: ZSAM 11423 (5:13-32), "South China Sea"; MNHN 1924.161 (82.0, holotype of *Petroscirtes annamensis*), "Coast of Annam", North Vietnam.

PHILIPPINE IS.: UW 7499 (5:38-55), Mindanao, Cotabato Pt.; SU 28450 (9:22-34), Negros, Dumaguete; USNM 137853 (1:67), Pandanon I., between Cebu and Bohol I.; USNM 137855 (1:86), USNM 137856 (2:71-87), USNM 137818 (1:60), Cebu Mkt.; USNM 138272 (1:52), Mindoro Strait, 12°47′15″N, 120°41′E; USNM 137829 (13:29-43), vicinity of Jolo I., 6°3′N, 121°1′E; USNM 137857 (1:68), Teomabal I., vicinity of Jolo; UW 7608 (7:25-36), Sulu Arch., 10°N, 120°21′E.

HONG KONG: ANSP 76749 (1:68), ANSP 77068 (2:60-84), ANSP 123735 (3:61-76), ANSP 128559 (2:56-71), CAS 28949 (7:38-71), CAS 28950 (3:69-84), SU 61156 (5:46-102), SU 61502 (1:38).

CHINA: RMNH 1838 (1:41), "China"; USNM 130446 (1:41), Foochow, Kukien, 26°9′N, 119°17′E; ZSZM 11236 (1:44), Fubien (Fokien) Prov.

TAIWAN: USNM 205285 (1:47), Wan-li.

RYUKYU IS.: HA 7301 (1:76), "Ryukyu Is."; USNM 111972 (1:25), Okinawa, Kimmu Wan Bay.

KOREA STRAIT: NMV 71804 (1:49), ZSZM 2122 (1:41), ZSZM 5465 (1:39), Museum Godeffroy.

JAPAN: USNM 122485 (1:86), "Japan"; USNM 150412 (1:43), Tsurikake, 31°38′30″N, 129°19′E; ZITU 5706 (1:70), Kashiwa jima, near Kochi; SU 40219 (5:17-27), Hamada, Matsubara Bay; CAS 28957 (1:20), Wakayama-Ken, Tanabe Bay, 34°41′34″N, 135°19′18″E; UMMZ 142765 (1:47), Wakanoura; UMMZ 187433 (1:36), Suruga Bay, Heta; SU 7114 (62.0, holotype of *Aspidontus trossulus*), SU 23531 (2:74-89), UMMZ 144778 (2:78-95), Misaki, 35°10′N, 139°37′E; SIO H53-355 (1:64), 35°4.5′N, 145°7.5′E.

YAP IS.: CAS 28968 (1:46), NMV 71762 (1:41).

PALAU IS.: ANSP 126031 (4:33-42), BPBM 7452 (2:45-54), BPBM 7455 (1:25), CAS 28958 (1:46), CAS 28959 (1:31), CAS 28962 (31:10-26), CAS 28963 (1:43), CAS 28966 (1:47), CAS 28967 (1:42).

NEW GUINEA: DASF FO2369 (1:31), USNM 210620 (1:42), Port Moresby; DASF FO2130 (1:77), found in bilge of pearl ship "Merindah Pearl", which had recently traveled from NW Australia to Port Moresby. (Continuous flow of saltwater in hold.)

Petroscirtes (Dasson) fallax, new species

Specimen Figures 13, 113; Tables 5-7

Petroscirtes lineatus (*non* De Vis) Whitley, 1926: 235 (misidentification, in part: specimen from Port Denison).

Petroscirtes variabilis (*non* Cantor) Grant, 1972: 313, pl. 70 (habits, distribution; color illustration).

Petroscirtes sp. Springer and Smith-Vaniz, 1972: 23, pl. 4. Allen *et al.*, 1975: 55 (illustration, suggested Batesian mimic of *Meiacanthus lineatus*).

Description. — Intercalar present. Dorsal fin XI, 18-20; total elements 29-31. Anal fin II, 18-20. Pectoral fin 13-15. Vertebrae: precaudal 11-12; caudal 21-24; total 33-36. Pleural ribs on vertebrae 3-12 or 13; epipleural ribs 13-15. Incisor teeth in specimens 29-73 mm, n = 13: lower jaw 28-41; upper jaw 28-39.

Symphyseal mandibular cirrus simple (minute in females); other cirri absent except for minute posttemporal cirrus. Posterior nostril a simple hole, without flap on anterior rim. Supratemporal pores consistently 3. Depth at anal-fin origin 14.8-21.5% SL. First dorsal-fin spine not elongate, shorter than second and fourth spines.

Sexual dimorphism slight in second dorsal-fin spine (Fig. 22) and pelvic fin, moderate in caudal fin in specimens > 30 mm SL. Second dorsal-fin spine: males 8.4-11.8% SL, females 8.1-11.1% SL; pelvic fin: males 13.0-19.4% SL, females 15.3-17.0% SL; caudal fin: males 19.6-27.1% SL, females 20.2-24.1% SL. Outer lobes of caudal fin elongate in adult males only. Anterior margin of first dorsal-fin spine with slight fleshy flap (distally) in adult males only.

Color pattern. — Head and body with two dark stripes, the pale interspace approximately equal width of dark stripes. Dorsal stripe extends from snout through eye, envelops gill-opening and ends on the proximal region of the caudal fin. The ventral stripe extends from lower jaw, onto ventral half of pectoral-fin base and ends at lower margin of caudal fin. A lighter stripe originates on top of head and follows the dorsal contour of the body extending slightly into the dorsal-fin base. The dorsal fin is otherwise immaculate except for a black spot, usually restricted to the distal margin of the interradial membranes of the first 3-4 spines. The other fins are also immaculate. Juveniles have vertical bands similar to those of *P. breviceps.*

Life colors, based on a color plate in Grant (1972) and color slides taken underwater of living specimens at the type locality, are as follows: Dark stripes brown, pale interspaces dirty yellow, especially anteriorly; Pelvic fins and body below ventralmost stripe white. Caudal and anal fins transparent or pale amber. Dorsal fin amber proximally followed by wide transparent area, submarginal amber stripe and narrow pale blue or green border.

Species comparisons. — Only two other species of *Petroscirtes* have alternating dark and pale stripes, and neither occurs sympatrically with *fallax.* The dorsal fin of *marginatus* differs from both in having a conspicuous dark border. *P. breviceps* differs in usually having orbital cirri, and in those individuals that are striped the ventral stripe is typically less well developed, especially on the head and base of the pectoral fin.

Etymology. — The Latin name *fallax,* meaning deceitful or deceptive, alludes to the presumed mimetic status of this species and *Meiacanthus lineatus.*

Biology. — This species usually occurs in the vicinity of beds of eel grass (*Zostera*) and often associates with *Meiacanthus lineatus,* which is similarly striped. See also comments under account of *M. lineatus.*

Distribution (Fig. 16). — Endemic to eastern Australia.

Material examined. — 57 specimens (11.0-73.2 mm SL) from 20 collections.

Holotype. — USNM 201365, male, 68.1 mm SL, Australia, Queensland, Great Barrier Reef, Capricorn Group, One Tree Island; 18 Nov. 1966; V. G. Springer; sta. VGS 66-4.

Paratypes (all from eastern Australia arranged from north to south). — BMNH 1933.8.12.45 (1:41), "Great Barrier Reef Expedition"; AMS IB.4318-19 (4:22-29), Green I., 16°43'S, 146°0'E; AMS IA.1788 (1:50), Bowen, Port Denison; QMB I. 10205-6 (2:28-29), Queensland, Whitsunday Group; AMS IA.6179 (2:22-54), Lindeman I.; AMS IB.6062 (1:53), AMS IB.6200 (9:43-63), Swains Reef Expedition, Gillet Cay; AMS I.11418 (1:59), Capricorn Group, Masthead I.; USNM 201797 (3:23-50), USNM 201798 (3:40-70), USNM 201799 (4:51-56), USNM 201800 (6:22-44), USNM 201801 (1:43), USNM 201802 (1:50), USNM 201803 (1:67), USNM 205290 (1:65), USNM 205643 (12:11-20), Capricorn Group, One Tree I.; USNM 205425 (1:73), Capricorn Group, Fairfax I.; AMS I.7188 (1:50), Gladstone, Port Curtis; AMS IB.3970 (1:73), New South Wales, Richmond River.

Petroscirtes (Dasson) marginatus, new species
Specimen Figure 112; Tables 5-7.

Petroscirtes anema (*non* Bleeker) Günther, 1861: 236 (misidentification in part; specimen from China).

Description. — Intercalar presumably present. Dorsal fin XI, 19. Anal fin II, 17. Pectoral fin 13.

Vertebrae: precaudal 13; caudal 22. Pleural ribs on vertebrae 3-12; epipleural ribs 15. Incisor teeth: 34 in both jaws.

Cirri absent. Posterior nostril a simple hole, without flap on anterior rim. Supratemporal pores apparently 4, including a pair of mid-dorsal ST pores. (Count not completely reliable because dorsum of head slightly damaged.) Depth at anal-fin origin 12.0% SL. First dorsal-fin spine not elongate, shorter than second and fourth spines.

Sexual dimorphism (female holotype only known specimen) probably as in *fallax*. Second dorsal-fin spine 9.8% SL; pelvic fin 6.4% SL; caudal fin 16.2% SL. Outer lobes of caudal fin not elongate. Anterior margin of first dorsal-fin spine without fleshy flap.

Color pattern. — Head and body with two dark stripes, the pale interspace approximately equals width of dorsal stripe. Dorsal stripe extends from snout through eye, envelops the gill-opening and ends at the caudal-fin base. The ventral stripe extends from lower jaw, just above anal fin to caudal-fin base. The inner margin of both dark stripes touches the pectoral-fin base, but the majority of pectoral-fin base is pale. The top of head and dorsum anteriorly is dark, but paler than stripes. Dorsal fin with narrow dusky area proximally, wider pale stripe above and distal half to two-thirds of fin very dark. Remainder of body and other fins pale or immaculate. Life coloration unknown.

Species comparisons. — Refer to discussion under account of *P. fallax*.

Etymology. — The Latin name *marginatus*, meaning to enclose with a border, alludes to the dark marginal stripe of the dorsal fin.

Distribution. — The only capture locality given for the holotype is "China." The specimen was probably collected during the voyage of H.M.S. SAMARANG. In the preface of the work treating the natural history of the voyage, Adams (1850) mentions a number of places visited by the vessel in the South China Sea.

Material examined. — Holotype, only known specimen.

Holotype. — BMNH 1848.3.16.151, female, 39.2 mm, "China"; Captain Sir Edward Belcher.

Petroscirtes (*Dasson*) *variabilis* Cantor
Specimen Figures 24a-c, 114-116; Tables 4-7, 10.

Petroscirtes variabilis Cantor, 1850: 1182 (type locality: Sea of Pinang, Singapore; neotype SU 67355).

Petroskirtes bankanensis Bleeker, 1852b: 727 (type locality: Bangka Island, near Koba; holotype RMNH 26669).

Petroskirtes amboinensis Bleeker, 1853a: 114 (type locality: Amboina; syntypes lost ?).

? *Petroskirtes solorensis* Bleeker, 1853c: 81 (type locality: Sea of Lawajong [= Solor, Indonesia]; syntypes lost ?).

Salarias viperidens De Vis, 1884: 697 (type locality: Somerset, Queensland, Australia; lectotype QMB I.11/104).

Petroscirtes eretes Jordan and Seale, 1905: 801, Fig. 19 (type locality: Negros, Philippine Islands; holotype USNM 51949).

Petroscirtes vulsus Jordan and Seale, 1907: 47, Fig. 20 (type locality: Manila; holotype SU 9253).

Description. — Intercalar present. Dorsal fin X-XI, 16-19; total elements 27-30. Anal fin II, 16-19 (rarely 19). Pectoral fin 13-15. Vertebrae: precaudal 11-12; caudal 20-23; total 32-35. Pleural ribs on vertebrae 3-12 or 13; epipleural ribs 12-14. Incisor teeth (Fig. 14) in specimens 23-105 mm, n = 86: lower jaw 24-48; upper jaw 24-45.

Symphyseal mandibular cirrus simple; other cirri absent except as follows: orbital and postorbital cirri present or absent, typically present; nuchal cirri present or absent, usually present; posttemporal cirrus present. Posterior nostril usually with short flap on anterior rim. Supratemporal pores 3-5, typically 5. Depth at anal-fin origin 14.8-21.5% SL. First dorsal-fin spine not elongate, shorter than second and fourth spines.

Sexual dimorphism slight in second dorsal-fin spine, pelvic and caudal fins in specimens > 30 mm SL. Second dorsal-fin spine: males 7.5-16.1% SL, females 8.1-11.9% SL; pelvic fin: males 10.7-21.1% SL, females 11.8-18.5% SL; caudal fin: males 19.6-27.1% SL, females 20.2-24.1% SL. Outer lobes of caudal fin not elongate. Anterior margin of first dorsal-fin spine without a fleshy flap.

Color pattern. — Body with 5 broad brownish or dusky bands on sides, usually indistinct and occasionally conjoined to form a stripe; head and body mottled with brown, speckled with small brownish dots and pale spots. Rays of dorsal and anal fins barred or spotted with brown, spots frequently form-

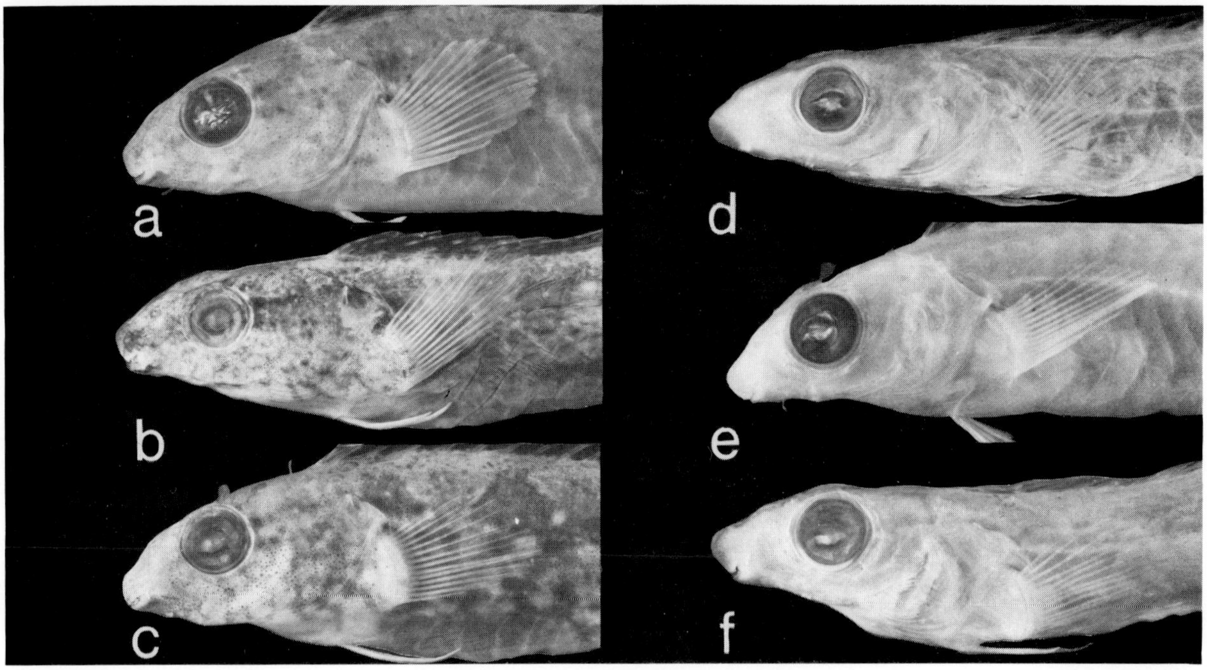

Figure 24. — Comparison of snout physiognomy in two closely related species of *Petroscirtes:* left *variabilis* (a) ANSP 126030, male, 46.9 mm, Philippines, Oriental Negros (b) USNM 205679, female, 54.8 mm, New Guinea, Kapa Kapa (c) ANSP 109865, male, 43.0 mm, Australia, Queensland; right *thepassi* (d) CAS 24697, male, 58.7 mm, Palau Islands (e) CAS 29014, female, 45.9 mm, Palau Is. (f) ANSP 126024, male, 45.1 mm, Palau Is.

ing diagonal rows posteriorly; the interradial membranes of the first 4-5 spines of the dorsal fin are the most heavily pigmented and often there is a small dark spot distally between the first two spines. The proximal region of the dorsal fin is not as heavily pigmented as in *breviceps.* The caudal fin is faintly barred; a dusky band, typically present at the base, extends slightly onto the fin centrally. Good illustrations of *variabilis* are given in Jordan and Seale (1905: 802) and Whitley (1932: 299).

According to field notes taken by Tyson R. Roberts, the life colors of Sri Lanka specimens were: "sea-grass green above, lighter below. Belly bright greenish yellow. Tiny sky blue flecks on posterodorsal half of body; white flecks somewhat larger below midline. Median fins with dull seaweed red striations." Jordan and Richardson (1908: 283) also stated that in life *variabilis* is mottled with white and dark green, and with numerous points of pale blue. See also color description in discussion of nomenclature below.

Geographic variation. — The Sri Lanka population of *P. variabilis* exhibited the highest average numbers of total vertebrae, dorsal-fin elements and segmented anal-fin rays. Populational variation in frequencies for these meristic characters is shown in Table 10. See also comments under species comparisons.

Species comparisons. — *P. variabilis* seems to be most closely related to *thepassi;* both have virtually identical color patterns, at least in preservation. The second dorsal-fin spine is usually longer (Fig. 25) and the body deeper (Fig. 26) in *variabilis,* but the character most useful in distinguishing these two species is the shape of the snout (Fig. 24). There is considerable overlap between the two species in snout/orbit measurements expressed as a percentage of SL or head length, but the elongate slightly overhung snout of *thepassi* is unmistakable in its "rodent-like" appearance. The frequency of specimens with a single pore on one or both sides of the mid-dorsal

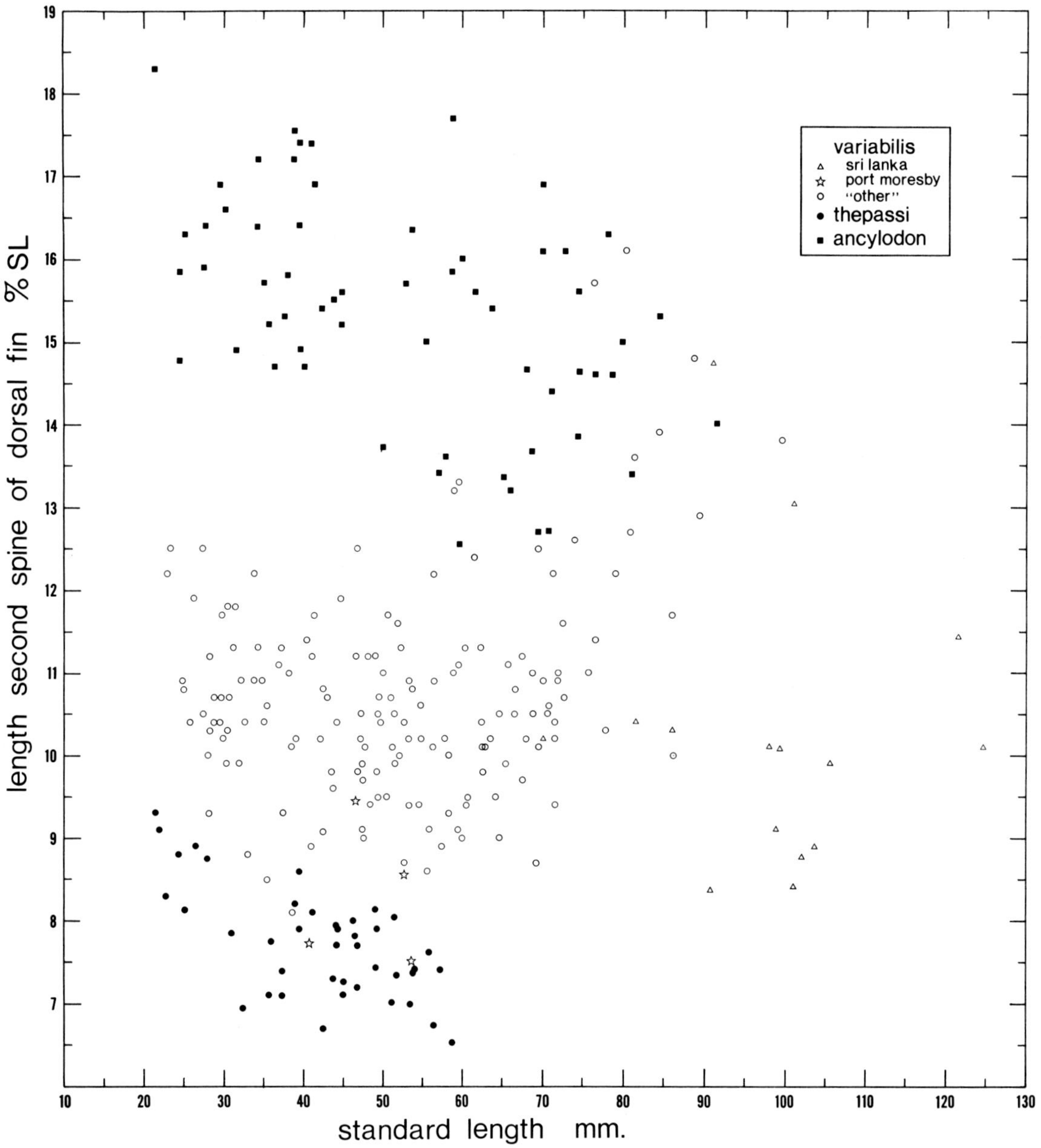

Figure 25. — Relationship between length of second spine of dorsal fin versus standard length in *Petroscirtes variabilis, thepassi* and *ancylodon.* (All *specimens* of *P. variabilis* with spines longer than 12.0 percent SL are males.)

ST pore is also much greater in *thepassi* (see Table 7).

The seven specimens of *variabilis* from Palau (see distributional comments) do not closely approach *thepassi* in their morphometric characters, but the four specimens from Port Moresby, New Guinea and a single questionable specimen, identified as *thepassi,* from the Trobriand Islands suggest the possibility of intergradation. This problem deserves additional consideration when more specimens from areas of sympatry become available.

Nomenclature. — The name *variabilis,* widely

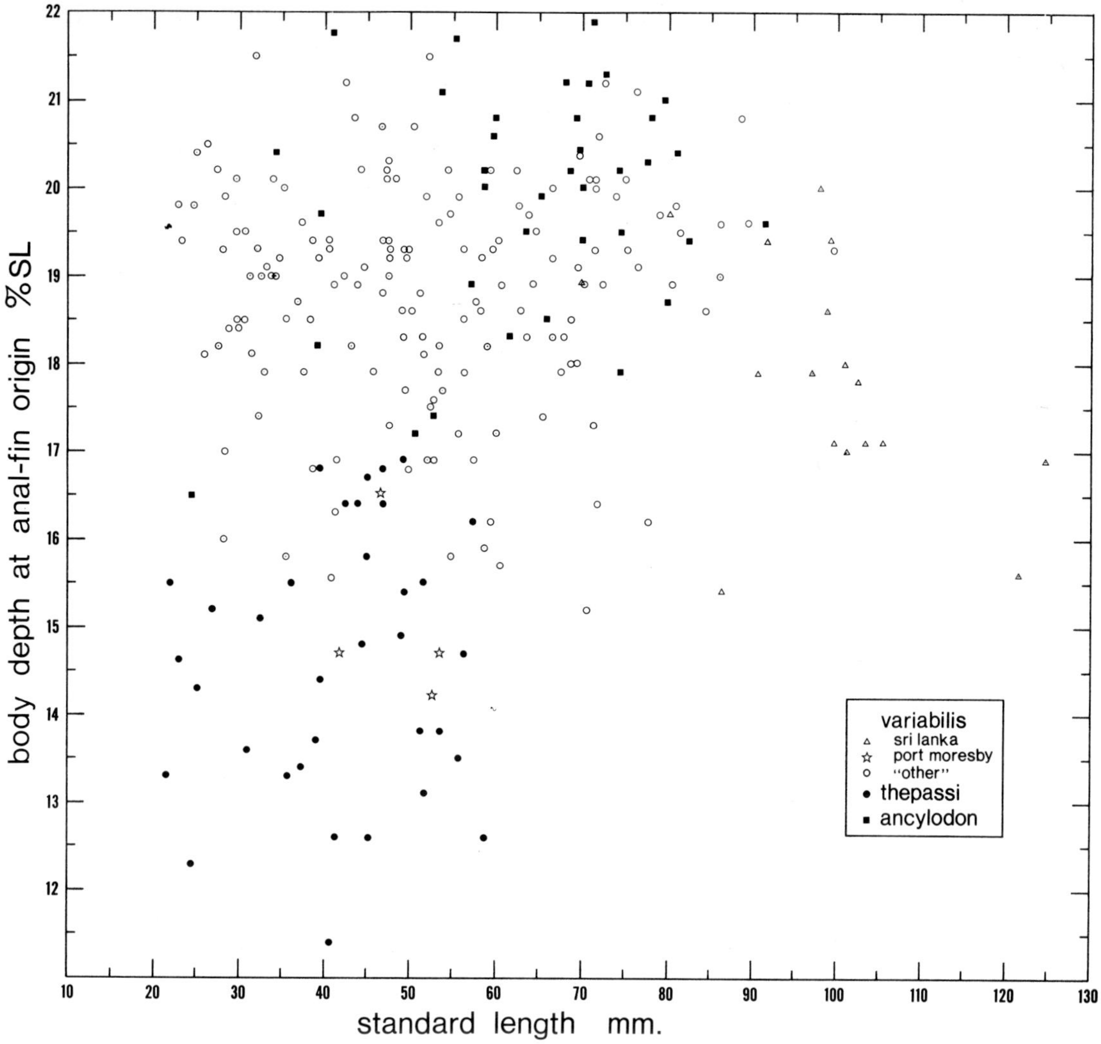

Figure 26. — Relationship between body depth at anal-fin origin versus standard length in *Petroscirtes variabilis, thepassi* and *ancylodon.*

used in the literature, has been applied to at least four species of *Petroscirtes.* The description, however, could apply to only two species that occur in the Indo-Malayan region — *breviceps* and the species to which the name is here applied. Both species have been taken in the same collection at Singapore, the type locality of *variabilis.* The color patterns of each are highly variable and until the diagnostic characters have been pointed out they are difficult to distinguish. Several museum collections, supposedly sorted to species, have contained both species. The count of 17 to 21 for the anal fin given in the original description suggests that Cantor also may have confused these two species (see Table 7). Cantor's description is otherwise remarkably thorough and detailed for his time.

Cantor's description of the caudal fin, quoted below, certainly does not apply to *breviceps* (which has the fin truncate in juveniles, with the outer lobes produced in adults). "The caudal equals the length of the head; when spread out, the posterior margin is rounded with the points of the rays exposed, but ordinarily the fin appears lanceolate." Likewise, the following complete quotation of the coloration

TABLE 10. — Geographic variation in frequencies of certain meristic characters in the *Petroscirtes variabilis* species group as indicated by specimens from selected localities.

	Total vertebrae							Total dorsal-fin elements							Segmented anal-fin rays					
Species	32	33	34	35	36	N	x̄	27	28	29	30	31	N	x̄	16	17	18	19	N	x̄
ancylodon																				
Red Sea			7	45	1	53	34.9			18	35		53	29.7			27	25	52	18.5
Persian Gulf			9	11	2	22	34.7		1	25	6	1	33	29.2		3	26	4	33	18.0
variabilis																				
Sri Lanka			17	2		19	34.1		1	15	3		19	29.1		4	13	2	19	17.9
Gulf of Thailand	1	20	5			26	33.2	3	17	6			26	28.1	5	20	1		26	16.8
Indonesia	1	21	4			26	33.1	1	21	5			27	28.1	3	21	3		27	17.0
Borneo	1	13	2			16	33.1		14	2			16	28.1	3	12	1		16	16.9
Philippines	6	75	12			93	33.1	4	70	22			96	28.2	7	72	17		96	17.1
Banda Sea		2	1			3	33.3		3				3	28.0		3			3	17.0
New Guinea		5	1			6	33.2		6				6	28.0		6			6	17.0
E. Australia		7	2			9	33.2		5	9			14	28.6	1	10	3		14	17.1
thepassi																				
Palau-Yap Is.	6	25	6			37	33.0	7	27	3			37	27.9	4	30	3		37	17.0
New Guinea	1	1				2	32.5	1	1				2	27.5	2				2	16.0
Solomon Is.	2					2	32.0	3					3	27.0	3				3	16.0

applies best to *variabilis* (especially diagnostic statements are given in italics):

"Head above and back light brownish olive; sides of the head and body yellowish; abdomen greenish yellow; from the orbit to the caudal a longitudinal blackish band, above which the head and body minutely dotted and clouded with black; *dorsal yellowish olive, dotted with black, rays with more or less distinct blackish spots forming two or three longitudinal series; in some individuals a black spot between the points of the first and second ray;* caudal yellow, base minutely dotted with black and in some with a central black spot; *anal yellow minutely dotted with black, in some with black spots on the rays forming two or three longitudinal series;* ventrals yellow; pectorals transparent. Iris bright yellow clouded with blackish; orbit margin olive."

The only statement in the above color description that might suggest *breviceps* is the blackish longitudinal band (stripe) from the orbit to caudal. Most preserved specimens of *variabilis* do not exhibit this feature. However, such a stripe is apparent in the specimen illustrated in Figure 114, and in Bleeker's unpublished color illustration of *Petroscirtes bankanensis* in the Rijksmuseum in Leiden. As noted by Jordan and Richardson (1908: 283), the lateral stripe apparently disappears readily. The character most useful in recognizing *breviceps,* the presence of a broad dark stripe along the entire base of the dorsal fin, is not mentioned in Cantor's color description.

The types of *variabilis* were stated by Günther (1861: 234) to consist of two dried skins in the British Museum. However, neither I nor Peter J. Whitehead (*in litt.*) have been able to locate these specimens, and they are presumed lost. Because the name *variabilis* has frequently been misapplied and the species is relatively common, it seems desirable to fix the name by designating a neotype. Accordingly, a 65.8 mm specimen (SU 67355) from Singapore, the type locality, is here designated as the neotype of *P. variabilis.* This specimen is illustrated in Figure 115.

I have examined the primary type specimens upon which the descriptions of *bankanensis, viperidens, eretes* and *vulsus* were based. None of these nominal species was compared with any other species in their original descriptions, and the descriptions and types apply unequivocally to *P. variabilis.* Jordan and Richardson (1908: 283) placed *vulsus* in the synonymy of *eretes* and questionably suggested that both might be synonyms of *variabilis.*

The types of *amboinensis* and *solorensis* (both named by Bleeker) are not recognizable among the Bleeker specimens that I have examined and may be lost. Bleeker (1857b: 215) synonymized *amboinensis* with *bankanensis* (a junior synonym of *variabilis*) which he had previously described. Because Bleeker frequently mixed later material with syntypical series (Whitehead *et al.*, 1966: 14), it is possible that the two syntypes of *amboinensis* are in the jar with the 19 specimens from which the holotype of *bankanensis* was segregated.

In the original description *P. solorensis* was compared only with *Meiacanthus grammistes,* but the color illustration of the species included among the unpublished plates of Bleeker's atlas of fishes (here reproduced as Fig. 109) suggests a species of *Petroscirtes* rather than a *Meiacanthus.* Günther (1861: 236) also was of the opinion that *solorensis* was more closely allied to a species of *Petroscirtes* rather than to *M. grammistes.* Bleeker's count of 20 anal-fin rays is too high for most species of *Meiacanthus* and excludes any that resemble Bleeker's illustration. His tooth count (14 premaxillary and 18 dentary, based on two specimens 27 and 30 mm total length) does suggest a species of *Meiacanthus,* but some of the teeth could easily have been overlooked. *P. variabilis* and *breviceps* are the only species of *Petroscirtes* known to occur in the general region of the type locality. In life the juveniles of both species sometimes have markings on the body that might resemble the two rows of spots shown on the illustration. *P. solorensis* is here tentatively regarded as a synonym of *variabilis* because the fin pigmentation depicted on the illustration suggests that species rather than *breviceps;* however if the name should later prove to have been based on *breviceps* or even *Meiacanthus grammistes* it would still remain in synonymy.

Etymology. — From the Latin *variabilis,* meaning variable. In view of the large synonymy and confusion in the literature, the specific name seems most appropriate for this species.

Distribution (Fig. 17). — Widely distributed in the western Pacific Ocean, but unknown from western Australia or the western Indian Ocean, where records of the species are based on *breviceps.* The two collections of *variabilis* from Palau, where it apparently occurs sympatrically with its close relative *thepassi,* are based on a single collection (now split between two museums) made by Herre on 18 October 1933.

Material examined. — 239 specimens (11.6-124.7 mm SL) from 72 collections.

SRI LANKA (CEYLON): MCZ 46982 (10:80-103), USNM 207578 (6:70-122), USNM 207579 (2:100-125), all from Jaffna vicinity.

INDIA: FMNH uncat. (1:26), sta. LW-3, Gulf of Mannar, Hare I., (Musal tivu) across channel from Mandapam fish camp.

INDONESIA: RMNH 26669 (99.6, holotype of *Petroskirtes bankanensis*), Sumatra, Bangka I., near Koba; RMNH 4776 (19:39-89), originally in same jar with holotype; RMNH 6294 (4:42-75), Bleeker collection, locality unknown; UMMZ 187434 (1:32), Java, Batavia Group, Hoarn I.; ZMA 109.335 (4:36-52), "Java Sea"; USNM 212003 (3:49-62), Seribu I., 5°51′S, 106°34′E; ZMA 109.343 (1:27), Kangean I., 6°59′S, 115°24′E; ZMA 109.344 (1:30), Sumbawa I., Bimba, 8°24′S, 118°44′E; ZMA 109.345 (1:17), Roti I., S of Timor; USNM 211899 (1:12), Kabaena I., 5°17′20″S, 122°4′E; USNM 214395 (14:19-37), Borneo, Darvel Bay; USNM uncat. (2:29-30), Borneo, 6°33′N, 115°51′E, Te Vega cr. 6, sta. 206; RMNH 20239 (1:14), Borneo, 2°0′N, 118°41.5′E; SU 27834 (1:86), USNM 138275 (2:44-58), Borneo, Sandakan Bay; RMNH 1840 (1:58), Seram (Ceram); NMV 71763 (1:78), NMV 71764 (1:72), Ambon I. (Amboina).

MALAYSIA: RMNH 17928 (1:79), Port Dickson; SU 67355 (65.8, neotype by present designation of *Petroscirtes variabilis*), Singapore, 12-15 March 1934, W. C. T. Herre; SU 30661 (6:47-59), same data as neotype; SU 32727 (2:59-70), NMV 71803 (1:70), Singapore.

GULF OF THAILAND: NMV 71795 (1:25), "Thailand"; CAS 29019 (7:23-35), CAS 29020 (1:47), CAS 29025 (3:52-75), CAS 29026 (1:43).

PHILIPPINE IS.: ANSP 48736 (2:45-50), CAS 29021 (1:40), CAS 29024 (7:23-60), "Philippines"; USNM 138274 (1:64), Pilas I., S of Zamboanga, 6°35′N, 121°38′E; USNM 138268 (2:44-48), Mindanao, Capunuy pugan; USNM 138257 (7:53-72), Pandanon I., between Cebu and Bohol; USNM 138249 (2:51-61), USNM 138250 (15:41-69), USNM 138251 (15:34-75), USNM 138252 (4:58-68), USNM 138253 (1:86), USNM 138254 (3:64-72), USNM 138255 (1:56), USNM 138256 (1:52), Cebu Mkt.; USNM 51949 (62.4, holotype of Petro*scirtes eretes*), Negros I.; USNM 210617 (2:31-62), SU 9132 (2:38-80), taken with holotype; USNM 138259 (6:30-56), Negros, Guijulugan; ANSP 126030 (1:47), CAS 29022 (4:28-33), CAS 29023 (2:31-35), SU 28450 (1:54), Negros, Dumaguete; USNM 122594 (2:28-32), USNM 148590 (18:38-71), Panay I., Iloilo; ZSZM 1492 (1:63), SU 9253 (46.3, holotype of *Petroscirtes vulsus*), Manila.

TAIWAN: NMV 71776 (1:74), Takao (Kaohsiung).

PALAU IS.: FMNH 47027 (3:45-67), SU 29194 (4:65-72).

NEW GUINEA: USNM 205679 (1:55), Kapa-Kapa, ca.

80 km SE Port Moresby; USNM 210619 (1:53), DASF FO1927 (3:41-53), Port Moresby, Fairfax Harbor; DASF FO1925 (1:42), Yule I., 8°45′S, 146°25′E.

E. AUSTRALIA: QMB I.1934 (2:14-25), Torres Strait, Raine I.; QMB I.11/104 (46.8, lectotype by present designation of *Salarias viperidens*), (3:33-46, in same jar with lectotype), AMS I.423 (3:42-48), Queensland, Cape York, Somerset; ANSP 109684 (1:38), ANSP 109685 (2:35-43), Queensland, Little Hope I.; AMS IA.4447 (1:49), AMS IA. 4896 (1:46), Low Is., Batt Reef, 16°18′S, 145°35′E; QMB I.4406 (1:64), Bowen; AMS IA.1792 (1:55), AMS IA.1794 (1:67), AMS IA.1795 (2:29-40), Port Denison, 20°0′S, 148°10′E; AMS IA.6622 (1:54), Hayman I., 20°2′S, 148°53′E.

Petroscirtes (Dasson) thepassi Bleeker
Specimen Figures 24d-f, 117-118; Tables 4-7, 10.

Petroskirtes thepassi Bleeker, 1853b: 136 (type locality: Ternate; holotype RMNH 4661).

Description. — Intercalar present. Dorsal fin X-XI (rarely X), 16-18; total elements 27-29. Anal fin II, 16-18. Pectoral fin 13-15. Vertebrae: precaudal 11-12; caudal 20-22; total 32-34. Pleural ribs on vertebrae 3-12; epipleural ribs 12-14. Incisor teeth in specimens 22-59 mm, n = 24: lower jaw 24-41; upper jaw 22-36.

Symphyseal mandibular cirrus simple; other cirri absent except as follows: orbital and postorbital cirri typically present; nuchal cirri present or absent, frequently absent; posttemporal cirrus present. Posterior nostril usually with a short flap on anterior rim. Supratemporal pores 3-5, highly variable. Depth at anal-fin origin 12.3-16.9% SL. First dorsal-fin spine not elongate, shorter than second and fourth spines.

Sexual dimorphism not apparent in specimens > 30 mm SL. Second dorsal-fin spine: males 6.6-8.6% SL, females 6.6-8.2% SL; pelvic fin: males 12.4-14.7% SL, females 12.4-14.5% SL; caudal fin: males 18.4-21.8% SL, females 19.2-22.5% SL. Outer lobes of caudal fin not elongate. Anterior margin of first dorsal-fin spine without fleshy flap.

Color pattern. — In preserved specimens, the color pattern of *thepassi* is virtually identical to that of *variabilis,* except that the small dark spot frequently present between the first two dorsal-fin spines is absent or indistinct. The life coloration, based on a transparency of a freshly collected specimen from Palau, is as follows: head and body brownish-olive, superimposed with small dark brown dots and pale spots; the belly is pale. The dorsal and anal fins are similarly mottled and spotted but there is some chestnut-salmon coloration proximally and the tips of the anal-fin rays are white. The base of the caudal fin is also salmon colored and the iris is dark yellow.

Species comparisons. — Refer to account of *P. variabilis*.

Nomenclature. — The holotype of *thepassi* is in poor condition, but in the original description the acute snout was specifically mentioned and Bleeker's unpublished illustration (here reproduced as Fig. 118) is reasonably diagnostic for the species. The notched dorsal fin, the result of mechanical injury, caused de Beaufort (1951: 367), who obviously had not seen the holotype, to include the name in the synonymy of *mitratus*.

Etymology. — Named for A. H. Thepass.

Distribution (Fig. 17). — Known only from the eastern periphery of the Indo-Australian Archipelago.

Material examined. — 43 specimens (21.5-58.7 mm SL) from 17 collections.

YAP IS.: CAS 29016 (1:37), CAS 29017 (1:36), CAS 29018 (1:23).

PALAU IS.: ANSP 126024 (2:22-45), BPBM 9550 (9: 26-56), BPBM 9600 (1:41), CAS 24697 (1:59), CAS 29013 (1:25), CAS 29014 (8:32-52), CAS 29015 (1:27), USNM 165710 (3:44-56), USNM 205296 (8:36-57).

INDONESIA: RMNH 4661 (40.7, holotype of *Petroskirtes thepassi*), Ternate.

NORTHERN NEW GUINEA: USNM 210618 (1:54), Sepik River area, Murik Lakes, Darapap; USNM 205678 (1:37), Trobriand Is., Kiriwina Boli Point.

SOLOMON IS.: USNM 123479 (2:22-24), Tulagi I.; USNM 123443 (1:31), Florida I.

Petroscirtes (Dasson) ancylodon Rüppell
Specimen Figures 119-120; Tables 5-7, 10.

Petroscirtes ancylodon Rüppell, 1838:1, pl. 1, Fig. 1 (type locality: Massawa, Red Sea; holotype NFIS 1818).

Description. — Intercalar present. Dorsal fin X-XII (typically XI), 17-19; total elements 28-31. Anal fin II, 17-19. Pectoral fin 13-15. Vertebrae: precaudal 11-12; caudal 22-24; total 34-36. Pleural ribs on vertebrae 3-12 or 13; epipleural ribs 13-15. Incisor teeth in specimens 17-91.5 mm, n = 15: lower jaw 18-41; upper jaw 18-40.

Symphyseal mandibular cirrus simple; other cirri absent except as follows: orbital, postorbital and nuchal cirri typically present; posttemporal cirrus present. Posterior nostril a simple hole, without flap on anterior rim. Supratemporal pores 3-5, usually 5. Depth at anal-fin origin 17.2-21.9% SL. First dorsal-fin spine not elongate, shorter than second and fourth spines.

Sexual dimorphism absent in second dorsal-fin spine, slight in pelvic and caudal fins in specimens > 30 mm SL. Second dorsal-fin spine: males 13.3-17.7% SL, females 12.6-17.7% SL; pelvic fin: males 14.0-20.9% SL, females 12.2-16.1% SL; caudal fin: males 21.7-27.5% SL, females 21.2-24.0% SL. Outer lobes of caudal fin not elongate. Anterior margin of first dorsal-fin spine sometimes with a fleshy flap (distally) in large adult males only.

Color pattern. — The color of *P. ancylodon* is difficult to characterize because of its complex and variable nature. The range of variation is shown in Figures 119-120. A good drawing of a 27 mm juvenile is given in Smith (1959: 233). Although difficult to describe, the pigmentation of the dorsal fin is relatively consistent and distinguishes *ancylodon* from other species of *Petroscirtes* except *lupus*. At least in preservation, the color pattern of juvenile *lupus* and some specimens of *ancylodon* is remarkably similar.

Smith (1959: 233) described the life colors as follows: "brownish, 6 irregular darker crossbars and spots, oval white spots on belly. D & A brownish, marbled, D edge with white spots. Cd with light margin."

Geographic variation. — Specimens from the Red Sea tend to have more dorsal- and anal-fin rays and total vertebrae than Persian Gulf specimens (see Table 10).

Species comparisons. — *P. lupus* and *variabilis* are the only species with which *ancylodon* might be confused. The frequency of specimens with a single pore on one or both sides of the mid-dorsal ST pore is much greater in *lupus* (see Table 7), and unlike *ancylodon* the nuchal cirri are frequently absent in adults. In large males of *lupus* (> 60 mm) the length of the second dorsal-fin spine is also longer than in *ancylodon.*

In addition to differences in color pattern, especially of the dorsal fin, between *ancylodon* and *variabilis,* the length of the second dorsal-fin spine is usually longer in *ancylodon* (Fig. 25), and *variabilis* usually has a short flap on the anterior rim of the posterior nostril (absent or minute in *ancylodon*).

Etymology. — The specific name is a combination of the Greek *ankylos* (bent or crooked) and *odon* (tooth).

Most authors cite 1835 as the date of publication for *Petroscirtes ancylodon.* However, Sawyer (1952) established the date of publication for that portion of Rüppell's work treating blenniids as 1838.

Distribution (Fig. 17). — Known only from the Red Sea, Gulf of Aden and the Persian Gulf.

Material examined. — 90 specimens (12.2-91.5 mm SL) from 20 collections.

PERSIAN GULF: USNM 147973 (20 of 100+:12-64), USNM 147974 (6:16-57), USNM 177975 (2:70-75), USNM 196489 (4:40-56), USNM 196490 (4:60-81), Tarut Bay.

GULF OF SUEZ: ZSZM 17010 (4:77-92), "Gulf of Suez"; BMNH 1915.10.25.8 (1:71), Suez; HUJ SLR2126 (3:42-57), El Tur, Sinai Peninsula; MNHN 99-286 (1:71), "Egypt," probably Gulf of Suez.

GULF OF AQABA: HUJ E57/281.2 (1:24), HUJ E62/451 (2:24-40), HUJ E62/649b (16 of 82:13-45), HUJ E62/1218b (1:45), HUJ E62/2001 (1:18), HUJ F3707 (1:40).

NORTHERN RED SEA: USNM 167027 (18:53-90), Hurghada, Egypt.

SOUTHERN RED SEA: NFIS 1818 (42.0, holotype of *Petroscirtes ancylodon*), Massawa; RMNH uncat. (2:27-44), off Ethiopia, 17°30′N, 40°39′E; USNM uncat. (1:51), off Ethiopia, 16°56′N, 40°58′E.

GULF OF ADEN: BMNH 1939.5.24.1436 (1:21), 12°13′42″N, 50°54′18″E.

Petroscirtes (*Dasson*) *lupus* De Vis
Specimen Figures 121-122; Tables 4-7.

Salarius lupus De Vis, 1886: 58 (type locality: Moreton Bay, Queensland, Australia; holotype QMB I.102).

Petroscirtes icelii Ogilby, 1894: 370, pl. 24, Fig. 3 (type locality: Lord Howe Island; holotype lost ?).

Petroscirtes (*Ostreoblennius*) *steadi* Whitley, 1930: 20 (type locality: Port Jackson, N. S. W., Australia; holotype AMS IA.4509).

Description. — Intercalar present. Dorsal fin X-XI, 19-21; total elements 30-32. Anal fin II, 18-20. Pectoral fin 13-15. Vertebrae: precaudal 11-12; caudal 23-25; total 35-36. Pleural ribs on vertebrae 3-12 or 13; epipleural ribs 12-15. Incisor teeth

(Fig. 14) in specimens 28-110 mm, n = 30: lower jaw 24-44; upper jaw 24-42.

Symphyseal mandibular cirrus simple; other cirri absent except as follows: orbital cirri present or absent, usually present but often minute in adults; postorbital cirri present or absent, usually present; nuchal cirri present or absent, frequently absent; posttemporal cirrus present. Posterior nostril a simple hole, without flap on anterior rim. Supratemporal pores 3-5, typically 3. Depth at anal-fin origin 14.2-22.3% SL. First dorsal-fin spine not elongate, shorter than second (which may be noticeably elongate in adult males) and fourth spines.

Sexual dimorphism well developed in second dorsal-fin spine (Fig. 22), slight in pelvic and caudal fins in specimens > 30 mm SL. Second dorsal-fin spine: males 10.9-31.9% SL, females 10.9-18.7% SL; pelvic fin: males 10.7-20.5% SL, females 12.7-17.4% SL; caudal fin: males 20.6-31.7% SL, females 19.9-24.6% SL. Outer caudal-fin lobes slightly elongate in males > 60 mm only. Anterior margin of first dorsal-fin spine with fleshy flap (distally) in some large males.

Color pattern. — The color pattern of *P. lupus* is difficult to characterize because of its complex and variable nature. The dorsal fin is heavily pigmented anteriorly and has conspicuous white spots; although difficult to describe, the pigmentation of the dorsal fin is remarkably similar in *lupus* and *ancylodon.* In some juveniles and, less frequently, adults the bands or blotches on the sides are distinctly U-shaped, enclosing a pale area dorsally. The sides have numerous pale spots, those of the dorsal half of the body are blue in life. In juveniles and small adults the caudal fin is faintly banded; in adults the fin is uniformly pale except that the interradial membranes are dusky proximally. A good illustration of an adult is given in Waite (1904: pl. 24). Ogilby (1894: 371) recorded the life colors as follows:

> "Head horn colour, lips black, throat and body sandy; broad, irregular black band traverses the whole length of the body, rather nearer to the dorsal than to the ventral surface; above the anal fin this band is crossed by five or six vertical bars, which extend more or less distinctly on to the dorsal and anal fins; a black spot at the base of the caudal fin. Incisor teeth in both jaws tipped with gold."

Species comparisons. — Color pattern, especially that of the dorsal fin, will distinguish *lupus* from all species of *Petroscirtes* with which it occurs sympatrically. See also comments under account of *P. ancylodon.*

Nomenclature. — The holotype of *P. icelii* could not be located, but two paratypes (AMS I.16591-2) were examined. *P. icelii* and *steadi* were compared only with *variabilis* in the original descriptions. The description of *steadi* is especially brief because Whitley stated his intention to figure and describe the species more fully at a later date (this he neglected to do).

Etymology. — The Latin specific name, *lupus,* means wolf.

Distribution (Fig. 16). — Known only from eastern Australia, New Caledonia and Lord Howe Island.

Material examined. — 131 specimens (11.9-110.2 mm SL) from 49 collections.

E. AUSTRALIA: AMS IA.1789-93 (4:66-79), Queensland, Port Denison; QMB I.4120 (8:28-59), Great Barrier Reef, Whitsunday Group; AMS IA.2200 (5:14-60), Kennedy Sound, between Shaw I. and Lindeman I.; AMS IA.6556 (1:71), AMS IA.6768 (1:54), Lindeman I.; BPBM 14334 (6:37-46), USNM 201881 (1:27), USNM 205643 (11:12-24), USNM 209413 (2:39-40), Great Barrier Reef, Capricorn Group, One Tree I.; AMS IB. 1277 (1:77), Queensland, Wide Bay; AMS IA.7987-8 (8:30-69), Queensland, Noosa River; QMB I.14/1704 (1:80), Queensland, Caloudra; QMB I.102 (54.3 holotype of *Salarius lupus*), AMS I.507 (1:68), AMS IB. 6347 (1:110), AMS IA.4939-40 (2:54-59), ANSP 98795 (1:84), MCZ 38543 (1:99), QMB I.4220-1 (2:66-107), QMB 3844-5 (2:69-82), QMB 4353 (1:105), USNM 209414 (1:62), Queensland, Moreton Bay; QMB I.7844 (1:47), Moreton I.; AMS IB.3418 (1:95), QMB I.4265 (1:93), North Stradbroke I.; AMS IA.7350-1 (2:52-62), QMB I.9814 (3:43-71), Queensland, South Port; AMS I.12645 (2:39-58), Queensland, Tweed River Heads; CSIRO A.2022 (1:88), New South Wales, Lake Macquarie; AMS IA.4509 (80.0, holotype of *Dasson steadi*), AMS IA.1546-7 (2:64-72), AMS IA.4147 (1:74), AMS I.16498-2 (1:53), N. S. W., Port Jackson; AMS IA.1834 (1:110), N. S. W., Woy Woy; AMS I.15965-1 (1:98), AMS I.16799-3 (2:27-56), AMS I.17662-1 (2:71-76), Sydney area.

NEW CALEDONIA: AMS IB.3453 (2:48-51), AMS IB. 3847 (1:62), ANSP 124405 (2:62-64), SU 47022 (16: 21-64), SU 47023 (9:27-62), all from Noumea.

LORD HOWE IS.: AMS I.6591-2 (2:71-77, paratypes of *Petroscirtes icelii*); AMS I.3123 (2:75-82), AMS IB.5708 (1:81), AMS I.5899 (1:89), AMS uncat. (1:88), sta. 73-

10, AMS uncat. (2:72-84), sta. 73-14, AMS uncat. (5:64-104), sta. 73-67, BPBM 14657 (3:99-107).

ASPIDONTUS Cuvier

Aspidontus Cuvier, *in* Quoy and Gaimard, 1834: 719 (type-species: *A. taeniatus* Quoy and Gaimard, 1834 by monotypy; the genus and species were described on the same page, but only the genus was credited to Cuvier).

Blennechis Valenciennes, *in* Cuvier and Valenciennes, 1836: 279 (type-species: *B. filamentosus* Valenciennes, 1836 [= *Aspidontus taeniatus* Quoy and Gaimard, 1834], by subsequent designation, in parentheses[1], of Jordan and Seale, 1906: 431).

Oncesthes Jordan and Hubbs, 1925: 319 (type-species: *Petroscirtes fluctuans* Weber, 1909 [=*Blennechis dussumieri* Valenciennes, 1836], by original designation).

Escadotus Smith, 1959: 235 [as a subgenus of *Aspidontus*] (type-species: *Petroscirtes fluctuans* Weber, 1909 [= *Blennechis dussumieri* Valenciennes, 1836], by original designation; a junior objective synonym of *Oncesthes* Jordan and Hubbs, 1925).

Diagnosis. — Major characters useful in distinguishing the genera of Nemophini are summarized in Tables 1, 3 and 4. *Aspidontus* is unique in having two distinct types of dentary canines (Fig. 29): strongly hooked in prejuveniles and "straight" in adults. It also differs from the other genera in usually having more incisors in the upper jaw (see Table 4). Superficially *Aspidontus* is most similar to *Petroscirtes* but differs most obviously in having more dorsal-fin elements, segmented anal-fin rays, and vertebrae, and a gill opening that is more vertically oriented and extends farther ventrally. The general physiognomy of the head of the two species of *Aspidontus* is shown in Figures 27-28.

Nomenclature. — Smith-Vaniz and Randall (1973) recently have shown that the type-species of *Aspidontus* and *Blennechis* are synonymous. *Blennechis* was based on the distinctive and generally unrecognized prejuveniles of *A. taeniatus*. *Oncesthes* and *Escadotus* were both based on *dussumieri,* the only other species of *Aspidontus*. Although there are important differences between the two species of *Aspidontus,* they agree in sharing at least one presumably apomorphic character — two distinct types of dentary canines during different stages of their ontogeny — and they are certainly no more distinct than some of the species united under *Meiacanthus* and *Plagiotremus*.

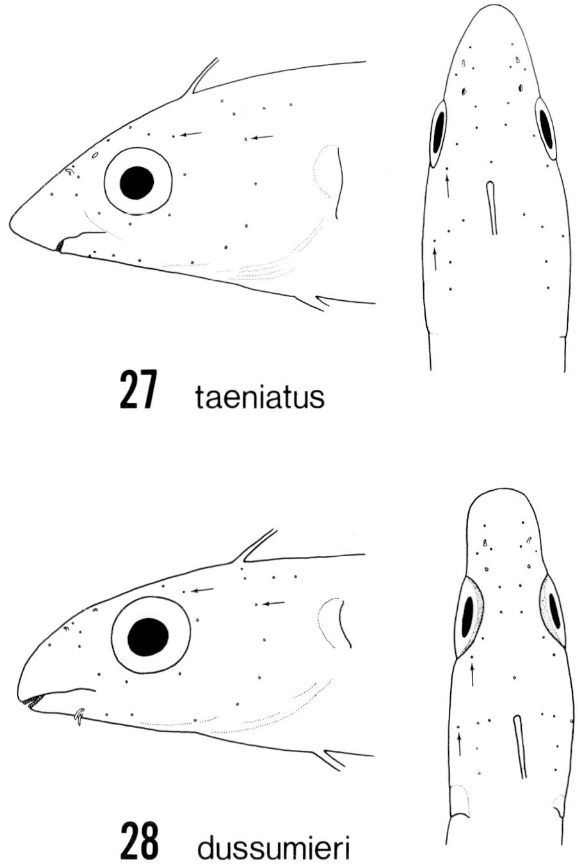

27 taeniatus

28 dussumieri

Figures 27-28. — General physiognomy and cephalic sensory pore distributions in species of *Aspidontus:* Fig. 27, *taeniatus,* ANSP 109672, male, 70.1 mm, Australia, Queensland; Fig. 28, *dussumieri,* USNM 201598, male, 64.1 mm, Marshall Islands, Kwajalein. Small arrows identify ventralmost supraorbital and dorsalmost preopercular pores in both lateral and dorsal views.

Etymology. — The generic name is derived from the Greek *aspis* (shield) and *odon* (tooth); gender masculine.

Biology. — *Aspidontus dussumieri* is a cryptic species that quickly backs into a worm tube when approached. *A. taeniatus,* a conspicuous species, is an aggressive mimic of *Labroides dimidiatus* (see general section on mimicry). Many experienced divers have failed to recognize the adults of *taeniatus* in the field despite their relative abundance because, when not in a worm tube, they are usually in the

[1] For a discussion of type-species designations in parentheses, see Hubbs *et al.* (1963).

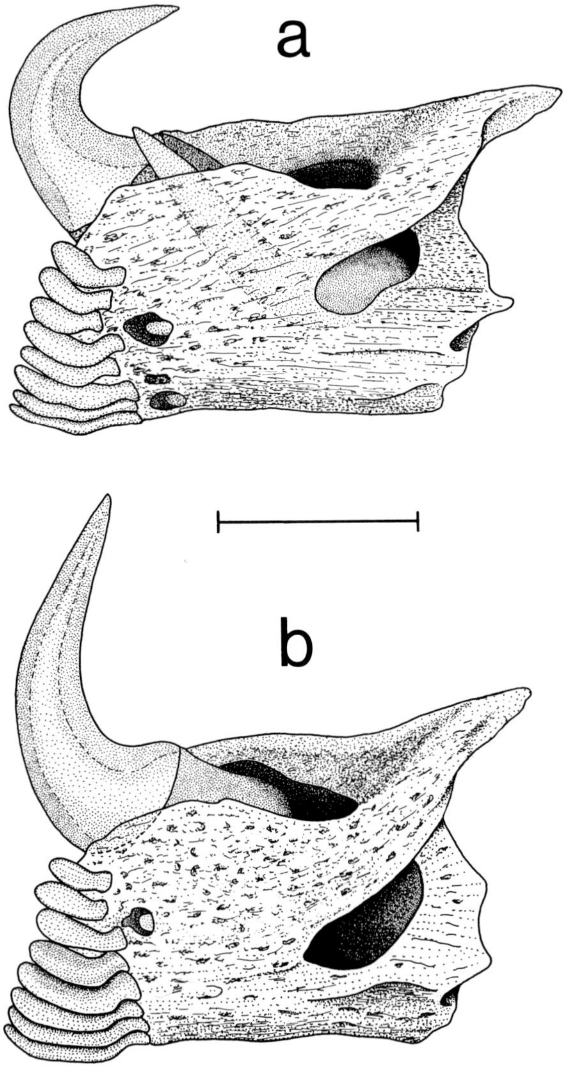

Figure 29. — Lateral view of left dentary of *Aspidontus taeniatus* showing canine of (a) prejuvenile, ANSP 122373 (note replacement canine immediately behind functional canine) and (b) recently metamorphosed juvenile, ANSP 122374. Both specimens 48.2 mm SL. Scale = 1.0 mm.

immediate vicinity of the ubiquitous cleaner wrasse which they mimic almost perfectly. When they are observed near a group of the wrasses they make no attempt to hide and are rather easy to collect with a dip net. The similarity of *A. taeniatus* to the cleaner wrasse enables it to approach naive prey (only young fishes are deceived) from which it bites off fin rays and associated membranes. Wickler (1963a) has filmed this aggressive feeding behavior. Unlike *Plagiotremus, A. taeniatus* usually does not take scales from its prey. The dentition (Fig. 69) is virtually identical in both species of *Aspidontus* and does not appear to be specialized. Immunity from predation may be the primary basis for the mimetic interaction between *A. taeniatus* and *Labroides,* with exploitation of an unusual food resource a secondary benefit. Other fishes that have been reported to feed on the fins of fishes are characoids (Roberts, 1970), and the blenniid *Haptogenys bipunctata* (Day) (Springer, 1972: 8, recorded as *H. quadripora*).

Randall and Randall (1960: 454) found fin rays and attached membranes in four of eight stomachs of *A. taeniatus* that they examined; sabellid worm tentacles and numerous elongate fish eggs (probably pomacentrid eggs) were also present in three stomachs respectively. The relatively few stomachs of *taeniatus* that I have examined also contained fish eggs. The feeding habits of *A. dussumieri* are obviously unlike those of its congener; it differs in having more teeth (Fig. 30) and a uniformly dark peritoneum (refer to section on evolutionary trends for discussion of significance of this character). The stomach contents of the three specimens of *dussumieri* that I examined were essentially identical to those of *Petroscirtes* (see p. 25); no fin rays or eggs were present.

No information is available on the reproductive biology of *Aspidontus.* The prejuveniles of both species are pelagic and apparently metamorphose very rapidly after they leave the plankton (Smith-Vaniz and Randall, 1973). At Christmas Island, Ashmole and Ashmole (1967: 38) found the pelagic prejuveniles of *taeniatus* (30 fish in 26 of 152 samples) to be a food item of the white tern, *Gygis alba.* Most prejuveniles have been collected in the vicinity of drifting "seaweed." The paucity of small specimens in collections (see Fig. 5) is a clear indication that the adults and juveniles occupy different habitats. At the Cocos-Keeling Islands prejuveniles of *dussumieri* were observed at the surface or just underneath a small boat on several occasions when it was anchored at a steep drop-off site in "blue water." The adults of both species have been observed or collected only from shallow reef habitats.

Zoogeography. — Both species of *Aspidontus* are distributed (Fig. 31) throughout much of the Indo-

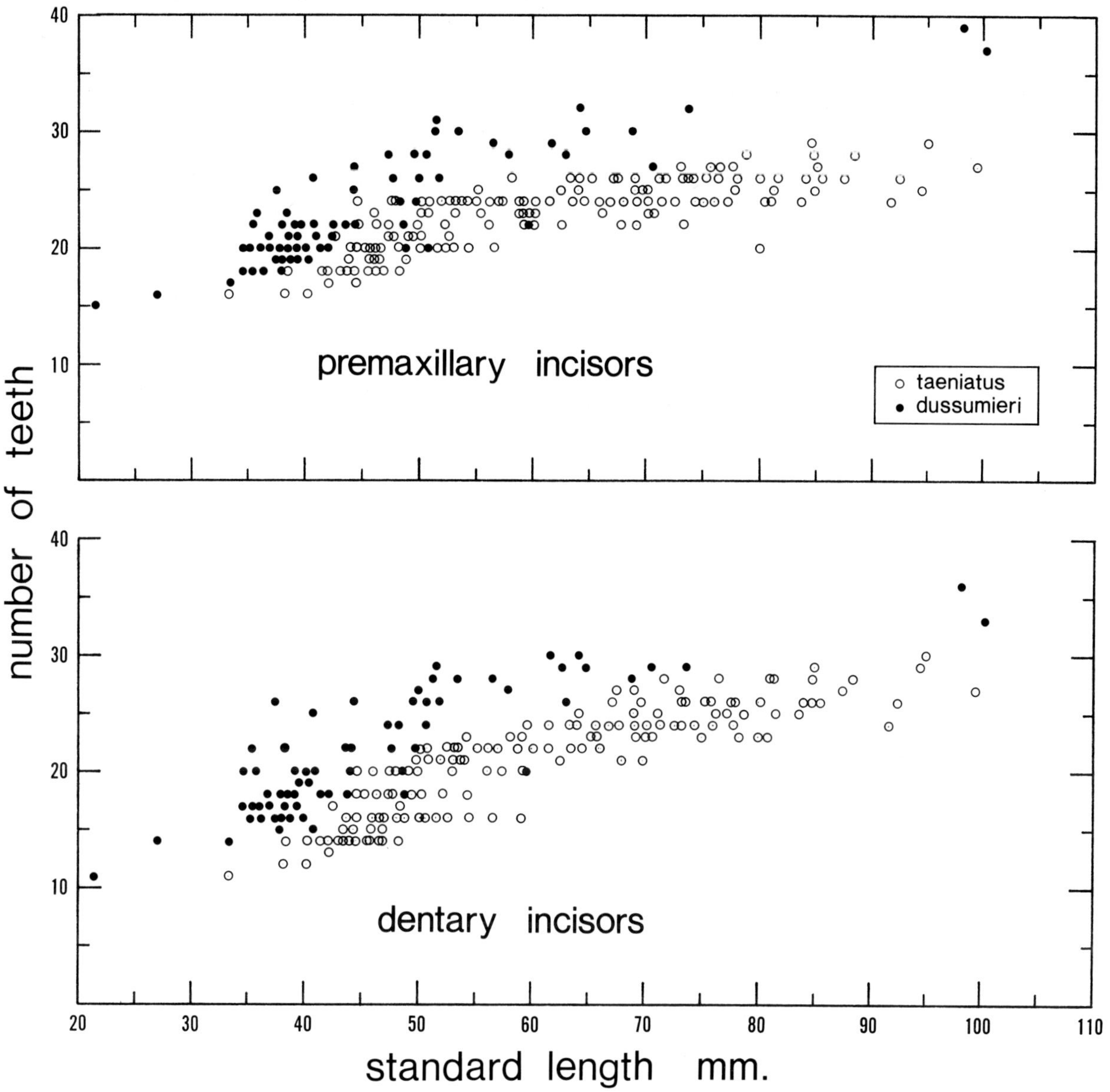

Figure 30. — Relationship between number of incisor teeth versus standard length in species of *Aspidontus.*

Pacific. The apparent adaptation of the early juveiles to a pelagic existence undoubtedly has facilitated their wide dispersal. The absence of specimens from the Persian Gulf-Arabian Sea only emphasizes the inadequacy of collections in that region.

Intrageneric relationships. — The two species of *Aspidontus* do not appear to be especially closely related. The lower number of dorsal-fin spines, synostosis of the lateral extrascapular with the parietal and the unforked posttemporal bones all point to *dussumieri* as the more osteologically specialized of the pair. *A. taeniatus* is more specialized than *dussumieri,* however, in its unusual feeding habits, fleshy conical snout (refer to discussion in section on evolutionary trends) and other characters correlated with its mimetic status.

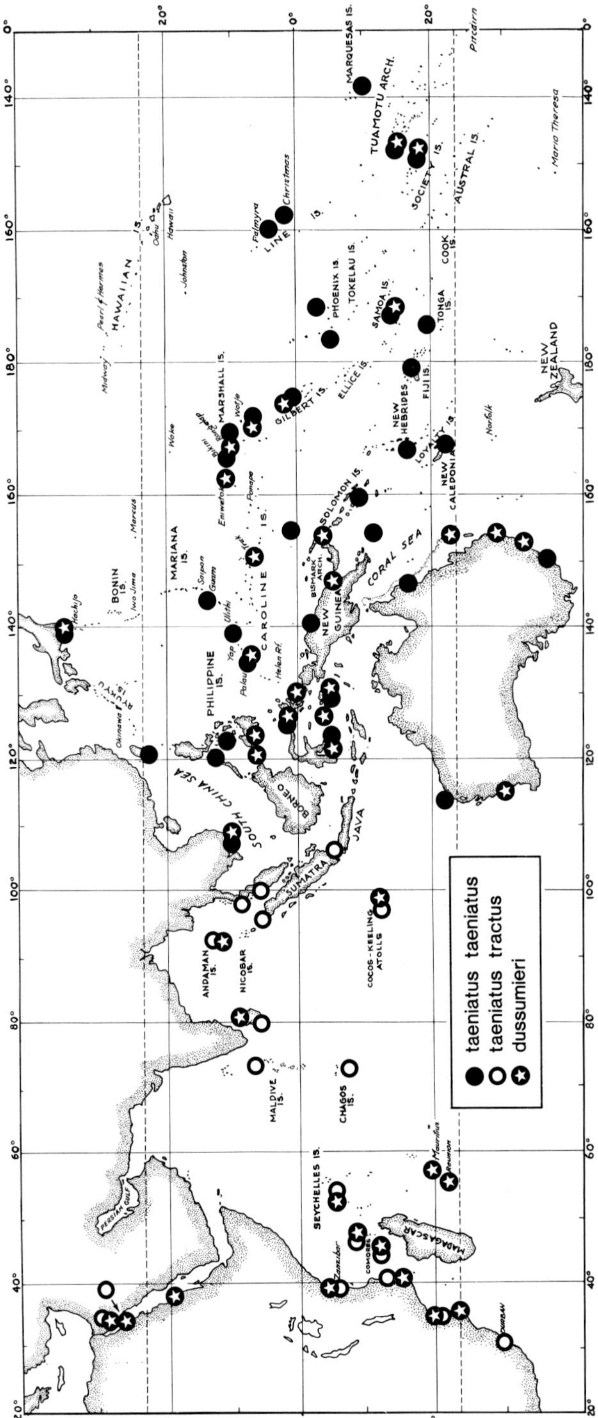

Figure 31.—Distributional records for species and subspecies of *Aspidontus.*

KEY TO SPECIES OF *ASPIDONTUS*

Lateral stripe, if present, widens posteriorly and extends to distal tip of median caudal-fin rays; first dorsal-fin spine frequently elongate in prejuveniles (Fig. 32); caudal fin of adults truncate, median rays not noticeably longer than outer rays; dorsal-fin spines 10-12, typically 11; total dorsal-fin elements 37-40 (rarely 40) .... *taeniatus*

Lateral stripe, if present, uniformly wide and only extends onto proximal 3rd of caudal fin; first dorsal-fin spine not elongate in prejuveniles; caudal fin of adults lanceolate, median rays longer than outer rays; dorsal-fin spines 9-11 (rarely 11); total dorsal-fin elements 38-44 .... *dussumieri*

Aspidontus taeniatus Quoy and Gaimard
Specimen Figures 27, 123-126; Tables 4, 11-13.

(For synonymy, see both subspecies)

Aspidontus taeniatus taeniatus Quoy and Gaimard

Aspidontus taeniatus Quoy and Gaimard, 1834: 719, pl. 19, Fig. 4 (type locality: Guam, "devant la ville d'Agagna"; holotype lost).

Blennechis filamentosus Valenciennes, *in* Cuvier and Valenciennes, 1836: 280, pl. 326 (type locality: Humboldt Bay, New Guinea; holotype MNHN A.2046).

Petroskirtes paradiseus Bleeker, 1854:495 (type locality: Amboina; holotype RMNH 26668).

Petroskirtes rostratus Bleeker, 1857a: 62 (type locality: Amboina; holotype RMNH 4453).

Petroscirtes azureus Jordan and Seale, 1906: 432, Fig. 109 (type locality: Apia, Samoa; holotype USNM 51796).

Description. — Posttemporal forked, with moderate ventral arm (except for one of 8 specimens which exhibited only a slight indication of forking). Dorsal fin X-XII, 26-29 (rarely 29); total elements 37-40. Anal fin II, 25-28. Pectoral fin 13-15. Caudal fin: procurrent rays 3-4 + 3-4 = 6-8 (typically 6); segmented rays 11. Vertebrae: precaudal 13-15; caudal 27-30; total 40-44. Pleural ribs on vertebrae 3-14 or 15; epipleural ribs 15-20. Incisor teeth (Fig. 30), in specimens 33-99.5 mm SL, n = 192: lower jaw 11-30; upper jaw 16-29. Each premaxilla with 0-2 posterior canines (canines usually absent in specimens larger than 55 mm SL).

Fleshy conical snout distinctly projects beyond tip of premaxillae. Caudal fin of adults truncate or nearly so. Anterior dorsal-fin spines noticeably elongate in prejuveniles (Figs. 32 and 123). Pelvic fin moderately elongate in specimens < 50 mm: 13.7-24.7% SL (Fig. 32); pelvic fin not sexually dimorphic, in specimens > 50 mm: males 8.0-19.2% SL, females 9.8-17.3% SL.

Color pattern. — Dark stripe from snout to caudal fin, widening posteriorly. Caudal fin mostly black, with broad pale dorsal and ventral margin. Occasional specimens exhibit some evidence of the black markings on the caudal fin forming a ventrally curving hook, with the fin pale centrally (as illustrated in Wickler, 1968: 161). Dorsal and anal fins dark, with narrow pale margin in some specimens. Pelvic and pectoral fins and remainder of body pale. In pelagic prejuveniles the dark mid-lateral stripe is not as distinct and the dorsal and anal fins are banded.

In life, adults with stripe brownish on head, black posteriorly; body grayish white anteriorly becoming pale blue posteriorly. Caudal fin black centrally with narrow blue border and white margin. Dorsal and anal fins brownish black with narrow white or hyaline margin, some specimens with fins pale blue proximally, especially posteriorly. Pelvic and pectoral fins pale. Monochrome photographs of preserved specimens of prejuveniles and adults of *taeniatus* are given in Springer and Smith-Vaniz (1972a: 36) and Smith-Vaniz and Randall (1973: 3). A good illustration of the prejuvenile is also given in Tomiyama (1956a), and excellent color photographs of the adult are given in Burgess and Axelrod (1974: 1333) and Masuda *et al.* (1975: 86).

Life color also may be strongly influenced by the coloration of certain wrasses, which *Aspidontus taeniatus* mimics. The blue of juveniles of *Aspidontus* is considerably deeper and more brilliant than the blue of adults; the change in color from young to adult is a gradual one. This color change closely duplicates that of *Labroides dimidiatus.* Refer to

TABLE 11.—Frequency distributions of certain meristic characters in species of *Aspidontus*.

Species	Dorsal-fin spines						Segmented dorsal-fin rays											Total dorsal-fin elements									
	9	10	11	12	N	x̄	26	27	28	29	30	31	32	33	34	N	x̄	37	38	39	40	41	42	43	44	N	x̄
taeniatus		18	160	2	180	10.9	17	92	66	5						180	27.3	18	100	60	2					180	38.3
dussumieri	16	73	5		94	9.9			1	5	32	40	12	3	1	94	30.7		2	6	33	42	7	3	1	94	40.6

Species	Precaudal vertebrae					Caudal vertebrae										Total vertebrae										
	13	14	15	N	x̄	27	28	29	30	31	32	33	34	N	x̄	40	41	42	43	44	45	46	47	48	N	x̄
taeniatus	2	137	34	173	14.2	10	88	70	5					173	28.4	1	4	67	94	7					173	42.6
dussumieri	19	70	1	90	13.8			5	20	47	14	3	1	90	30.9			1	5	30	41	9	3	1	90	44.7

Species	Segmented anal-fin rays								Total pectoral-fin rays							Procurrent caudal-fin rays								
																Dorsal		Ventral		Total				
	25	26	27	28	29	30	N	x̄	26	27	28	29	30	N	x̄	3	4	3	4	6	7	8	N	x̄
taeniatus	23	114	40	2			179	26.1	2	5	155	6	4	172	28.0	155	6	152	9	149	9	3	161	6.1
dussumieri	1	5	30	46	8	4	94	27.7	3	6	76	2	2	89	27.9	54	3	55	2	54	1	2	57	6.1

TABLE 12. — Change in number of premaxillary canines with growth in species of *Aspidontus*.

Species	standard length class (mm)	Premaxillary canines 0-0	0-I	I-I	I-II	II-II
taeniatus				–	–	1
dussumieri	32-37.9			12	4	5
taeniatus		2	1	8	2	10
dussumieri	38-43.9	–	1	13	7	12
taeniatus		6	2	16	10	8
dussumieri	44-49.9	–	–	11	–	4
taeniatus		16	7	12	–	1
dussumieri	50-55.9	–	–	9	–	1
taeniatus		17	–	2	1	1
dussumieri	56-61.9	–	1	3	–	–
taeniatus		19	1	1		
dussumieri	62-67.9	–	–	4		
taeniatus		64	–	–	–	–
dussumieri	>68	–	–	4	–	1

TABLE 13. — Change in extent of gill-opening with growth in species of *Aspidontus*.

Species	standard length class (mm)	Pectoral-fin ray (counting ventrally) opposite ventral margin of gill opening 3	4	5	6	7	8	9	10	11	12	13	14	N	x̄
dussumieri	32-37.9								2	3	3			8	11.1
taeniatus	38-43.9									2	–	1	40	43	13.8
dussumieri					2	–	2	–	2	3	7	4	3	23	11.3
taeniatus	44-49.9			1	–	1	–	2	1	–	2	3	38	48	13.2
dussumieri						2	2	2	1	–	–	1		8	8.9
taeniatus	50-55.9				2	3	5	2	2	2	3	2	8	29	10.6
dussumieri				1	2	3	–	1						7	6.7
taeniatus	56-61.9	1	1	1	2	2	1	2	–	–	–	1	5	16	9.2
dussumieri		1	1	–	–	1								3	4.7
taeniatus	62-67.9	1	1	–	3	3	4	2	–	1	1			16	7.4
dussumieri			1	–	–	–	1	–	–	1				3	7.7
taeniatus	68-73.9			4	5	2	2	4	1					18	7.0
taeniatus	74-79.9		1	3	2	4	1	3						14	6.7
taeniatus	80-85.9	1	1	1	2	1	1	1						8	6.0
taeniatus	>68	2	–	2	–	1								5	4.6
dussumieri						1								1	—

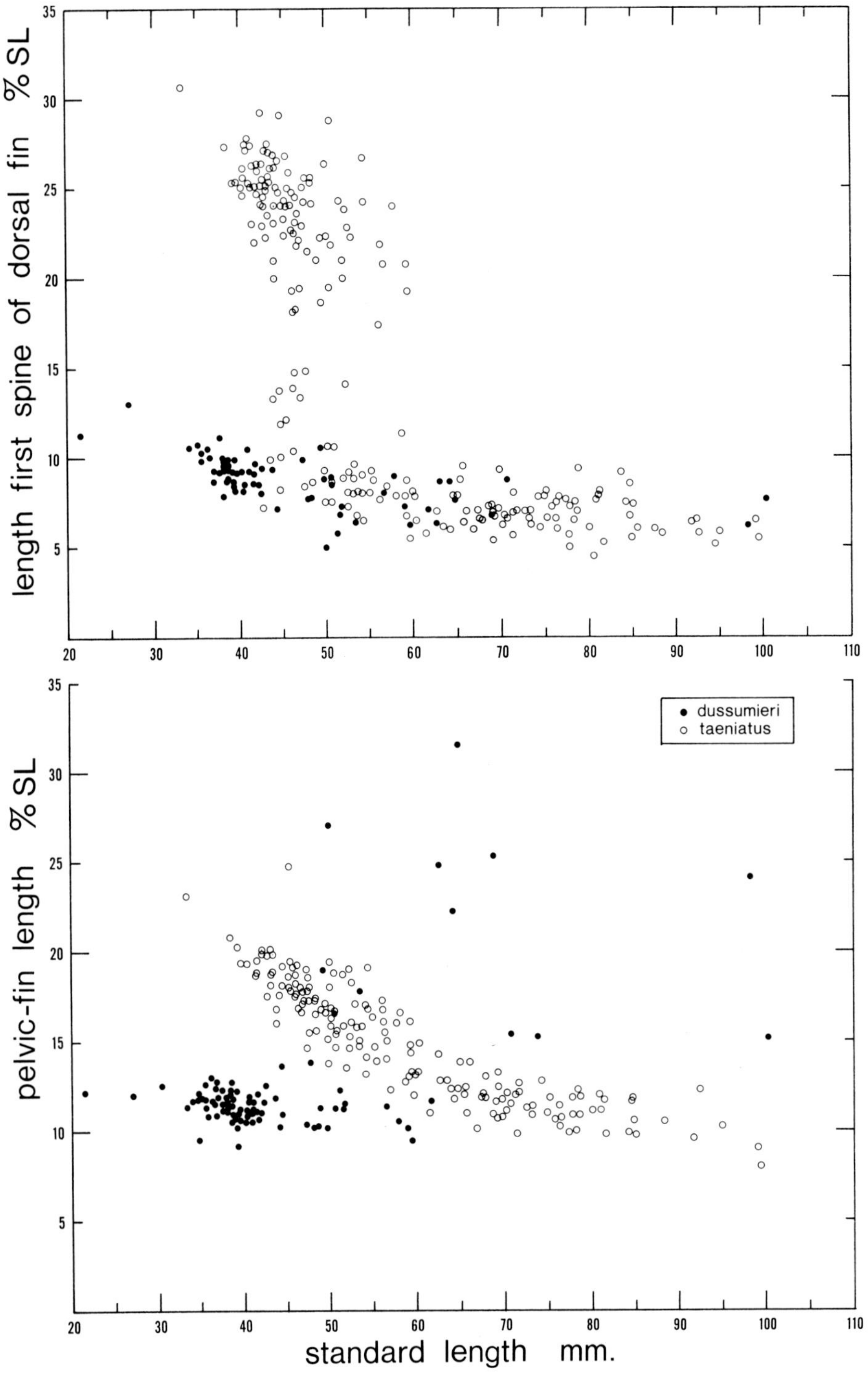

Figure 32. — Relationship between pelvic-fin and caudal-fin lengths versus standard length in species of *Aspidontus*. (All specimens of *A. dussumieri* with pelvic fins longer than 12.5 percent SL are males, except for single 100.3 mm female with pelvic fin 15.2 percent SL.).

discussion of geographic variation below for description of color variants.

Geographic variation. — The only significant geographic variation observed in *Aspidontus taeniatus* involves coloration. Randall and Randall (1960: 453) observed a pair of *Labroides dimidiatus* at Makatea in the Tuamotus with a dull orange-red area in the middle of the body just below the black lateral stripe. In the immediate vicinity, two *Aspidontus* of about the same size displayed precisely the same variation in color pattern. Observations of the same color variation in *L. dimidiatus* and its duplication by *A. taeniatus* were also made on two occasions in the Society Islands.

The following quote from Springer and Smith-Vaniz (1972a: 26) discusses color variants of *A. taeniatus* from the Marquesas Islands. One of the atypically pigmented specimens referred to below is shown in Figure 125.

"While our manuscript was in press, J. E. Randall sent us eight uniquely pigmented specimens of *Aspidontus taeniatus* from the Marquesas Islands. Based on a color transparency of a freshly collected specimen, their fins, except for the transparent pectorals, were pale gray. A dark lateral stripe, in the same position as in normally pigmented *A. taeniatus,* is present anteriorly, but is less intense and somewhat brown posteriorly and does not extend far onto the caudal fin. The background color of the body is pale gray, almost white ventrally. Randall observed the pale *A. taeniatus* in the vicinity of a group of similarly pigmented wrasses belonging to an apparently undescribed, endemic species of *Coris. Labroides dimidiatus* and typically pigmented *A. taeniatus* were also present in the area, but it is not known if the typical *A. taeniatus* also associated with the pale *Coris.*

Wickler (1968: 169) noted that blennies have a pronounced ability to change color pattern with change in emotional state. One of four such patterns of *A. taeniatus* illustrated by Wickler (1968, Fig. 37c) closely approximates that of the pale Marquesas individuals. This suggests that the pale pattern is derived from the typical pattern. The apparent restriction of the pale *Coris* to the Marquesas Islands may explain the absence of the pale form of *A. taeniatus* elsewhere. Randall noted that *A. taeniatus* appeared to be more abundant in the Marquesas than elsewhere, and suggested that this might, in part, account for the apparent switch-over of some individuals to another model."

Species comparisons. — In addition to the characters given in the key to distinguish the two species of *Aspidontus, taeniatus* has a more acute snout, fewer incisor teeth (Fig. 30) and epipleural ribs (15-20 versus 31-36).

Nomenclature. — All recent workers have agreed in regarding *P. paradiseus, rostratus* and *azureus* as junior synonyms of *Aspidontus taeniatus.* Some workers have continued to recognize *Blennechis filamentosus,* but Smith-Vaniz and Randall (1973) have shown that all of the distinguishing characters alter during metamorphosis, thus revealing *filamentosus* as the pelagic prejuvenile of *taeniatus.* Losey (1974: 434) reported that after being in an aquarium for only one day two of four individuals of *"filamentosus"* were living in holes and had assumed the striped *"taeniatus"* coloration. The other two fish, which were prevented from descending from the surface, retained their banded color pattern.

Etymology. — The Latin specific name, *taenia* (ribbon, stripe), refers to the prominent mid-lateral stripe of adults.

Mimicry. — The most widely reported example of mimicry in fishes involves *A. taeniatus* and the cleaner wrasse, *Labroides dimidiatus* (refer to discussion of biology and general section on mimicry). See also preceding comments on geographic variation.

Distribution (Fig. 31). — Widely distributed in the western and central Pacific Oceans. The only record of *taeniatus* from Japan is based on the detailed description (Tomiyama, 1956) of a single specimen from the bay of Aburatsubo, Misaki, Kanagawa Prefecture.

Not listed in the material examined section is a single prejuvenile specimen (ANSP 16666) of *taeniatus* collected by W. H. Jones from the "Sandwich Islands." On the basis of this specimen Fowler (1928: 428) credits the species to the Hawaiian Islands, although Strasburg (1956: 243) regarded the record as dubious. Since I know of no other record of the species from Hawaii and fishes collected by Jones from other Pacific localities are deposited at the Academy, I concur with Strasburg in treating the record as suspect.

Material examined. — 226 specimens (40.0-99.5 mm SL) from 92 collections.

INDONESIA: SU 28841 (1:42), Celebes, Lembeh Strait; USNM 211910 (3:46-55), Celebes, Kabaena I.; SU 30088 (1:46), Sangihe Is., 3°33′N, 125°32′E; RMNH 26668 (96.2, holotype of *Petroskirtes paradiseus*), RMNH

4453 (45.0, holotype of *Petroskirtes rostratus*), NMV 71772 (1:50), Amboina; USNM 195695 (2:46-47), USNM 195696 (4:53-82), "Longley-Dutch East Indies" — presumably Banda Sea; RMNH 4772 (4:60-85), USNM 210881 (1:45), ZMA 109.346 (1:49), Banda Sea.

W. AUSTRALIA: WAM P25370-19 (1:77), North West Cape, off Tantabiddi Creek.

SOUTH CHINA SEA: CAS 28898 (1:78), 10°29′N, 108°57′E; USNM 205288 (1:53), Taiwan.

PHILIPPINE IS.: SU 29570 (1:93), Mindoro; USNM 137841 (1:47), Panay.

MARIANA IS.: CAS 28893 (1:45), Guam.

YAP IS.: ANSP 124871 (1:72), CAS 28895 (3:66-76), CAS 28896 (1:77); ANSP 122373 (1:48), Ulithi Atoll.

PALAU IS.: CAS 28889 (3:70-76), CAS 28892 (1:71), CAS 28894 (1:43), CAS 28897 (2:52-75), CAS GVF524 (1:84), CAS GVF551 (6:73-88), CAS GVF798 (1:71), CAS GVF843 (1:80), CAS GVF1043 (1:73), CAS GVF1869 (2:75-80), USNM 154301 (1:78).

W. CAROLINE IS.: CAS 28882 (1:51), CAS 28883 (1:65), Ifaluk Atoll.

KAPINGAMARANGI ATOLL: CAS 28884 (3:69-73), CAS 28885 (1:65), CAS 28886 (1:52), CAS 28887 (1:76), CAS 28888 (2:52-74).

MARSHALL IS.: USNM 113595 (2:58-85), USNM 142314 (1:46), USNM 142315 (1:57), Bikini Atoll; USNM 142228 (1:81), Rongerik Atoll; USNM 211131 (1:85), Taka Atoll; USNM 200339 (1:69), Kwajalein Atoll; USNM 166785 (1:73), Arno Atoll.

GILBERT IS.: USNM 205548 (3:45-49), Kuria Atoll; USNM 167323 (1:53), USNM 167360 (1:69), Onotoa; AMS uncat. (1:64), sta. DFH 73-27, AMS uncat. (1:53), sta. DFH 73-48, AMS uncat. (1:64), sta. DFH 73-51, Apaiang Atoll.

PHOENIX IS.: NMFS 1098 (1:47), Canton I.; NMFS 1113 (2:50-52), Enderbury Is.; USNM 207571 (1:53), Gardner I.

LINE IS.: ANSP 75827 (65:40-49), ANSP 122374 (12:47-60), ANSP 124870 (2:45-50), ANSP 128567 (2:48-57), ANSP 128568 (1:48), Christmas I.; ANSP 75800 (2:46-60), BPBM 13454 (3:43-47), Fanning I.

MARQUESAS IS.: CAS 15796 (3:51-57), BPBM 10870 (1:68), BPBM 10871 (3:48-65), BPBM 10881 (7:52-66), BPBM 10894 (4:53-64), BPBM 13233 (2:52-60).

TUAMOTU IS.: USNM 122657 (1:56), "Wilkes Exped."; CAS 28881 (1:92), Raroia; ANSP 124232 (1:47), Rangiroa; BPBM 13471 (1:58), Takaroa.

SOCIETY IS.: CAS 28890 (5:54-92), CAS 28891 (1:60), USNM 211128 (1:68), USNM 211130 (1:66).

SAMOA IS.: USNM 51796 (80.2, holotype of *Petroscirtes azureus*), Apia.

TONGA IS.: ISZZ 10140 (1:56), USNM uncat. (3:88-100), Vavau Group.

FIJI IS.: CAS uncat. (1:45), ISZZ 9601 (1:74), USNM 176618 (1:94), USNM uncat. (2:63-69), Te Vega Cr. 7 sta. 278, ZSZM 12287 (1:77).

NEW HEBRIDES IS.: AMS uncat. (1:69), Efate I., sta. GRA-19.

SOLOMON IS.: CAS 5651 (1:44), Rennell I.; AMS uncat. (1:44), off Florida I., sta. GRA-63; USNM 195782 (1:60), New Georgia I.; USNM 212030 (2:64-65), Guadalcanal Is.

NEW GUINEA: MNHN A.2046 (48.9, holotype of *Blennechis filamentosus*), Humboldt Bay; RMNH uncat. (2:45-50), Hollandia Bay; DASF FO1931 (1:46), Louisiade Is., Rossel I.

E. AUSTRALIA: AMS IB.5749-50 (2:62-72), Manly, N. S. W.; ANSP 109672 (1:70), Queensland, Endeavour Reef.

NEW CALEDONIA: ANSP 124237 (1:59), stomach contents of *Thunnus albacares*.

Aspidontus taeniatus tractus Fowler

Aspidontus tractus Fowler, 1903: 170, pl. 7 (type locality: Zanzibar; holotype ANSP 24207).

Color pattern. — *A. taeniatus tractus* differs from the nominal subspecies only in having a black bar on the fleshy pectoral-fin base. Only a faint indication of the bar exists in the single specimen (tentatively assigned to *tractus*) available from the Cocos-Keeling Islands; however the single specimen available from the Seribu Islands, Indonesia has a well-developed bar. The fleshy pectoral-fin base is immacuate in all other specimens of *taeniatus* from the western and central Pacific oceans, except for two of 20 specimens from the Palau Islands that have a minute spot present.

The presence or absence of the bar may have biological significance because *taeniatus* mimics *Labroids dimidiatus* and all adults of this wrasse examined from within the range of *tractus* (specimens examined from the Red Sea, Aldabra, Seychelles, Sri Lanka and Cocos-Keeling Islands) also have a bar in the region of the pectoral-fin base. See also comments in Springer and Smith-Vaniz (1972a: 25-26) regarding the presence or absence of the bar in *Aspidontus* and *Labroides*.

The same color photograph of *A. taeniatus tractus* appears in Springer (1970: 62) and Burgess and Axelrod (1973a: 471). Monochrome photographs of a juvenile (identified as *A. filamentosus*) and an adult are also given in Springer and Smith-Vaniz (1972a: 36).

Etymology. — In the original description the etymology of the Latin word *tractus* was given as "a streak", obviously in reference to the dark bar on the pectoral-fin base. The correct translation is "a space drawn out, district or region."

Distribution (Fig. 31). — Essentially restricted to the Red Sea and Indian Ocean.

Material examined. — 54 specimens (33.3-99.1 mm SL) from 40 collections.

GULF OF AQABA: USNM 204553 (1:44), USNM 204554 (1:45), USNM 207643 (1:47), USNM 211132 (1:63), USNM 211133 (1:57), USNM uncat. (1:48), sta. VGS 71-29.
RED SEA: USNM 204552 (2:54-61), Strait of Jubal.
SEYCHELLES: ANSP 111002 (1:78), Curieuse I.; ANSP 111016 (4:47-76), USNM 201562 (4:45-81), Amirante Is.
ALDABRA IS.: USNM uncat. (1:51), sta. HA 67-62, USNM uncat. (1:56), sta. HA 67-66.
AGALEGA IS.: USNM 215441 (3:44-66), USNM 215442 (1:47).
COMORO IS.: USNM 199432 (1:51), USNM 201563 (3: 71-85), USNM 207572 (1:50).
E. AFRICA: RUSI uncat. (1:47), Mombasa; RUSI uncat. (1:33), Pemba I.; ANSP 24207 (76.7, holotype of *Aspidontus tractus*), Zanzibr; RUSI uncat. (1:63), Dar-es-Salaam; RUSI uncat. (1:52), Ibo; RUSI BP490 (1:52), RUSI BP285 (1:56), RUSI BP661 (1:99), Pinda; USNM uncat. (1:55), Chesterfield I.; SAM 9897 (1:54), St. Lucia.
DIEGO GARCIA ATOLL: USNM 214390 (2:84-85).
MALDIVE IS.: FMNH 71347 (1:54), FMNH 71348 (1: 67), NFIS 9760 (2:74-75).
SRI LANKA (CEYLON): USNM 214391 (1:50), USNM 214392 (2:59-73), USNM 214393 (1:70), USNM uncat. (1:59), sta. TI 70-349.
THAILAND: USNM 211129 (1:64), Similan I., 8°29′N, 97°39′E; NMV 71794 (1:42), Penang.
COCOS-KEELING IS.: ANSP 128048 (1:69).
INDONESIA: USNM 212001 (1:64), Java Sea, Seribu Is., 5°51′S, 106°34′E; ZMA 109.354 (1:46), N. Sumatra, "W coast Atjeh."

Aspidontus dussumieri (Valenciennes)

Specimen Figures 28, 127-130; Tables 4, 11-13.

Blennechis dussumieri Valenciennes, *in* Cuvier and Valenciennes, 1836: 282 (type locality: l'ile de Bourbon [= Reunion]; holotype MNHN A.2084).
Petroscirtes striatus Day, 1888: 262 (type locality: Ceylon; holotype BMNH 1889.2.1.3587).
Petroscirtes fluctuans Weber, 1909: 146 (type locality: Siboga Expedition Station 149, Fau anchorage, West Coast Gebe Island; lectotype ZMA 109.338).
Petroscirtes quadrimaculatus Kendall and Goldsborough, 1911: 329, pl. 7, Fig. 1 (type lqocality: Arno Atoll, Marshall Islands; holotype USNM 65980).
Petroscirtes gorrorensis Herre, 1936a: 285 (type locality: Koror, Palau Islands; holotype SU 29102).
Aspidontus (Escadotus) wamiziensis Smith, 1959: 235, pl. 17, Fig. 10 (type locality: Wamizi Island, Mozambique; holotype RUSI 961c).

Description. — Posttemporal not forked. Dorsal fin IX-XI, 28-34 (rarely 28); total dorsal elements 38-44. Anal fin II, 25-30. Pectoral fin 13-15. Caudal fin: procurrent rays 3-4 + 3-4 = 6-8 (typically 6); segmented rays 11. Vertebrae: precaudal 13-15; caudal 29-34; total 42-48. Pleural ribs on vertebrae 3-14 or 15; epipleural ribs 31-36 (based on 3 cleared and stained specimens). Incisor teeth (Fig. 30) in specimens 21-100 mm SL, n = 76: lower jaw 11-36; upper jaw 15-39; each premaxilla with 1 or 2 posterior canines.

Broadly rounded snout scarcely projects beyond tip of premaxillae. Caudal fin of adults lanceolate, median rays longer than outer rays. Usually the middle rays of the caudal fin are only slightly longer than the outer rays (13.5-22.4% SL), but in two specimens 40.9 mm SL and 47.3 mm the longest ray is 61.9 and 92.5% SL respectively. Anterior dorsal-fin spines not elongate in prejuveniles. Pelvic fin relatively short in specimens < 50 mm: 9.2-13.8% SL (Fig. 32); pelvic fin sexually dimorphic in specimens > 50 mm: males 11.3-31.5% SL, females 9.4-15.2% SL.

Color pattern. — Adults with dark stripe extending from eye to caudal fin, ventral margin of which follows mid-line of body; stripe ends at caudal-fin base or extends onto proximal portion of fin. Body and head pale, except snout and dorsum slightly darker than other areas. Dorsal and anal fins dusky with narrow, dark, submarginal stripe and very narrow, pale margin. Only females have a black spot in the anterior portion of the dorsal fin. Prejuveniles differ from adults primarily in having the lateral stripe absent or indistinct, the body and the dorsal and anal fins banded.

In life the lateral stripe is dark brown, ventral half of body white, snout light brown or orange and dorsum light brown. The dorsal and anal fins are uniform olive-yellow, with a dark brown submarginal stripe and narrow white margin. Caudal fin white proximally, except for brown stripe, followed by dirty yellow or light brown area, and with dorsal and ventral margins of fin broadly transparent. Pelvic fins white; pectorals transparent. In prejuveniles the markings on the dorsal and anal fins and body consist of dark brown bands with narrow, pale brown interspaces.

Tomiyama (1955) gives an excellent illustration and detailed description of a prejuvenile from Misaki, Japan. Smith-Vaniz and Randall (1973: 8) also present monochrome photographs of a prejuvenile and adult. Masuda *et al.* (1975: 86) give a color photograph of a specimen from southern Japan.

Geographic variation. — The single Red Sea specimen and the three from the Gulf of Aqaba have the highest counts for segmented dorsal- and anal-fin rays and caudal vertebrae. The two largest specimens from the Gulf of Aqaba (Figs. 129-130) also differ in having an additional stripe present ventrally. The Gulf of Aqaba and northern Red Sea population of *Plagiotremus tapeinosoma* also is distinctive in having a well-developed ventral stripe.

Species comparisons. — Refer to discussion under account of *A. taeniatus.*

Nomenclature. — The primary types of four of the nominal species here referred to *A. dussumieri* (*P. striatus, fluctuans, quadrimaculatus* and *gorrorensis*) were based on the banded prejuveniles of *dussumieri,* and were not compared with *dussumieri* or with any other species in the original description. *Aspidontus wamiziensis* (based on a single adult) was compared only with *P. fluctuans* and *gorrorensis* and considered distinct, primarily on the basis of color pattern.

Etymology. — This species was named for Mr. L. I. Dussumier, who collected the holotype.

Distribution (*Fig. 31*). — Widely distributed throughout most of the Indo-Pacific.

Material examined. — 99 specimens (21.4-100.3 mm SL) from 62 collections.

GULF OF AQABA: USNM 204549 (3:60-100).

RED SEA: ISZZ 10507 (1:35), Kosseir (Quseir); BPBM 19578 (1:55), Port Sudan; BPBM 19579 (3:44-52), Suakin.

SEYCHELLES: ANSP 111000 (1:41), Curieuse I.; ANSP 109806 (2:51-63), Amirantes Is.

ALDABRA IS.: MNHN 54-136 (6:30-37).

GRANDE COMORE IS.: CAS 34380 (1:50), CAS 34387 (2:46-52).

E. AFRICA: USNM 201564 (1:35), off Kenya, 3°7'S, 40°39'E; RUSI 961c (73.8, holotype of *Aspidontus wamiziensis,* Wamizi I.; RUSI BP2144 (1:59), RUSI 961b (2:39-42), RUSI uncat. (1:71), Pinda; RUSI uncat. (1:57), Mozambique, Bazaruto I.

REUNION IS.: MNHN A.2084 (40.2, holotype of *Blennechis dussumieri*).

MAURITIUS: BPBM 15951 (2:58-62).

ST. BRANDON SHOALS (CARGADOS CARAJOS): USNM 215439 (6:66-73).

SRI LANKA (CEYLON): BMNH 1889.2.1.3587 (31.3, holotype of *Aspidontus striatus*).

ANDAMAN IS.: ZSIC 64/2 (1:38), Port Blair.

COCOS-KEELING IS.: ANSP 128040 (1:69), ANSP 128041 (1:53), ANSP 128042 (1:44), ANSP 128043 (2:36), ANSP 128044 (1:36), ANSP 128107 (1:27).

INDONESIA: ZMA 109.355 (1:42), Celebes, Gulf of Bone (Boni); SU 29497 (1:39), Celebes, Kerma; RMNH 21150 (1:33), Sulu Sea, Tawitawi I.; USNM 137842 (1:37), Sulu Sea, Teomabal I.; ZMA 109.338 (39.4, lectotype by present designation of *Petroscirtes fluctuans*), Gebe I., Siboga Exped. sta. 149; ZMA 109.339 (1:38), Banda Sea, Taam I.; ZMA 109.340 (1:38), Banda Sea, 3°50'S, 128°20'E; USNM 122337 (1:36), Buru I.

W. AUSTRALIA: WAM P11702-3 (2:37-40), between Cervantes and Green I., 30°44'S, 115°8'E; WAM P8753 (1:40), Beagle I.

SOUTH CHINA SEA: CAS 28880 (8:21-52), 10°49'N, 109°5'E; MNHN A.2162 (1:50), "China."

PHILIPPINE IS.: USNM 137830 (2:35-38), SU 32586 (1:39), Mindanao; USNM 111970 (1:36), Panay.

JAPAN: SU 23503 (1:41), Wakanoura.

PALAU IS.: SU 29102 (44.2, holotype of *Petroscirtes gorrorensis*), Koror I.

TRUK ATOLL: BPBM 7449 (1:52).

MARSHALL IS.: BPBM 12567 (1:41), Eniwetok Atoll; USNM 142219 (1:42), USNM 142220 (2:40-42), Bikini Atoll; USNM 142221 (2:41-42), Rongerik Atoll; USNM 201598 (1:64), Kwajalein Atoll; USNM 65981 (1:34), USNM 65980 (41.0, holotype of *Petroscirtes quadrimaculatus*), Arno Atoll.

GIILBERT IS.: AMS uncat. (1:47), Apaiang Atoll, sta. DFH 73-36.

TUAMOTU IS.: ANSP 124256 (1:41), Rangiroa Atoll.

SOCIETY IS.: MCZ 29393 (1:39), FMNH 77010 (1:65), Tahiti.

SAMOA IS.: BPBM 11337 (1:50), BPBM 13231 (1:38), Tutuila I.

NEW BRITAIN: USNM 201595 (1:49), Dawapia Rocks.

NEW GUINEA: BPBM 15816 (1:51), SU 28234 (1:39), Madang.

E. AUSTRALIA: AMS IB.6690 (3:44-49), AMS I.17057 (1:58), Forster, N. S. W., 23°11'S, 152°30'E; AMS IB.2585 (1:44), off Byron Bay, N. S. W.; USNM 201596 (1:63), Queensland, Capricorn Gr., One Tree I.

XIPHASIA Swainson

Xiphasia Swainson, 1839: 259 (type-species: *X. setifer* Swainson, 1839, by monotypy).

Nemophis Kaup, 1858: 168 (type-species: *N. lessoni* Kaup, 1858 [= *Xiphasia setifer* Swainson, 1839], by monotypy).

Xiphogadus Günther, 1862: 374 (type-species: *Xiphasia setifer* Swainson, 1839, by monotypy; a junior objective synonym of *Xiphasia* Swainson, 1839).

Diagnosis. — Major characters useful in distinguishing the genera of Nemophini are summarized in Tables 1, 3 and 4. The high meristic values, forward position of the dorsal fin in adults and the membranous attachment of the last dorsal- and anal-fin rays (Fig. 36) easily distinguish *Xiphasia* from the other genera of Nemophini. The general physiognomy of the head of the two species of *Xiphasia* is shown in Figures 33-34.

Nomenclature. — Putnam (1874) gives a detailed and accurate account of the synonymy and nomenclatural history of *Xiphasia.* In his original description Kaup did not compare *Nemophis* with *Xiphasia.* Günther adopted the specific name of *setifer* for his new genus and cited *Xiphasia* with an exclamation mark, probably intending thereby to indicate that Swainson's name was not admissible on account of its construction. Although *Xiphasia* is indeed improperly constructed, it still has nomenclature validity and priority.

Etymology. — Although Swainson gave no statement of etymology, *Xiphasia* is probably derived from the Greek neuter noun *Xiphos,* meaning sword, in allusion to the slender body and lanceolate tail. The gender of *Xiphasia* is considered to be feminine in accordance with the rules outlined by Emiliani (1952).

Biology. — The available data indicate that both species of *Xiphasia* are semipelagic and probably nocturnal. The capture data for the 38 collections for which such information is available (data for both species combined) is as follows: trawl (17); fish market purchase (2); night light and dip net (7); stomach contents of tuna (6); washed up on beach (2); captured by hand (see below) (1). Most of the trawl collections were in depths shallower than 75 meters, usually between 25-50 m. The two exceptions are a 31 mm juvenile *setifer* from R/V METEOR sta. 95 (Nellen, 1973) taken in a 55 m oblique tow over a depth of 1923 m, and the type of *matsubarai,* which was reported to have come from a depth of "about 150 hiro" (178 m). The method of capture for the latter collection was not stated but, assuming it was a trawl, the specimen may have been captured near the surface.

The hand collection of an adult *setifer* from the Gulf of Aqaba (Fridman and Masry, 1971) may provide valuable insight on the daytime habits of *Xiphasia.* The following observations are especially relevant:

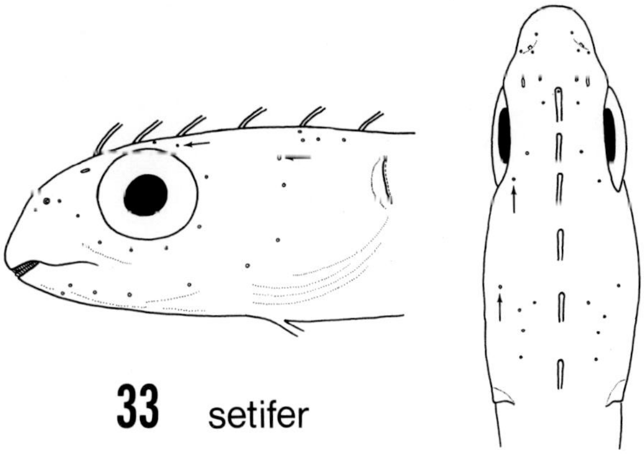

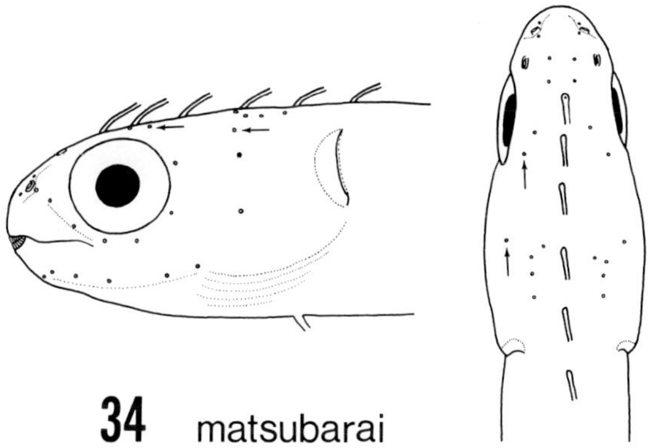

Figures 33-34. — General physiognomy and cephalic sensory pore distributions in species of *Xiphasia:* Fig. 33, *setifer,* USNM 205292, female, 290.4 mm, Gulf of Thailand; Fig. 34, *matsubarai,* USNM 203276, female, 244.2 mm, Solomon Islands, Rennell Is. Small arrows identify ventralmost supraorbital and dorsalmost preopercular pores in both lateral and dorsal views.

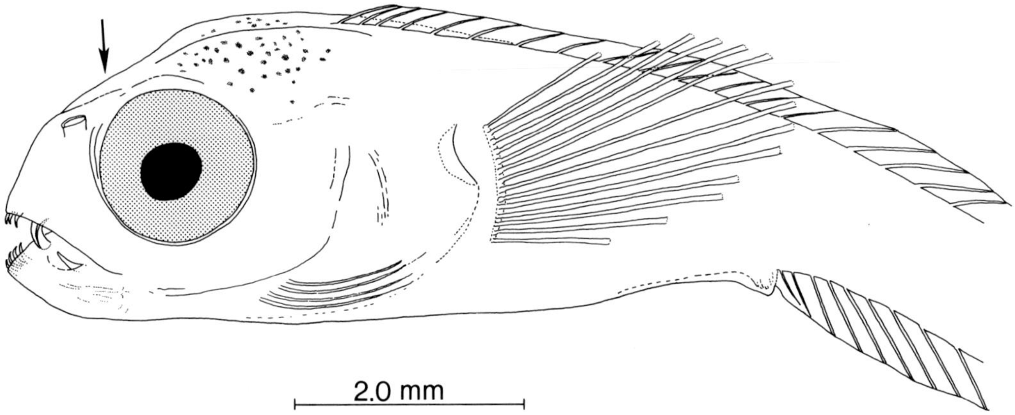

Figure 35. — Prejuvenile *Xiphasia setifer,* IMUK uncataloged, 45.9 mm, from Persian Gulf, R/V METEOR sta. 287. Arrow indicates origin of dorsal fin in juveniles and adults.

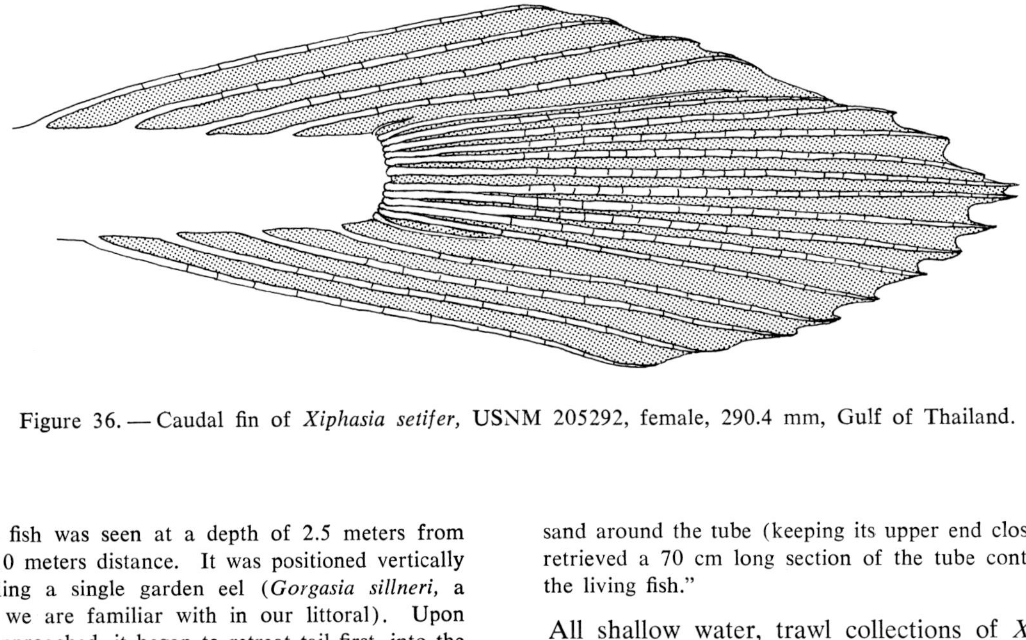

Figure 36. — Caudal fin of *Xiphasia setifer,* USNM 205292, female, 290.4 mm, Gulf of Thailand.

"The fish was seen at a depth of 2.5 meters from about 10 meters distance. It was positioned vertically resembling a single garden eel (*Gorgasia sillneri,* a species we are familiar with in our littoral). Upon being approached, it began to retreat tail first, into the sand. From a distance of 1 m, not more than the head of the fish was seen protruding. Around the head the clear outline of a tube opening was observed. The fish's head and eyes kept moving in a jerky manner, very much resembling that of a blenny. I (D. F.) was skindiving and when I actually reached out for the fish, it quickly disappeared in the tube.

No other fish or tube openings were observed anywhere in the vicinity. After carefully digging into the sand around the tube (keeping its upper end closed), I retrieved a 70 cm long section of the tube containing the living fish."

All shallow water, trawl collections of *Xiphasia* were made over sand or mud bottoms, and the hand captured specimen came from an area of calcareous sand covered with the sea grass *Halophila. Xiphasia* differs from all other genera of Blenniidae in having the gill opening protected by both a lateral and mesial fleshy flap, which undoubtedly functions as a one-way valve to prevent sand and fine sediment from entering the gill cavity. This adaptation to-

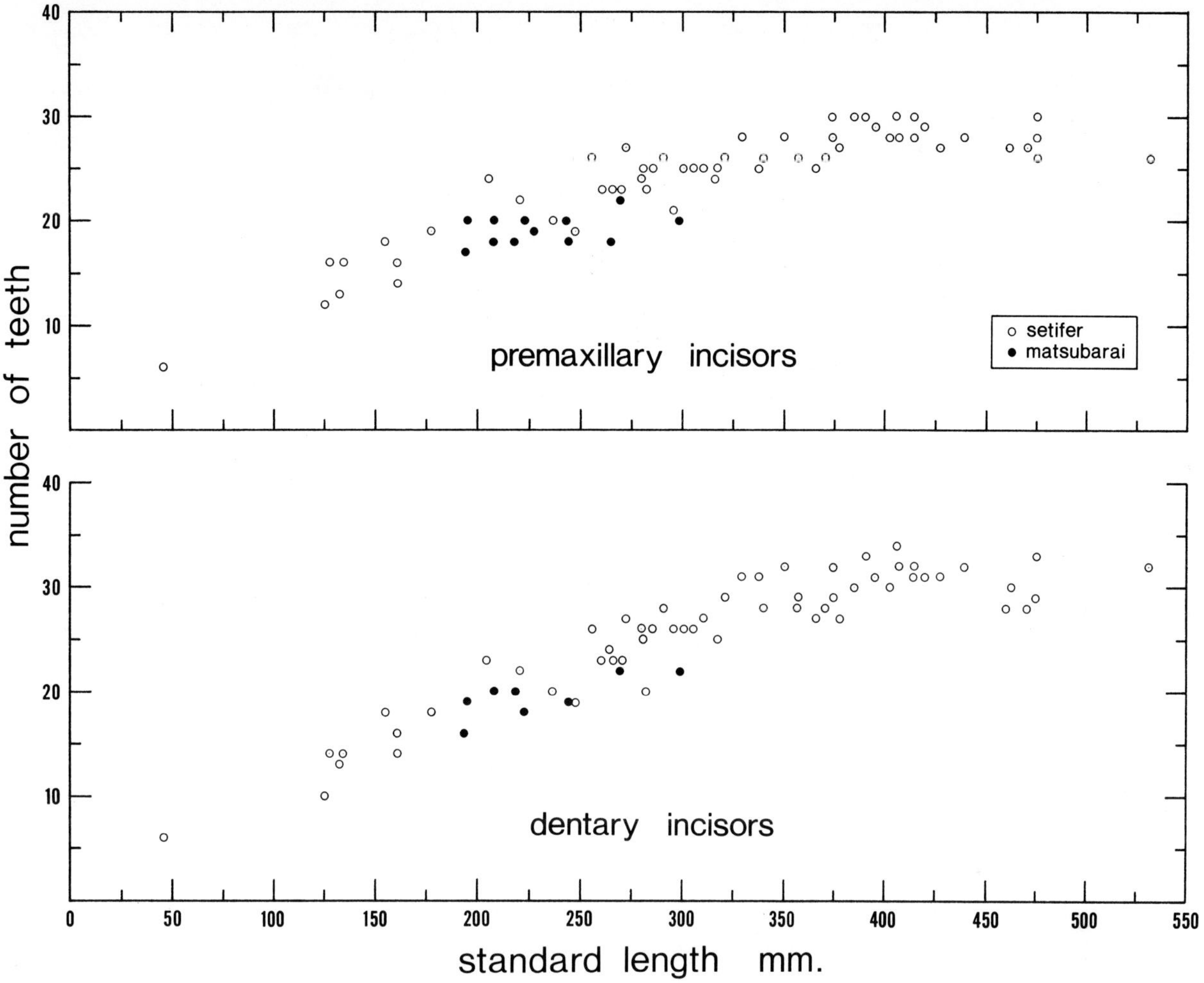

Figure 37. — Relationship between number of incisor teeth versus standard length in species of Xiphasia.

gether with the above observations suggest that *Xiphasia* typically inhabits burrows or tubes in relatively soft-bottom habitats. Masuda et al. (1975: 265) reported that *X. setifer* "dives into sand when threatened." The lanceolate caudal fin (Fig. 36) is also characteristic of many burrow-dwelling fishes. The collection of adults of *Xiphasia* in several surface, night-light stations may indicate that it is most active at night, behavior apparently unique among the Blenniidae. The extremely elongate body and generally shallow-water habitat would seem to make it especially vulnerable to predation, at least during the daylight hours.

Kuthalingam and Menon (1965) reported that the majority of the identifiable stomach contents (excluding sand and mud) of two trawl caught *X. setifer* consisted of copepods, crustacean and polychaete remains. The number of incisors (Fig. 37) suggests an unspecialized, carnivorous feeding habit. The gut (Fig. 99d), and speckled peritoneum is most similar to that of *Aspidontus*.

Almost nothing is known of the early life history of *Xiphasia*. The three smallest specimens of *setifer* (16-46 mm SL) were taken in separate trawl stations. They differ from the adults in having the origin of the dorsal fin more posteriorly positioned

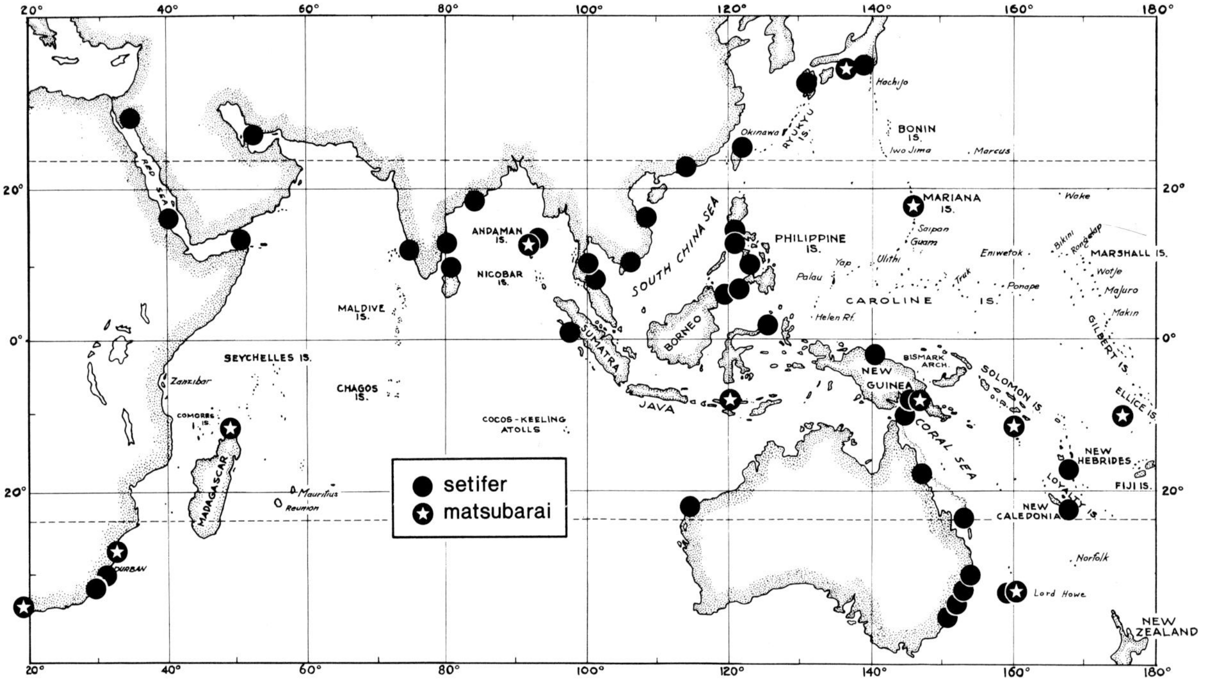

Figure 38. — Distributional records for species of *Xiphasia.*

(Fig. 35), the anterior nostrils very large, the gill opening extending farther ventrally (opposite 5th to 7th pectoral-fin ray versus 1st to 3rd), larger head and abdomen and the dentary canines relatively small. The illustrated prejuvenile has a normal complement of dorsal-fin spines and segmented rays, so the shift in position of the dorsal fin in adults can not be attributed to incomplete development of dorsal-fin elements in the prejuveniles.

Zoogeography. — Both species of *Xiphasia* are broadly distributed (Fig. 38), but only the adults of *matsubarai* have been collected in oceanic waters far from major land masses.

Intrageneric relationships. — The lower number of fin rays (especially pectoral-fin rays and dorsal-fin spines) and total vertebrae, and the relatively blunt snout and large anterior nostrils suggest that *matsubarai* is the most specialized species of *Xiphasia.*

KEY TO SPECIES OF *XIPHASIA*

Total pectoral-fin rays 24-28; dorsal-fin spines 13 or 14; total dorsal-fin elements 118-133; anterior two-thirds to one-half of anal fin not white distally .... *setifer*

Total pectoral-fin rays 20-22; dorsal-fin spines 11; total dorsal-fin elements 110-115; anterior two-thirds to one-half of anal fin white distally, the width of the white area variable .... *matsubarai*

TABLE 14. — Frequency distributions of certain meristic characters in species of *Xiphasia*.

Species	Dorsal-fin spines					
	11	12	13	14	N	x̄
setifer			31	27	58	13.5
matsubarai	17				17	11.0

Species	Segmented dorsal-fin rays																						
	99	100	101	102	103	104	105	106	107	108	109	110	111	112	113	114	115	116	117	118	119	N	x̄
setifer							1	–	1	2	8	8	7	8	5	5	3	2	3	2	2	57	112.1
matsubarai	2	8	4	–	1	2																17	100.8

Species	Total dorsal-fin elements																									
	110	111	112	113	114	115	116	117	118	119	120	121	122	123	124	125	126	127	128	129	130	131	132	133	N	x̄
setifer									1	–	–	2	6	6	9	7	7	4	4	4	2	2	1	2	57	125.6
matsubarai	2	8	4	–	1	2																			17	111.8

Species	Segmented anal-fin rays																								
	97	98	99	100	101	102	103	104	105	106	107	108	109	110	111	112	113	114	115	116	117	118	119	N	x̄
setifer											2	4	5	7	8	7	7	4	1	5	2	3	2	57	112.3
matsubarai	1	2	7	3	2	–	1	1																17	99.7

Species	Precaudal vertebrae					
	13	14	15	16	N	x̄
setifer			16	40	56	15.7
matsubarai	18				18	13.0

Species	Caudal vertebrae																					
	100	101	102	103	104	105	106	107	108	109	110	111	112	113	114	115	116	117	118	119	N	x̄
setifer						1	–	1	6	7	7	6	6	2	6	5	3	1	4	1	56	112.0
matsubarai	5	9	1	1	2																18	101.2

Species	Total vertebrae																								
	113	114	115	116	117	118	119	120	121	122	123	124	125	126	127	128	129	130	131	132	133	134	135	N	x̄
setifer									1	–	1	8	7	9	5	3	2	6	7	1	4	1	1	56	127.7
matsubarai	5	9	1	1	2																			18	114.2

Species	Total pectoral-fin rays											Procurrent caudal-fin rays: Dorsal		Ventral
	20	21	22	23	24	25	26	27	28	N	x̄	1	2	1
setifer					4	5	45	3	1	58	25.9	4	47	51
matsubarai	3	1	11							15	21.5		10	10

Xiphasia setifer Swainson

Specimen Figures 33, 35, 131; Tables 4, 14.

Xiphasia setifer Swainson, 1839: 259 (description based on "tonkah talawaree" Russell, 1803: 28, pl. 39, Vizagapatam).

Nemophis lessoni Kaup, 1858: 168, unnumbered Fig. (type locality: not stated, "from M. Lesson and Garnot, obtained in the expedition of M. Duperrey"; holotype MNHN 2123).

Xiphasia trachypareia Bleeker, 1864: 196; Bleeker, 1865: 196, unnumbered plate (type locality: not stated, "Hab ?"; holotype lost ?).

Xiphogadus madagascariensis Playfair, 1868: 11 (type locality: from stomach of fish from Maurounda River, Madagascar; holotype MNHN 5545).

Description. — Dorsal fin XIII-XIV, 105-119; total elements 118-133. Anal fin II, 107-119. Pectoral fin 12-14. Caudal fin: procurrent rays 1-2+1; segmented rays 10 (one of 51 with 11). Vertebrae: precaudal 15-16; caudal 105-119; total 121-135. Pleural ribs on vertebrae 3-15 or 16; epipleural ribs 11-13. Incisor teeth (Fig. 37) in specimens 125-532 mm SL, n = 58: lower jaw 10-34; upper jaw 12-30. Caudal fin 2.7-21.8% SL, sexually dimorphic (Fig. 39); two innermost segmented rays elongate in large males.

Color pattern. — Body with approximately 28 alternating dark and pale bands, which extend into dorsal fin (often indistinct in preserved specimens). Segmented portion of dorsal fin with a narrow, dark submarginal stripe and very pale margin (similar to that of *Aspidontus dussumieri*). Black spot or ocellus, slightly smaller than pupil diameter, usually present distally between 5th to 7th dorsal-fin spines, and a large black blotch between 10th to 14th spines sometimes with a pale stripe ventrally; dorsal fin otherwise brownish with faint trace of bands. Caudal fin dark brown. Anal fin dark with very narrow pale margin.

According to Shen and Ting (1972: 29) the life colors are as follows:

> "Color in fresh specimens is bright yellow with 28 broad dark brown vertical bands on the dorsal and body, the first behind head; the dorsal yellow at its origin and with an ovate black spot on the upper membrane between 4th to 6th spines, both dorsal and anal fins with a broad marginal blackish band; the caudal filaments blackish; the head dusky yellowish, the pelvic and pectoral fins yellow."

Good illustrations of the species are given in Day (1878: pl. 73), de Beaufort (1951: 382) and Menon and Varma (1967). Color illustrations of *X. setifer* appear in Tanaka (1908: pl. 4), Herre (1926: pl. 1), Masuda *et al.* (1975: 86), and Burgess and Axelrod (1973a: 466), the latter misidentified as *X. matsubarai.*

Geographic variation. — There are insufficient numbers of specimens to evaluate geographic variation, but a range of 8-9 was recorded for segmented dorsal-fin rays and precaudal vertebrae in 20 specimens from the Gulf of Thailand. Specimens with the highest counts are from the Red Sea, Hong Kong and Japan.

Species comparisons. — Except for the caudal-fin ray count, there is no overlap in meristic values between *setifer* and *matsubarai* (see Table 14). The pigmentation of the anal fin is also useful in distinguishing these two species, and the snout is blunter and the origin of the dorsal fin more posteriorly positioned in adults of *matsubarai.*

Nomenclature. — Russell's description and illustration of "tonkah talawaree", upon which the name *Xiphasia setifer* was established, is sufficiently accurate for positive identification. In the original description, *Nemophis lessoni* was compared with no other species; examination of the holotype leaves no doubt that it is a synonym of *setifer.* I have been unable to locate the type of *X. trachypareia,* but the description could apply only to *setifer.* The holotype of *X. madagascariensis* (MNHN 5545), a poorly preserved specimen, was not included in Bauchot's (1967) catalog of blennioid types in the Museum National d'Histoire Naturelle. The total dorsal-fin element count of 130 that I obtained for the holotype positively establishes it as a junior synonym of *setifer.* Putnam (1874) also correctly relegated the preceding nominal species to the synonymy of *setifer.*

Etymology. — The Latin name *setifer,* meaning bristle bearer, apparently refers to the caudal fin which Russell described as terminating in two short setaceous cirri. The name is here treated as a substantive.

Remarks. — Under the account of *Xiphasia setifer* Smith (1949: 346) remarked: "Ordinarily stated to grow to 24 ins., but pearl divers of the tropical

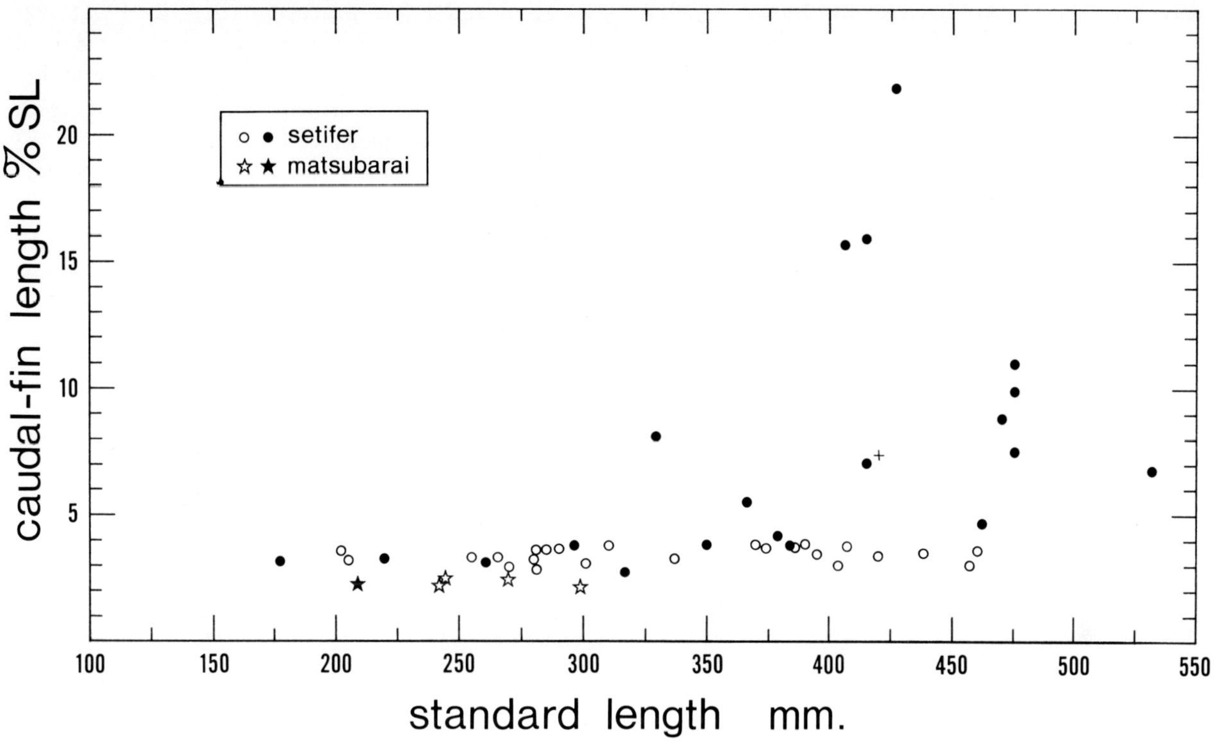

Figure 39. — Relationship between caudal-fin length versus standard length in species of *Xiphasia.* Closed symbols = males; open symbols = females. (One male *X. setifer,* 415 mm SL, that had the tip of caudal fin partially broken off, indicated by plus sign, was plotted on the figure.)

Pacific describe an eel-like creature several yards long much like this species. This they fear greatly, as it is said to attack with great ferocity, and once the great fangs are buried in a man he cannot escape, and drowns." The above observation, if indeed not based entirely on superstition, certainly does not apply to *X. setifer* for it has never been known to attack man.

Distribution (*Fig. 38*). — Widely distributed throughout the Indo-Pacific, but unknown east of the Andesite Line (refer to general discussion of Zoogeography) and largely confined to continental habitats.

The following acceptable distributional records for *X. setifer,* for which no specimens were available, are plotted on the distributional map: Vishakhapatnam, India (Russell, 1803: 28); Mangalore, India (Kuthalingam and Menon, 1965); Sri Lanka (De Silva, 1958); Dumaguete, Negros Is. (Herre, 1926: 225); Celebes, Lembeh Strait (Herre, 1939: 368); and the following Indonesia localities: Sibutu Passage, between Basilan and Jolo Islands; east of Cagayan Sulu Island; and Batangas Mkt., Luzon (Herre and Herald, 1951: 336-337).

The record of *setifer* from False Bay, South Africa (Penrith and Penrith, 1972: 84) is based on a misidentified specimen of *X. matsubarai.* The specimen has 10 pectoral-fin rays on each side, not 13 as they reported.

Material examined. — 76 specimens (16-532 mm SL) from 54 collections.

PERSIAN GULF: IMUK uncat. (1:46), R/V METEOR sta. 287, 26°45′N, 53°7′E; IMUK uncat. (1:16), R/V METEOR sta. 360, 27°4′N, 52°35′E.

GULF OF SUEZ: MNHN 1966-684 (2:405-423).

GULF OF AQABA: HUJ F5309 (1:415), Sinai Peninsula, bay at El Himeira.

RED SEA: MNHN 1966-685 (1:427), MNHN 1966-686 1:476), HUJ E58/326 (1:475), Ethiopia, "Eritrea."

GULF OF ADEN: IMUK uncat. (1:31), R/V METEOR sta. 95, 11°41′N, 48°52′E.

E. AFRICA: BMNH 1920.7.23.64 (1:316), Durban; RUSI

uncat. (1:283), "Natal Fisheries"; RUSI uncat. (1:301), small river E of Great Fish River, 33°30′S.

MADAGASCAR: MNHN 5545 (289, holotype of *Xiphogadus madagascariensis*), Maurounda River, from fish stomach.

INDIA: ZSIC F4730/2 (1:272), Madras, Royapuram Mkt.

ANDAMAN IS.: BMNH 1869.3.7.4 (1:238).

INDONESIA: ZMA 113.302 (2:255-321), Sumatra, Nias Is.

W. AUSTRALIA: CSIRO A1405 (1:330), WAM P5114 (1:338), WAM P11106 (1:249), Exmouth Gulf.

GULF OF THAILAND: CAS 29080 (5:350-438), CAS 29081 (1:471), CAS 29082 (1:462), CAS 29083 (1: 395), CAS 29084 (1:404), CAS 29085 (1:371), CAS 29086 (1:391), ANSP 124869 (4:285-310), ANSP 128699 (2:267-282), USNM 205292 (3:280-374).

SOUTH CHINA SEA: CAS 24700 (1:358), South Viet Nam, Da Nang; USNM 8293 (1:422), Hong Kong.

PHILIPPINE IS.: USNM 211134 (1:406), Manila Bay.

TAIWAN: CSC 70-143 (1:340), SIO 70-274 (1:408), USNM 192548 (1:460), USNM 205289 (1:384).

JAPAN: MCZ 36424 (1:378), Hondo, 32°28′N, 130°12′E; SU 26514 (1:366), Sagami Sea; USNM 149556 (1:532), Surugu Bay, Shimizu, 35°1′N, 138°29′E.

NEW HEBRIDES IS.: AMS I.11323 (1:221), Efate I.

NEW GUINEA: CSIRO C214 (1:221), Port Moresby; RMNH 26746 (2:204-205), Hollandia Bay.

E. AUSTRALIA: BMNH 1933.8.12.46 (1:173), Torres Strait, Maer I.; AMS I.12500 (1:475), Queensland, Dunk I., 17°55′S, 146°29′E; AMS IB.7455 (1:224), Queensland, Heron I.; AMS I.11488 (1:357), AMS I.14584 (1:237), AMS IB.634 (1:317), AMS IB.2172 (1:244), AMS IB. 2959 (1:203), AMS IB.3689 (1:242), New South Wales; AMS IB. 5231 (1:173), N. S. W., Bermagui, 36°28′S, 150°3′E, from tuna stomach.

NEW CALEDONIA: ANSP 124235 (1:265), USNM 126231 (1:177).

LOYALTY IS.: BMNH 1900.5.29.2 (1:261), Lifu I. (Lifou); ANSP 124555 (6:125-162), from stomach of *Thunnus albacares.*

LORD HOWE IS.: AMS IA.941 (1:270).

LOCALITY UNKNOWN: MCZ 12457 (1:155), MNHN 2123 (248.5, holotype of *Nemophis lessoni*).

Xiphasia matsubarai Okada and Suzuki

Specimen Figures 34, 132; Tables 4, 14.

Xiphasia matsubarai Okada and Suzuki, 1952: 75, unnumbered Fig. (type locality: Sea of Japan, off coast of Owashi; holotype FFUM 4052 [not examined]).

Description. — Dorsal fin XI, 99-104; total elements 110-115. Anal fin II, 97-104. Pectoral fin 10-11. Caudal fin: procurrent rays 2 + 1; segmented rays 10. Vertebrae: precaudal 13; caudal 100-104; total 113-117. Pleural ribs on vertebrae 3-12 or 13; epipleural ribs 8-11. Incisor teeth (Fig. 37) in specimens 194-299 mm SL, n = 8: lower jaw 16-22; upper jaw 17-20. Caudal fin 2.1-2.5% SL (Fig. 39).

Color pattern. — Head and body uniformly dark (pale brown in preservation) or with faint bands similar to those of *setifer.* Dorsal and caudal fins uniformly dark brown or blackish. Anterior two-thirds of anal fin white at least distally, the width of the white area variable, uniformly dark posteriorly. A good illustration of the species accompanied the original description. Life coloration unknown.

Remarks. — There is no evidence of secondary sexual dimorphism in *matsubarai* but, if the species attains a larger size than is suggested by the available specimens, this may not be valid. The largest male is only 265 mm SL, a size at which secondary sexually dimorphic characters are also not apparent in males of *setifer* (see Fig. 39).

Etymology. — Named for the late Dr. Kiyomatsu Matsubara, a famous Japanese ichthyologist.

Distribution (Fig. 38). — This rare species is widely distributed in the Indo-Pacific Region.

Material examined. — 22 specimens (140-298.5 mm SL) from 14 collections.

E. AFRICA: ANSP 53066 (1:193), S. Natal coast, between Pondoland and Durban; SAM 23692 (1:265), Muizenberg, False Bay, 34°10′S, 18°50′E.

MADAGASCAR: RUSI 964a (1:218), SW of Nossi Be, from stomach of *Thunnus albacares;* MNHN 2211 (1: 228), Nossi Be.

ANDAMAN IS.: BMNH 1869.3.7.4 (1:239).

INDONESIA: ZMA 113.304 (1:269), S coast of Flores I.

MARIANA IS.: NMFS 2765 (1:208), USNM 200562 (1: 224), Pagan I.; ANSP 123722 (4:140-195), ANSP 128537 (2:195-208), BPBM 15072 (3:194-205), between Alamagan and Guguan Is., from stomach of *Thunnus albacares.*

NORTH FIJI BASIN: SIO 64-239 (1:243), Alexa Bank, 11°40′S, 175°10′E.

SOLOMON IS.: USNM 203276 (1:244), Rennell I.

NEW GUINEA: ZMA 113.303 (1:229), S coast of New Guinea, coll. J. D. F. Hardenberg.

LORD HOWE IS.: AMS I.5345 (1:299).

LOCALITY UNKNOWN: MCZ 12456 (1:166).

MEIACANTHUS Norman

Meiacanthus Norman, 1943: 805 (type-species: *Petroscirtes oualanensis* Günther, 1880 [= *Meiacanthus atrodorsalis ovalauensis* (Günther)], by original designation).

Diagnosis. — Major characters useful in distinguishing the genera of Nemophini are summarized in Tables 1, 3 and 4. The grooved dentary canines (Fig. 43) and associated glands (Fig. 44) easily distinguish *Meiacanthus* from the other genera of Nemophini. The general physiognomy of the head of representative species of *Meiacanthus* is shown in Figures 40-42.

Etymology. — The generic name, a combination of the Greek *meion* (less) and *akantha* (thorn), alludes to the relatively few dorsal-fin spines in most species; gender masculine.

Biology. — *Meiacanthus* is unique among fishes in having highly specialized toxic buccal glands. Springer and Smith-Vaniz (1972a) and Losey (1972) give fairly detailed accounts of the general biology of *M. nigrolineatus* and *atrodorsalis* respectively, with primary emphasis devoted to the value of these glands as defense organs. Predation experiments show that *Meiacanthus* is generally unacceptable as prey and that naive predators rapidly learn to ignore it. Especially important is the fact that individuals of *Meiacanthus* are frequently rejected unharmed after being ingested by a naive predator. After taking a *Meiacanthus* into their mouths, some predators exhibit violent quivering of the head with distension of the jaws and operculi. The relatively large number of fishes that mimic species of *Meiacanthus* (see Fig. 9) attest to the effectiveness of the toxic glands in providing immunity from predation. A brief description of the glands and their arrangement is given below.

Each gland is positioned at the base of a large, elaborately grooved dentary canine (Fig. 43). The dentary bones are also specialized to accommodate the gland and canine (Fig. 44). It is especially interesting to note that in each of the three basic gland and dentary types a flange of bone guards the base of the canine anteriorly (best seen in the mesial views in the figure). Fishelson (1974) gave a detailed description of the histology and ultrastructure of the gland in *M. nigrolineatus.* Slight pressure on the glands, which lack ducts, exudes a

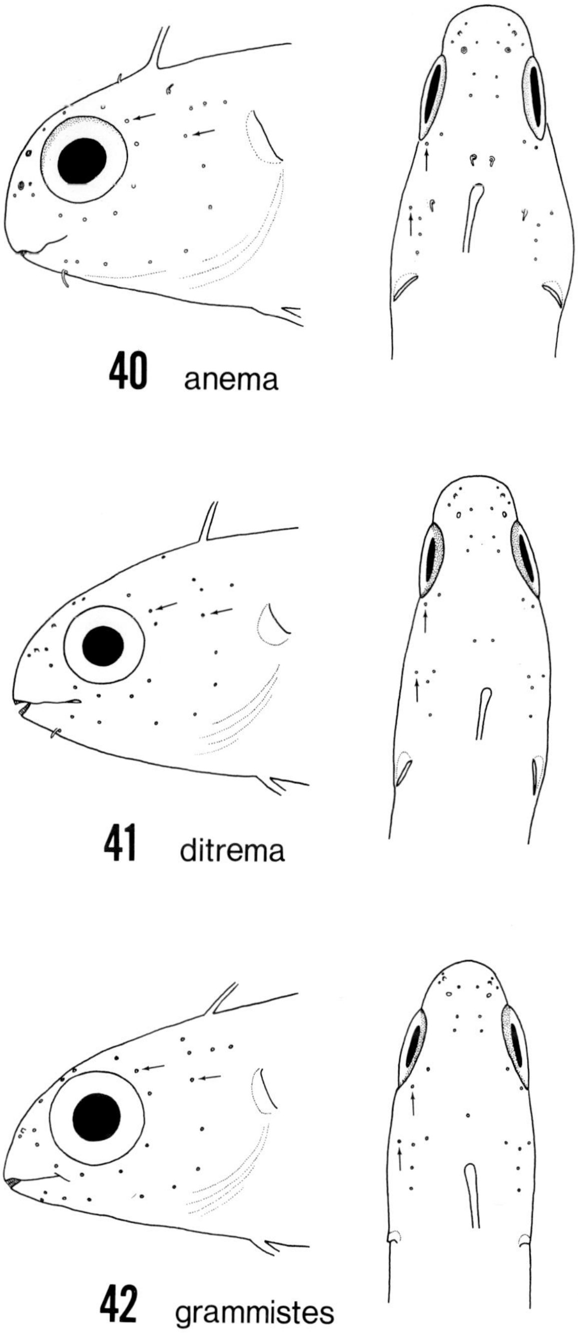

Figures 40-42. — General physiognomy and cephalic sensory pore distributions in selected species of *Meiacanthus:* Fig. 40, *anema,* USNM 211959, male, 42.0 mm, Indonesia, Banda Sea; Fig. 41, *ditrema,* BPBM 7453, male, 39.3 mm, Ryukyu Islands; Fig. 42, *grammistes,* ANSP 124879, female, 43.2 mm, Palau Islands. Small arrows identify ventralmost supraorbital and dorsalmost preopercular pores in both lateral and dorsal views.

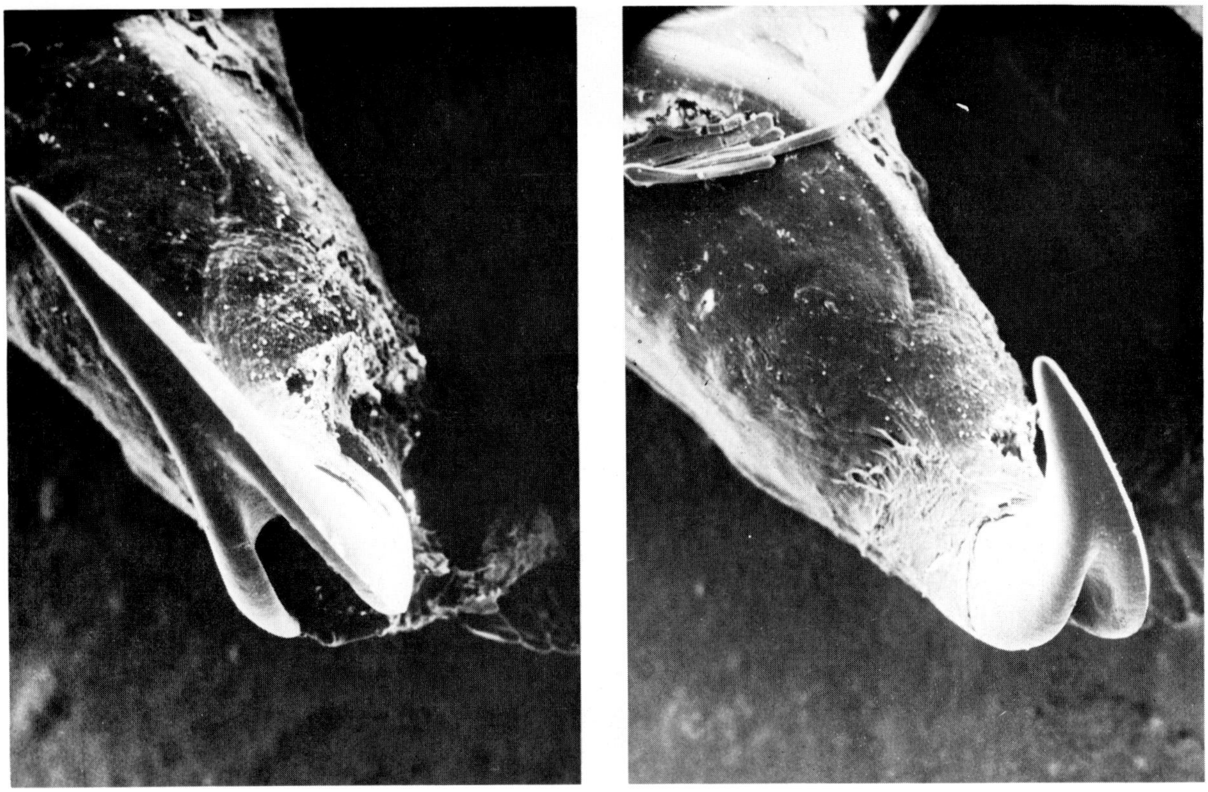

Figure 43. — Anterodorsal views of right dentary canine of *Meiacanthus atrodorsalis* showing deep groove along anterior surface (Scanning Electron Photomicrograph, X 50).

pale secretion that tends to be channeled along the groove of the canine. The general morphology of the glands is similar to that of various poison glands in the dermis of other fishes. The gland is formed by invagination of the buccal epithelium. As in other poison glands of fishes, it has a holocrine secretory function, in which the ripe cells exude their products, then degenerate and are replaced by new cells from the peripheral germinative layer.

The behavior and ecology of *Meiacanthus* differs fundamentally from that of most blennies. The combination of a well-developed swimbladder and reduced vulnerability to predation has allowed *Meiacanthus* to feed in the water column and forage over a wide area. The deepest sighting of a *Meiacanthus* (*nigrolineatus*) is 37 meters; but most collections have come from shallow reef habitats seldom exceeding 20 meters. *M. anema* has been collected only in estuarine or freshwater habitats, often in depths shallower than one meter. Limited observations of *M. ditrema* suggest that its biology is also different from that of other species. On two occasions groups of at least 40 individuals have been observed in back reef areas hovering in the water column. Bruce A. Carlson (*in litt.*) reported sighting a compact and relatively stationary "school" of about 50 individuals feeding in about 3 meters of water at Suva, Fiji. Differences in the relative numbers of teeth, especially dentary incisors, in some species (Fig. 45) also suggest slightly different feeding habits.

A recent study by Fishelson (1975) gives an excellent account of the territorial behavior, courtship and reproduction of *Meiacanthus nigrolineatus*. This paper is especially valuable because it demonstrates that *Meiacanthus* can be successfully spawned and reared from the egg in the laboratory. The following is a brief summary of some of Fishelson's obser-

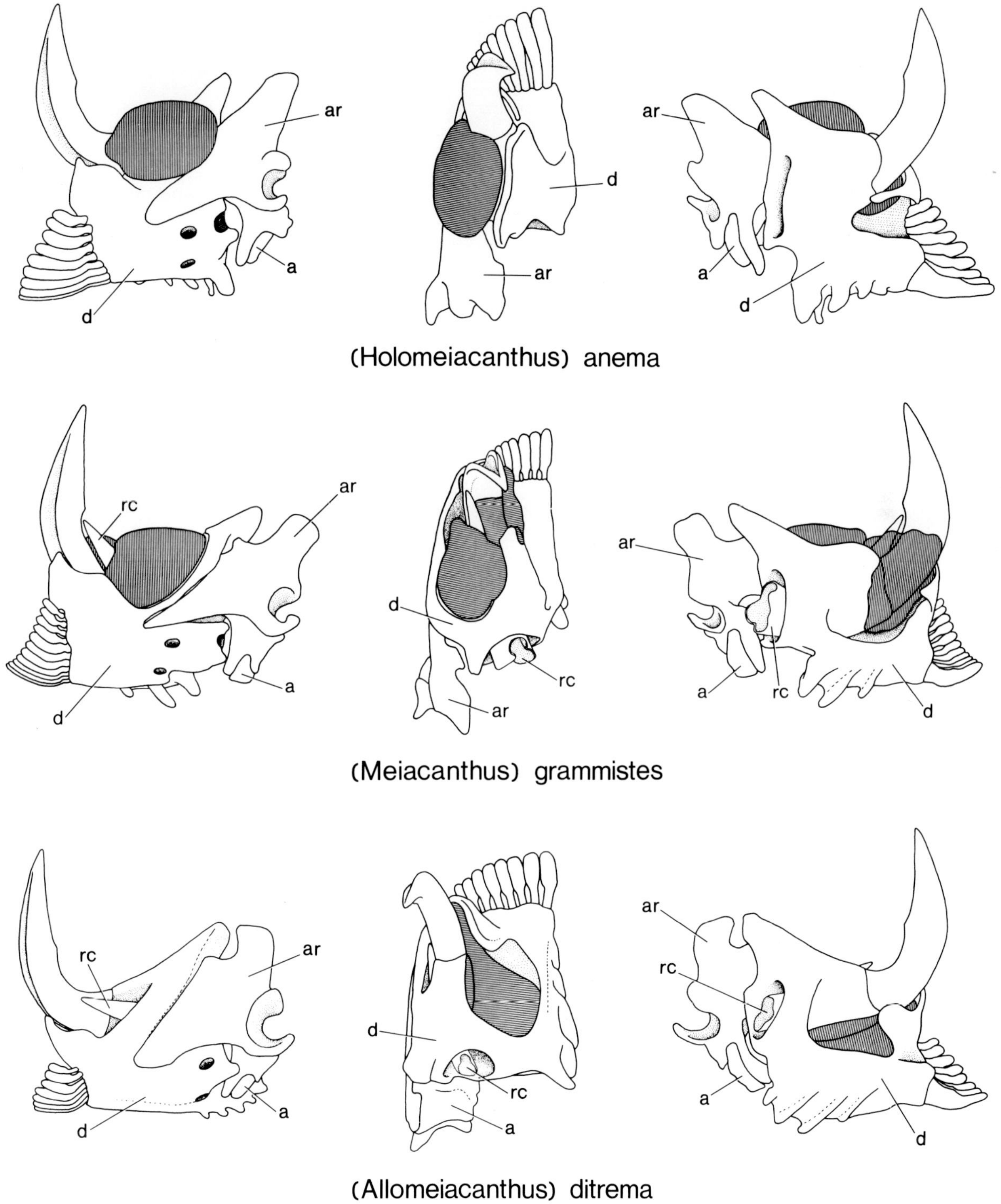

Figure 44. — Lower jaws (lateral, dorsal and mesial views) of three species of *Meiacanthus: M.* (*Holomeiacanthus*) *anema,* ANSP 124867, 40.3 mm; *M.* (*Meiacanthus*) *grammistes,* ANSP 128536, 42.9 mm; *M.* (*Allomeiacanthus*) *ditrema,* ANSP 124891, 26.2 mm. Dentary gland shaded. Abbreviations: d, dentary; ar, articular; a, angular; rc, replacement canine.

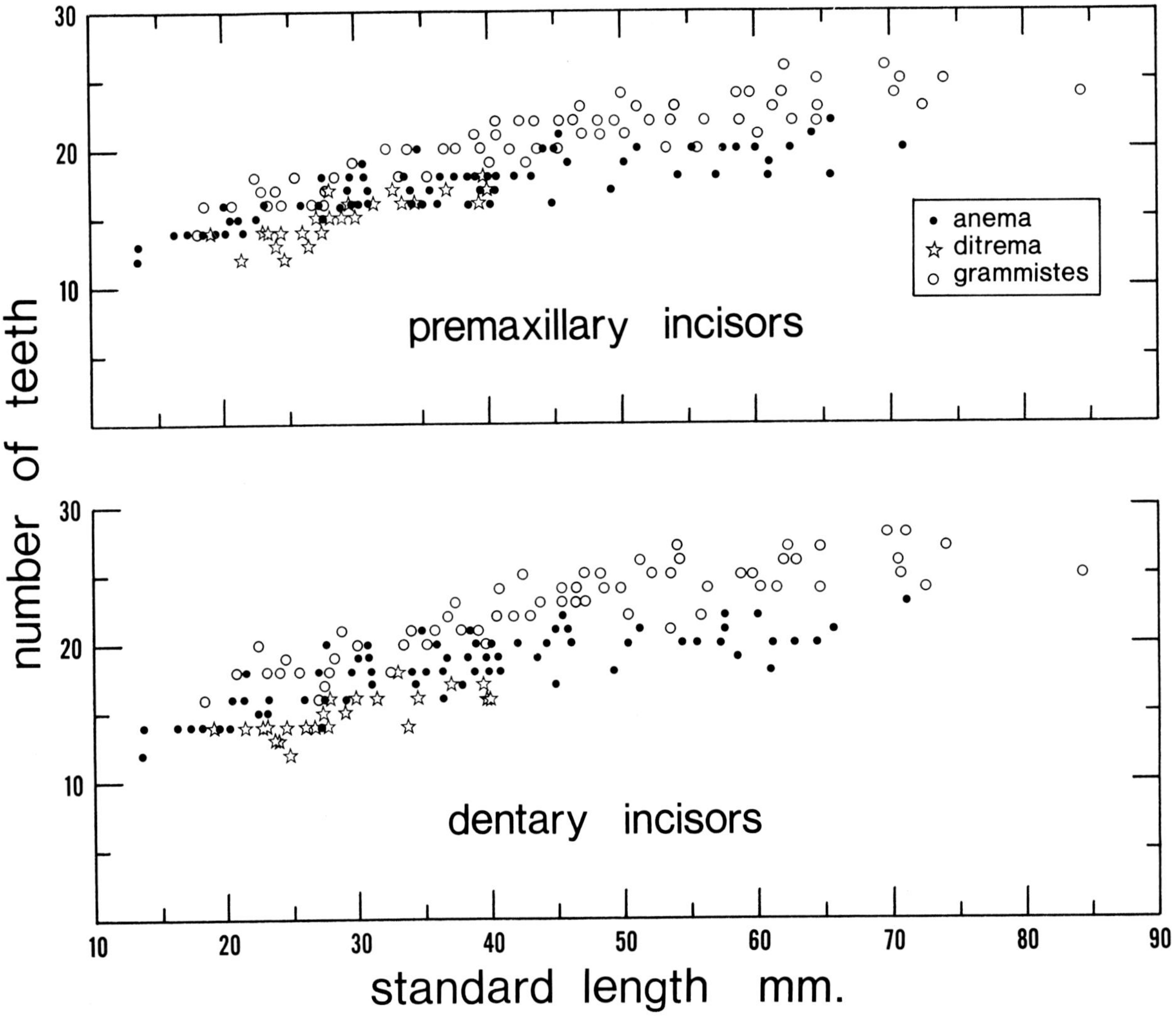

Figure 45. — Relationship between number of teeth versus standard length in selected species of *Meiacanthus*.

vations, which I believe are probably true for most species of *Meiacanthus*. *M. nigrolineatus* is usually observed swimming in the open, rarely resting on the substrate. During the day the activities of the fish are mostly concerned with feeding, with peak feeding times just after sunrise and early in the afternoon. The fish usually swim up into the water column or hover head down when searching for food. They tend to jerk from spot to spot and, hovering over one place, detect prey visually and seize it with a short strike. Benthic invertebrates, such as amphipods, isopods and young stages of crabs and small polychaetes are major food items. They also feed on plankton, while facing into the current. On such occasions, they tend to orient vertically, head up, focusing their mobile eyes on particles passing by. Detecting a prey, they thrust forward, seize, and then immediately turn back to the substrate.

Tubes of sponges and empty shells, especially vermetid tube-mollusks are utilized as shelters. They prefer vertical tubes and usually enter the tubes tail first. Groups of 10 to 15 juveniles or sub-adult fish sometimes inhabit tube-forming colonies of sponges, but adult males and females occupy separate terri-

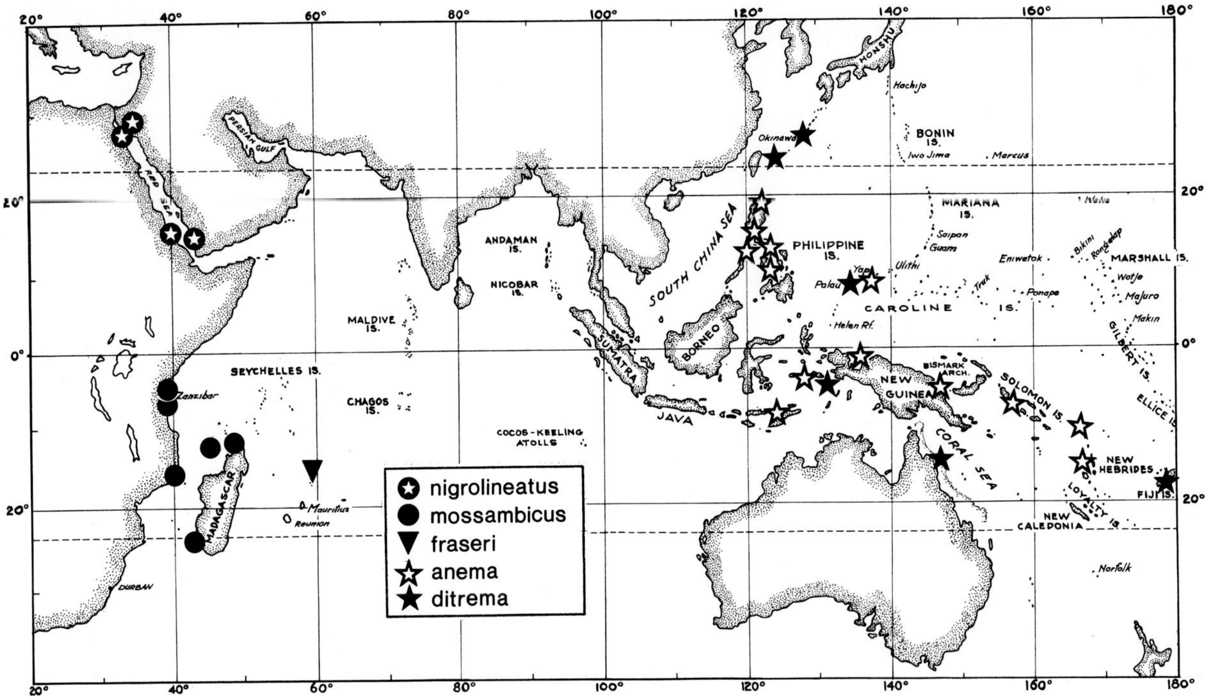

Figure 46. — Distributional records for selected species of *Meiacanthus*.

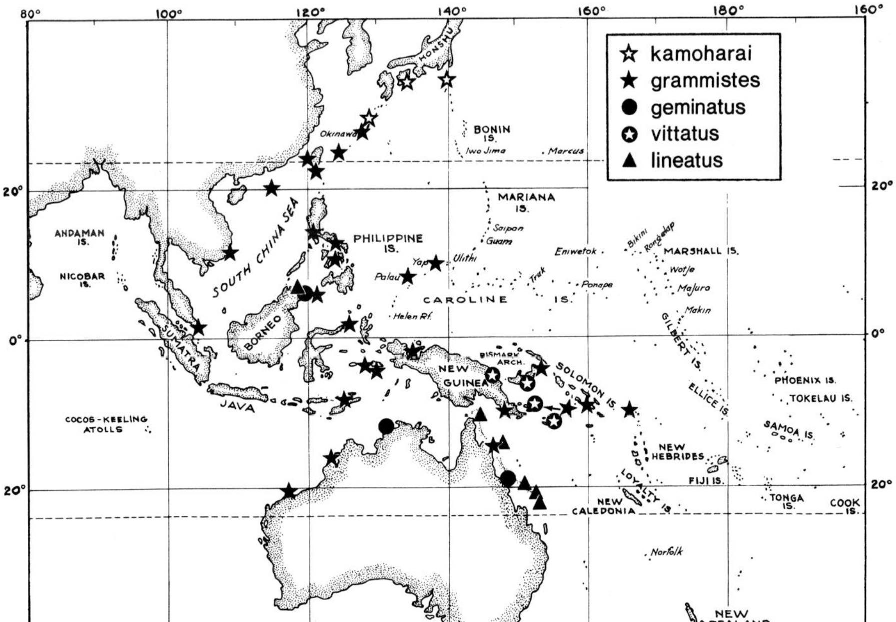

Figure 47. — Distributional records for selected species of *Meiacanthus*.

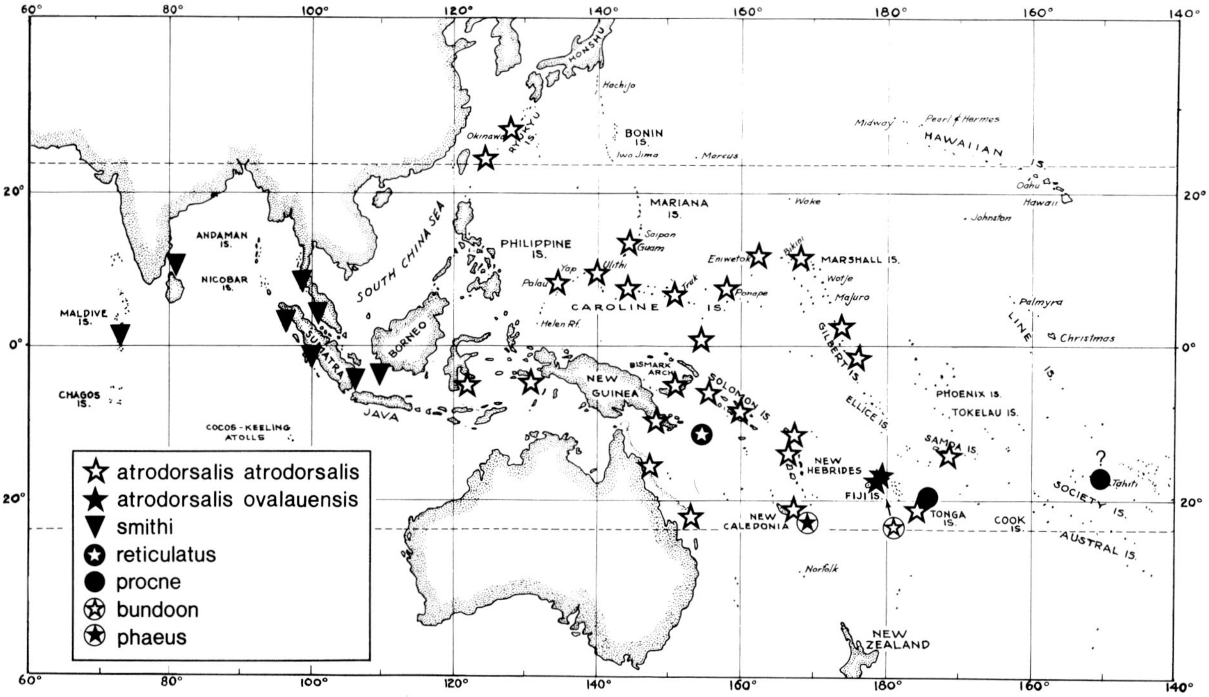

Figure 48. — Distributional records for selected species of *Meiacanthus.*

tories. The boundaries of such territories have a radius of one or two meters. Adults over 50 mm TL usually behave agonistically towards intruders, especially conspecific males.

Reproduction occurs in the spring and summer in fish maintained in aquaria and kept on an essentially natural day-night cycle. Following courtship, the male leads receptive females to their nesting tubes where spawning occurs. The female deposits only 5 to 10 eggs at a time, with 100 to 160 total eggs deposited during approximately 1½ days of spawning activity. Several females may spawn with a single male, who cares for the eggs until they hatch. At a temperature of 24°C (± 1°), the eggs hatch in 10 to 11 days. The larvae are at first planktonic, but after about 20 days they begin to show signs of sedentary behavior.

Several species of *Meiacanthus* have spectacularly beautiful life colors, which may be aposematic in function, and all would probably adapt well to an aquarium. At least one species, the "canary blenny", *M. atrodorsalis ovalauensis,* is already a popular marine tropical aquarium fish.

Zoogeography. — The distribution of *Meiacanthus* (Figs. 46-48) extends from the northern Red Sea and southern Madagascar in the western Indian Ocean and to the Society Islands in the central Pacific Ocean. Four species of *Meiacanthus* are known from the western Caroline Islands, but only *M. atrodorsalis* occurs throughout the remainder of Micronesia. No species of *Meiacanthus* is known from the Persian Gulf-Arabian Sea, perhaps as a result of inadequate collecting, and three species are known from single collections. The distributional hiatus between New Guinea and eastern Australia is surprisingly strong, only two of seven species of *Meiacanthus* found in the region occur on both sides of Torres Strait.

Intrageneric relationships. — The three subgenera of *Meiacanthus* are compared in Table 15. The following phylogenetic character analysis indicates: the monophyletic nature of the 14 species assigned to the nominal subgenus, the sister-group relationship between the subgenus *Meiacanthus* and *M.* (*Allomeiacanthus*) *ditrema,* and establishes *M.* (*Holomeiacanthus*) *anema* as the most primitive species.

M. ditrema is the most derived species of *Meiacanthus* and is extremely specialized in several of its characters, although sharing at least two plesiomorphic characters exclusively with *Holomeiacanthus*. *M.* (*Allomeiacanthus*) *ditrema* is accorded separate subgeneric status to call attention to its evolutionary divergence. Since it is phylogenetically most closely related to the nominal subgenus (Fig. 49), it was also necessary to erect a new, monotypic subgenus (*Holomeiacanthus*) for *M. anema*.

Briggs (1974: 253) has noted that one way for a "primitive" species to persist in major evolutionary centers is to shift into a different niche. *M. anema* appears to be a good example of a species that has avoided competition from its more specialized relatives by this device. It is the only species of *Meiacanthus* known to occur in estuarine or freshwater habitats, and most collections of *anema* have come from such habitats.

The remaining 14 species of *Meiacanthus* appear to be the product of recent speciation, judging from the high incidence of endemism and their relatively trivial distinguishing characters. Within this group the relative development of secondary sexual dimorphism (lengths of anal- and caudal-fin rays and degree to which inner caudal-fin rays are exserted beyond the margin of the interradial membranes) and color pattern seem to be the most important taxonomic characters. Although several species-pairs are identifiable (refer to individual species accounts), their interrelationships are difficult to assess.

Numbers given below correspond to those shown in Figure 49. For each character (or character complex) the plesiomorphic condition is given a lower case letter and the apomorphic condition a capital letter.

1. Dentary gland positioned dorsally with little or no modification of the dentary or articular bones to accommodate the gland (a). Dentary gland positioned either ventrally or dorsally, but with modifications of the surrounding or supporting bones (Fig. 44) (A). It should be stressed here that synapomorphy does not require that the shared apomorphic characters be expressed identically.

2. Bony interorbital width relatively narrow (b). Bony interorbital width relatively wide (Fig. 91) (B).

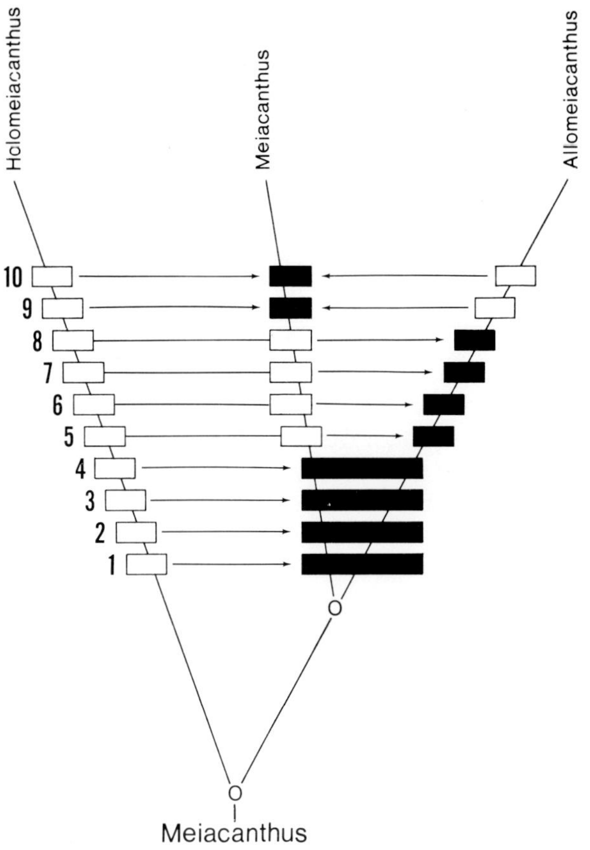

Figure 49.—A phylogenetic analysis of the three subgenera of *Meiacanthus*. Open squares are primitive (plesiomorph) character expressions and shaded squares are derived (apomorph) character expressions. Numbers correspond to characters discussed in text. Arrows are inferred directions of evolutionary change and do not imply ancestor-descendent relationships.

3. Segmented caudal-fin rays usually 13 (see Table 17) (c). Segmented caudal-fin rays 11 or 12 (rarely 13) (C). *M. anema* is the only species of Nemophini that typically has as many as 13 segmented caudal-fin rays.

4. Dorsal-fin spines 6-10, typically 8 (d). Dorsal-fin spines 3-6, rarely 6 (D). No other blenniid genus has fewer than 6 dorsal-fin spines, and most species of Nemophini have 8 or more.

5. Major portion of dentary gland positioned dorsally (e). Major portion of dentary gland positioned ventrally (E). It would be relatively easy to derive the slightly more complex gland arrange-

ment of *Meiacanthus* from that found in *Holomeiacanthus,* while the ventrally encapsulated gland of *Allomeiacanthus* would require much greater structural modification of the dentary bone.

6. Three mandibular pores (f). This is the number of pores found in most species of Nemophini (see Table 1), and in the Phenablenniini and most genera of the Omobranchini. The two mandibular pores (F) of *Allomeiacanthus* are probably correlated with the ventrally encapsulated gland and robust nature of the dentary bone in that subgenus.

7. Three lateral temporal pores and well-developed posttemporal bone (g). Two lateral temporal pores and reduced posttemporal bone (Fig. 50b) (G). The number of lateral temporal pores is determined by the shape and position of the bones associated with the temporal canal. *M.* (*Allomeiacanthus*) *ditrema* is the only species of Nemophini with two lateral temporal pores.

8. Lateral line present (h). Lateral line absent (H).

9. Four infraorbital bones (i). Three infraorbital bones (Fig. 94) (I). Four is the number of infraorbital bones found in all other genera of Nemophini, and in the Phenablenniini; many species of Omobranchini have 4 or 5.

10. Number of mid-dorsal supratemporal pores and alignment of the two branches of the supratemporal canal. These two characters are correlated in the genus *Meiacanthus* because the alignment of the two branches of the ST canal determines the number of mid-dorsal ST pores. A pair of mid-dorsal ST pores and the two branches of the ST canal aligned parallel to each other anteriorly (j). A single ST pore and the two branches of the ST canal forming an obtuse angle with the apex positioned anteriorly (J). In all blenniids (except *Xiphasia*) the ST canal terminates in front of the dorsal-fin origin. When the dorsal fin is positioned far forward, the two branches of the ST canal tend to become aligned parallel to the dorsal-fin pterygiophores (Fig. 91); this arrangement typically results in a separate pore near the anterior termination of each branch of the ST canal. The high number of dorsal-fin spines in proportion to the total number of dorsal-fin elements (see Table 16) accounts for the more anteriorly positioned dorsal-fin of *Holomeiacanthus.* In the Nemophini the spacing between adjacent dorsal-fin spines (especially those whose supporting pterygiophores rest on the skull) is wider than the spacing between segmented dorsal-fin rays (Fig. 94). Thus, in *Holomeiacanthus* the high number of dorsal-fin spines (a plesiomorphic condition for the genus, see no. 4 above) accounts for character state 10j. The retention of the same character state in *Allomeiacanthus* (Fig. 91), which has an intermediate number of dorsal-fin spines, suggests that reduction in number of dorsal-fin spines may have been a relatively recent evolutionary event in this subgenus.

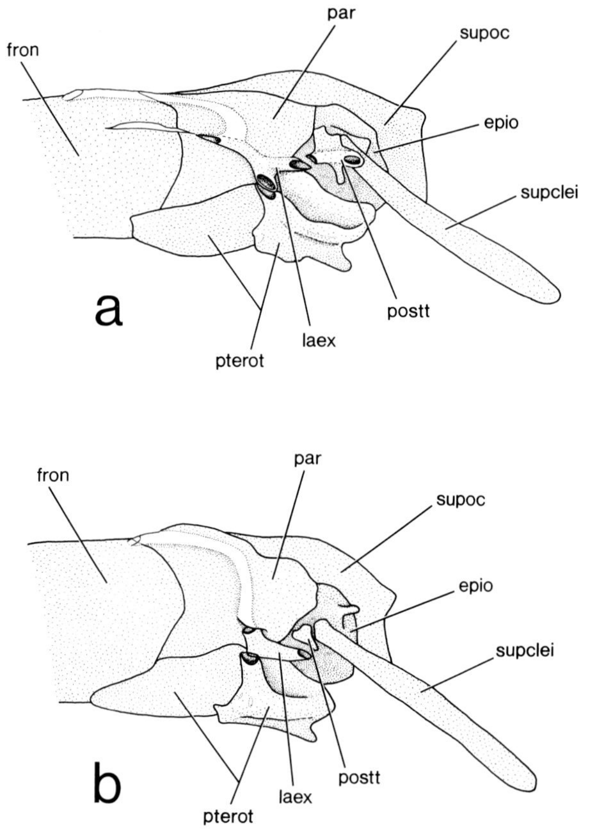

Figure 50. — Lateral view of skull in two species of *Meiacanthus* showing relative development of posttemporal bone: a, *M.* (*Holomeiacanthus*) *anema,* ANSP 128535, 38.8 mm; b, *M.* (*Allomeiacanthus*) *ditrema,* USNM 195706, 37.1 mm. Abbreviations: fron = frontal; par = parietal; supoc = supraoccipital; epio = epiotic; supclei = supracleithrum; postt = posttemporal; laex = lateral extrascapular; pterot = pterotic.

TABLE 15. — Comparison of the subgenera of *Meiacanthus*.

Characters	*Holomeiacanthus*	*Meiacanthus*	*Allomeiacanthus*
Dentary gland (major portion) see Fig. 44	positioned dorsolaterally; supported by dentary and articular, held in place laterally by flange of articular	positioned dorsolaterally; supported solely by dentary, held in place laterally by flange of dentary	positioned ventrally; encapsulated in dentary
Posttemporal bone see Fig. 50	well developed; moderate ventral arm	well developed; moderate ventral arm	reduced; no ventral arm
Infraorbital bones	4	3	4
Supratemporal pores	4 (1-2-1)	3 (1-1-1)	4 (1-2-1)
Lateral temporal pores	3	3	2
Mandibular pores	3	3	2
Lateral line	present	present	absent
Dorsal-fin spines	6-10 (usually 8)	3-6 (rarely 6)	5-6
Segmented caudal-fin rays	11-13 (usually 13)	11-13 (infrequently 13)	11

KEY TO SPECIES OF *MEIACANTHUS*

1. Dentary gland encapsulated in dentary (Fig. 44); mandibular pores 2; lateral line absent; lateral stripe forks on head, dorsal arm of fork extends through eye and ventral arm extends to gape; (western and central Pacific) .... *ditrema*

Dentary gland not encapsulated in dentary; mandibular pores 3; lateral line present; color pattern not as above **2**

2. Mid-point of supratemporal canal typically ending in a pair of pores (Fig. 41); Dorsal-fin spines 6-10 (rarely 6); segmented dorsal-fin rays 20-24; (western and central Pacific) .... *anema*

Mid-point of supratemporal canal typically ending in a single pore; Dorsal-fin spines 3-6 (rarely 6); segmented dorsal-fin rays 23-28 .... **3**

3. Adults with two or more dark stripes on sides of body and/or ventral surface of head and belly dark and strongly vermiculated **4**

Color pattern of adults not as above **6**

4. Caudal fin immaculate or with dark stripes restricted to proximal fourth of fin; dark stripes on sides clearly defined posteriorly; (eastern Australia and Sulu Archipelago) .... *lineatus*

Caudal fin heavily pigmented on dorsal and ventral margins and/or middle rays distinctly spotted; dark stripes on body interrupted posteriorly or replaced by conspicuous dark spots **5**

5. Ventral surface of head and belly uniformly pale (throat with 2 or 3 spots in some specimens); dark stripes on head approximately equal in width to pale interspaces; (western and central Pacific) .... *grammistes*

Ventral surface of head and belly dark or vermiculated; dark stripes on head, if present, approximately twice the width of pale interspaces; (Japan and Ryukyu Islands) .... *kamoharai*

6. Dark lateral stripe extends from snout to caudal fin **7**
Color pattern not as above **8**

7. Dark lateral stripe extends onto dorsal part of pectoral-fin base; greatest width of lateral stripe equal to eye diameter; dorsal fin uniformly pale; (New Guinea and New Britain) .... *vittatus*
Dark lateral stripe does not extend onto pectoral-fin base; greatest width of lateral stripe narrower than eye diameter; dorsal fin dark proximally with narrow pale margin; (Borneo and northern Australia) .... *geminatus*

8. Axil of pectoral fin with dark spot approximately equal to eye diameter (Figs. 154-155); (western and central Pacific) .... *atrodorsalis*
Axil of pectoral fin without dark spot **9**

9. Anal fin pale **10**
Anal fin dark **13**

10. Outer lobes of caudal fin not elongated, middle rays longest; dark stripe extends from eye to origin of dorsal fin; (eastern Indian Ocean and Java Sea) .... *smithi*
Outer lobes of caudal fin elongated, at least in adults; color pattern not as above **11**

11. Narrow dark stripe extends obliquely from eye to below posterior third of dorsal fin; (Red Sea) .... *nigrolineatus*
Color pattern not as above **12**

12. Dorsal fin heavily and uniformly pigmented, except for pale tips of rays; posterior third of body uniformly pale; (western Indian Ocean) .... *fraseri*
Dorsal fin dark with central pale stripe; posterior third of body dark dorsally, spotted or speckled in large specimens; (Tonga and Society Islands) .. *procne*

13. Caudal fin immaculate; (western Indian Ocean) .... *mossambicus*
Outer lobes of caudal fin heavily pigmented **14**

14. Body distinctly reticulated; pectoral-fin rays 14; (Solomon Sea) .... *reticulatus*
Color pattern not as above; pectoral-fin rays 15 or 16 .... **15**

15. Outer lobes of caudal fin not distinctly darker than posterior fifth of body; head and body very dark except for pale wedge-shaped area (yellow-orange in life), base of which covers postorbital region of head; (Fiji Islands) .... *bundoon*
Outer lobes of caudal fin distinctly darker than posterior fifth of body; head and body uniformly light brown (in preservation); (New Caledonia) .... *phaeus*

Holomeiacanthus, new subgenus

Type-species. — *Meiacanthus anema* (Bleeker)

Diagnosis. — A subgenus of *Meiacanthus* with major portion of dentary gland positioned dorsolaterally and supported by dentary and articular (Fig. 44). Posttemporal bone well-developed, with short ventral arm (Fig. 50a). Infraorbital bones 4. Mid-point of supratemporal canal with a pair of pores; lateral temporal pores 3; mandibular pores 3. Lateral line present, consisting of a series of simple disconnected tubes that may terminate anywhere from below the base of fifth spine to second segmented ray of dorsal fin. Dorsal-fin spines 6-10 (rarely 6).

Comparisons. — Refer to Table 15 and general discussion of relationships.

Etymology. — A combination of the Greek *holos* (whole, entire) and *Meiacanthus* (a genus of blenniid fishes), in reference to the hypothesis that the type-species of the new subgenus most closely approximates the progenitor that gave rise to *Meiacanthus*.

Meiacanthus (*Holomeiacanthus*) *anema* (Bleeker)

Specimen Figures 40, 51, 133-134; Tables 4, 15-17.

Petroskirtes anema Bleeker, 1852: 273 (type locality: Amboina; holotype RMNH 26468).

Petroscirtes kulambangrae Herre, 1931: 9 (name only); Herre, 1935: 436 (type locality: Solomon Islands, Kulambangra Island; holotype FMNH 17392 [not examined]).

Description. — Characters of the subgenus. Dorsal fin VI-X (usually VIII), 20-24; total elements 28-31. Anal fin II, 16-19. Pectoral fin 13-16. Caudal fin: procurrent rays 5-7 + 5-7 = 10-14; segmented rays 11-13 (usually 13). Vertebrae: precaudal 11-12 (13 in 1 of 250 specimens); caudal 20-24; total 32-35. Pleural ribs on vertebrae 3-11 or 12; epipleural ribs 11-13. Incisor teeth (Fig. 45) in specimens 21-71 mm, n = 74: lower jaw 14-23; upper jaw 14-21.

Sexual dimorphism well developed in pelvic fins, moderate in caudal fin in specimens > 30 mm SL (Fig. 52). Pelvic fin: males 11.9-34.4% SL, females 9.9-13.9% SL; caudal fin: males 20.3-31.7% SL, females 19.5-22.4% SL. Outer lobes of caudal fin not elongated, interradial membranes of inner rays weakly incised in adults. Large males (only) occasionally have cirri (Fig. 51).

Color pattern. — Females and juvenile males have three dark stripes. The dorsal stripe, widest anteriorly, extends from top of head along dorsal contour of body and dorsal-fin base to end of fin. The posterior half of the stripe is restricted to the fin base. The middle stripe extends from the tip of upper lip through the middle of eye and above the pectoral fin to the dorsal half of caudal-fin base, tapering to a point on the proximal half of fin. The ventral-most stripe extends from the middle of the lower jaw, below the eye and through the base of pectoral fin to the ventral half of caudal-fin base, tapering to a point as in dorsal counterpart. The pale area (white in life) between the lower stripes is slightly wider than the bordering stripes, which are usually

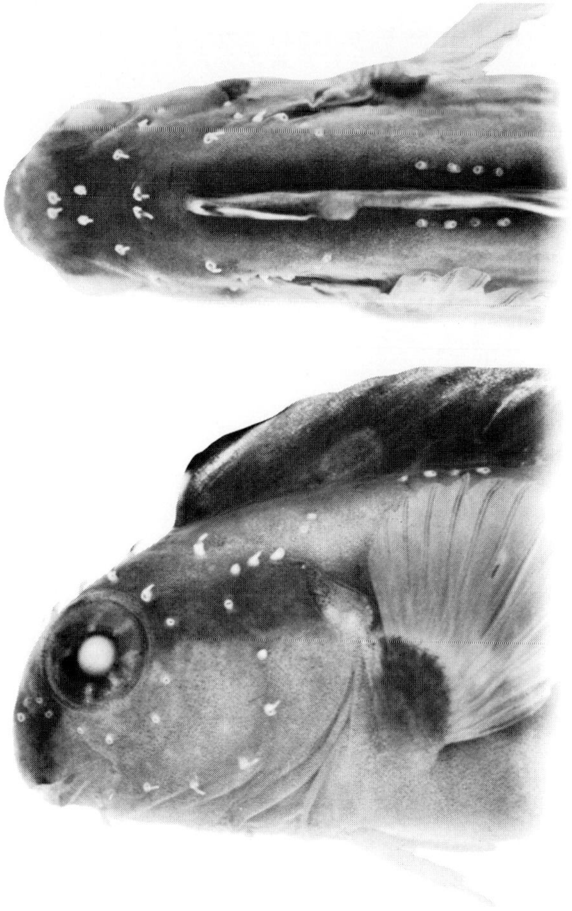

Figure 51. — Dorsal and lateral views of head of *Meiacanthus anema,* RMNH 25232, male, 43.6 mm, New Guinea.

scalloped on their inner margins. The ventral surface of the head and throat has a small median spot followed by a broad V-shaped marking, the apex of which is directed posteriorly. The anal fin is also pale except for narrow dusky margin.

In adult males, the ventral stripe is very diffuse, apparent only on the base of the pectoral fin, caudal peduncle and caudal fin. The head is very dark (except for the pale pores and, in the case of large males, associated cirri — see Fig. 51) and there are no markings on the ventral surface of the head. The dark pigment in the proximal region of the dorsal fin is more extensive, especially anteriorly, and the interradial membrane of the first and second spines has an intense black spot distally (Fig. 51). The border of the anal fin is also darker in adult males.

Table 16. — Frequency distributons of dorsal-fin rays and vertebrae in species of *Meiacanthus*.[1]

Species	Dorsal-fin spines 3	4	5	6	7	8	9	10	N	x̄
anema				1	10	223	17	1	252	8.0
ditrema			35*	2					37	5.1
lineatus		89							89	4.0
grammistes	9	155	3						167	4.0
kamoharai		10							10	4.0
vittatus	1	19*							20	4.0
geminatus		4							4	4.0
nigrolineatus		100	7						107	4.1
mossambicus		3	22						25	4.9
fraseri		3	24*						27	4.9
smithi	1	33	1						35	4.0
a. atrodorsalis	1	134	5	1					141	4.0
a. ovalauensis		41	1						42	4.0
procne		1	14*						15	4.9
reticulatus			1						1	5.0
bundoon		1*	4						5	4.8
phaeus		1							1	4.0

Species	Segmented dorsal-fin rays 20	21	22	23	24	25	26	27	28	N	x̄
anema	13	146	88	4	1					252	21.3
ditrema			1	6	24*	6				37	23.9
lineatus						2	58	26	3	89	26.3
grammistes						51	103	12	1	167	25.8
kamoharai						1	5	4		10	26.3
vittatus						5	9*	6		20	26.1
geminatus							3*	1		4	26.3
nigrolineatus				2	—	65				107	24.7
mossambicus					1	14	9	1		25	25.4
fraseri					9	18*				27	24.7
smithi					1	12	22			35	25.6
a. atrodorsalis						15	68	51	7	141	26.4
a. ovalauensis						1	4	29	8	42	27.1
procne						9	6*			15	25.4
reticulatus							1			1	26.0
bundoon							2	3*		5	26.6
phaeus								1		1	27.0

Species	Total dorsal-fin elements 28	29	30	31	32	N	x̄
anema	9	147	88	8		252	29.4
ditrema	7	23*	7			37	29.0
lineatus		2	58	26	3	89	30.3
grammistes		55	100	12		167	29.7
kamoharai		1	5	4		10	30.3
vittatus		5	10*	5		20	30.0
geminatus			3*	1		4	30.3
nigrolineatus	31	67	9			107	28.8
mossambicus		2	14	9		25	30.3
fraseri		12	15*			27	29.6
smithi	1	12	22			35	29.6
a. atrodorsalis		13	68	52	8	141	30.4
a. ovalauensis			5	29	8	42	31.1
procne			10	5*		15	30.3
reticulatus				1		1	31.0
bundoon				3*	2	5	31.4
phaeus				1		1	31.0

Species	Precaudal vertebrae 11	12	13	14	N	x̄
anema	141	108	1		250	11.4
ditrema	11	25*			36	11.7
lineatus		2	60	26	88	13.3
grammistes			23	142	165	13.9
kamoharai				9	9	14.0
vittatus			17*	1	18	13.1
geminatus			4		4	13.0
nigrolineatus		42	60		102	12.6
mossambicus			16	9	25	13.4
fraseri		9	18*		27	12.7
smithi		11	23	1	35	12.7
a. atrodorsalis		17	113	5	135	12.9
a. ovalauensis			32	8	40	13.2
procne			2*	13	15	13.9
reticulatus			1		1	13.0
bundoon			5		5	13.0
phaeus				1	1	14.0

Species	Caudal vertebrae 19	20	21	22	23	24	25	N	x̄
anema		5	56	115	68	6		250	22.1
ditrema			9*	22	4	1		36	21.9
lineatus		2	22	52	12			88	21.8
grammistes	1	67	81	15	1			165	20.7
kamoharai			4	5				9	21.6
vittatus		6	7*	5				18	20.9
geminatus			2*	1	1			4	21.8
nigrolineatus		22	57	23				102	21.0
mossambicus			5	15	5			25	22.0
fraseri			4	20*	3			27	22.0
smithi			26	9				35	21.3
a. atrodorsalis		5	28	75	25	1	1	135	21.9
a. ovalauensis			4	23	13			40	22.2
procne			7	8*				15	21.5
reticulatus					1			1	23.0
bundoon					1*	3	1	5	24.0
phaeus				1				1	22.0

Species	Total vertebrae 32	33	34	35	36	37	38	N	x̄
anema	8	121	108	13				250	33.5
ditrema		15*	20	1				36	33.6
lineatus			8	61	19			88	35.1
grammistes		2	82	71	10			165	34.5
kamoharai				4	5			9	35.6
vittatus		5	8*	5				18	34.0
geminatus			2*	1	1			4	34.8
nigrolineatus	1	41	59	1				102	33.6
mossambicus			2	13	10			25	35.3
fraseri			10	17*				27	34.6
smithi		7	22	6				35	34.0
a. atrodorsalis		5	34	74	20	2		135	34.9
a. ovalauensis			2	21	15	2		40	35.4
procne				9*	6			15	35.4
reticulatus					1			1	36.0
bundoon					1*	3	1	5	37.0
phaeus					1			1	36.0

[1] When counts vary, those for holotypes of new species are indicated by an asterisk.

TABLE 17. — Frequency distributions of certain meristic characters in species of *Meiacanthus*.[1]

Species	Segmented anal-fin rays 14	15	16	17	18	19	N	x̄
anema			42	153	56	1	252	17.1
ditrema		2	16*	17	2		37	16.5
lineatus	1	37	45	5			88	15.6
grammistes	27	111	29				167	15.0
kamoharai		5	5				10	15.5
vittatus	3	13*	4				20	15.1
geminatus		1*	1	1	1		4	16.5
nigrolineatus	12	78	15				105	15.0
mossambicus		6	16	3			25	15.9
fraseri		11	16*				27	15.6
smithi		10	22	3			35	15.8
a. atrodorsalis		11	61	59	9		140	16.5
a. ovalauensis			4	28	10		42	17.1
procne		11*	4				15	15.3
reticulatus				1			1	17.0
bundoon			2*	3			5	16.6
phaeus				1			1	17.0

Species	Total pectoral-fin rays 24	25	26	27	28	29	30	31	32	N	x̄
anema			10	15	213	8	5	1		252	27.9
ditrema			4	8	24*	1				37	27.6
lineatus					13	4	63	5	4	89	29.8
grammistes			2	2	61	26	65	5		161	29.0
kamoharai							10			10	30.0
vittatus			17*	2	1					20	26.2
geminatus	1*	–	1	–	–	–	1	–	1	4	28.0
nigrolineatus	1	1	45	17	39	1	1			105	26.9
mossambicus			1	1	20	2				24	28.0
fraseri					22*	3	2			27	28.3
smithi			1	2	28	2	2			35	28.1
a. atrodorsalis			2	5	81	18	32			138	28.5
a. ovalauensis			1	3	32	1	1			38	27.9
procne				1	14*					15	27.9
reticulatus					1					1	28.0
bundoon							3*	2		5	30.4
phaeus									1	1	32.0

Species	Segmented caudal-fin rays[2] 11	12	13	N	x̄
anema	5	9	98	112	12.8
ditrema	24*			24	11.0
lineatus	46	8	2	56	11.2
grammistes	104	5		109	11.0
kamoharai	3	4		7	11.6
vittatus	3*	7	2	12	11.9
geminatus	4			4	11.0
nigrolineatus	40	30	3	73	11.5
mossambicus	13	7	1	21	11.4
fraseri	20*	5	1	26	11.3
smithi	23	8	2	33	11.4
a. atrodorsalis	83	7	2	92	11.1
a. ovalauensis	35	3		38	11.1
procne	13*	2		15	11.1
reticulatus	1			1	11.0
bundoon	2	2	1*	5	11.8
phaeus	1			1	11.0

"Procurrent" caudal-fin rays[3]

Species	Dorsal 4	5	6	7	8	Ventral 5	6	7	8	Total 9	10	11	12	13	14	15	N	x̄
anema		84	116	3		104	97	2			71	46	82	3	1		203	11.1
ditrema		11	24*			17	18*				8	11	16*				35	11.2
lineatus	2	55	6			42	20	1		2	40	14	7				63	10.4
grammistes		27	113	4	1	16	112	17			10	23	93	16	2	1	145	11.9
kamoharai		2	3			2	3				2	–	3				5	11.2
vittatus	1	13*				8	6*			1	7	6*					14	10.4
geminatus		2*	2			1	2*	1				3*	–	1			4	11.5
nigrolineatus	4	68	22			37	57			4	33	36	21				94	10.8
mossambicus		5	15			2	18				1	6	13				20	11.6
fraseri		12*	14			1	24	1			1	11*	13	1			26	11.5
smithi		16	15	1		4	23	5			4	12	11	4	1		32	11.6
a. atrodorsalis		49	52	2		6	62	33	2		6	38	29	27	2	1	103	11.8
a. ovalauensis	2	23	8			5	22	6		1	4	18	7	3			33	11.2
procne		11*	2			3*	10				3*	8	2				13	10.9
reticulatus				1				1							1		1	14.0
bundoon		2	3*				3	2*				2	1	2*			5	12.0
phaeus			1					1						1			1	13.0

[1] When counts vary, those for holotypes of new species are indicated by an asterisk.
[2] Caudal-fin rays with a single segmental joint were not tabulated.
[3] Refer to methods section for discussion of "procurrent" caudal-fin rays.

Based on a color transparency of a freshly collected male from New Guinea, life colors are as follows: stripes and dark fin markings black; diffuse ventral stripe brownish. Head and body above mid-lateral stripe brown or grayish-brown. Belly and pale area of body below mid-lateral stripe white. Pale areas of fins bluish-gray or white.

Herre (1936b: 401) gives an illustration of a small adult male (as *Petroscirtes kulambangrae*), and good illustrations of both sexes are given in de Beaufort (1951: 356), although misidentified as *Petroscirtes grammistes.*

Species comparisons. — *M. anema* has more dorsal-fin spines than any other species of *Meiacanthus* and is the only striped species without produced caudal-fin lobes. It and *M. ditrema* are the only species of *Meiacanthus* with typically two mid-dorsal supratemporal pores (3 of 205 specimens with a single pore).

Nomenclature. — In the original description, *Petroscirtes kulambangrae* was compared with no other species. The description and subsequent illustration (Herre, 1936b: 401) could apply only to *Meiacanthus anema.* See also nomenclatural remarks under *M. grammistes.*

Etymology. — The specific name, a combination of the Greek *nema* (thread) and prefix *a* (without), alludes to the simple caudal fin without filamentous rays.

Mimicry. — Springer and Smith-Vaniz (1972a: 24) noted that two species of Gobiidae were superficially similar to *M. anema* and should be considered as possible candidates for mimicry. Now that I have observed one of these gobies in the field and examined well-preserved material of the other, the possibility that either is a mimic of the *Meiacanthus* seems highly unlikely.

Distribution (*Fig. 46*). — *Meiacanthus anema* is the only member of the tribe that typically occurs in estuarine or apparently freshwater habitats, often where mangroves are abundant. Some collections have come from as far as 1.5 km upstream in small rivers where the water was fresh to the taste.

Since *M. anema* was not found in any of the extensive collections from the western Caroline Islands housed at CAS, the record from Yap Island based on a single Museum Godeffroy specimen may be erroneous.

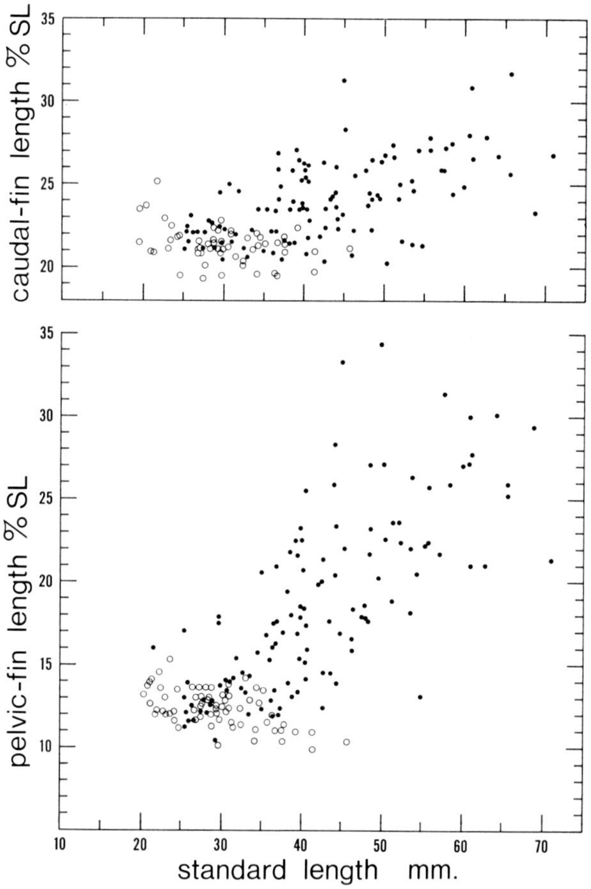

Figure 52. — Relationship between pelvic-fin and caudal-fin lengths versus standard length in *Meiacanthus anema.* Closed symbols = males; open symbols = females.

Material examined. — 328 specimens (13.4-71.0 mm SL) from 43 collections.

INDONESIA: RMNH 744 (2:51-61), Java; NMV 71759 (1:60), Celebes; RMNH 1889 (2:56-58), Timor; RMNH 1837 (4:46-64), ZMA 109.351 (2:17-39), Ceram; RMNH 26468 (48.5, holotype of *Petroskirtes anema*), Ambon (Amboina); RMNH 4774 (14 of 23:34-71, originally in same jar with holotype); ISZZ 7551 (2:47-56), NMV 12679 (1:36), RMNH 20718 (4:52-61), RMNH 20759 (2:56-69), RMNH 20809 (1:51), ZMA 109.336 (25:17-54), ZMA 109.348 (12:34-61), ZMA 109.350 (18:26-43), USNM 211954 (2:28-31), USNM 211956 (33:21-43), USNM 211959 (5:22-42), Banda Sea, Ambon I.

PHILIPPINE IS.: SU 38222 (16:23-55), Mindanao, Cagayan, in mangrove swamp; SU 29571 (2:38-45), Negros, Dumaguete; SU 32936 (1:30), Mindoro, Puerto Galera;

SU 29572 (2:39-49), Luzon, Legaspi; USNM 137851 (1:36), Luzon, Mariveles, Pucot River; USNM 137852 (3:27-30), Luzon, Ragay River; USNM 137854 (2:23-30), Luzon, Cabugo, Catanduanes I.; USNM 137850 (1:34), Luzon, Palaui I., 18°34′N, 122°9′E.

YAP IS.: RMNH 11232 (1:43), "Museum Godeffroy."

NEW GUINEA: BPBM 15832 (4:20-44), Madang, Meiro River; RMNH 25232 (54:17-48), RMNH 25364 (10:18-49), RMNH uncat. (18:19-45), Biak I., "creek"; AMS I.17064-6 (1:31), AMS I.16672-6 (2:40-43), AMS I.16670-14 (1:37), Madang Harbor; USNM 205673 (4:21-34), USNM 205674 (5:13-33), Madang Harbor, behind Nui I., "mangroves."

SOLOMON IS.: CAS 5647 (9:25-48), Malaita I., Auki; SU 24441 (1:33), Kulambangra I., "freshwater creek"; SU 47016 (1:25), New Georgia I., Enogai River.

SANTA CRUZ IS.: ANSP 124867 (5:27-43), ANSP 128535 (5:30-41), CAS 28900 (11:14-35), CAS 28901 (24:14-44), Vanikoro I., Sundi River.

NEW HEBRIDES IS.: AMS uncat. (3:22-29), Espiritu Santo, sta. GRA 27; USNM 122852 (10:25-36), Espiritu Santo, mouth Brigstock lagoon.

Allomeiacanthus, new subgenus

Type-species. — *Meiacanthus* (*Allomeiacanthus*) *ditrema,* new species

Diagnosis. — A subgenus of *Meiacanthus* with dentary gland positioned ventrally and encapsulated in dentary bone (Fig. 44). Posttemporal bone reduced, without ventral arm (Fig. 50b). Infraorbital bones 4. Mid-point of supratemporal canal with a pair of pores; lateral temporal pores 2; mandibular pores 2. Lateral line absent. Dorsal-fin spines 5 or 6.

Comparisons. — Refer to Table 15 and general discussion of relationships.

Etymology. — A combination of the Greek *allos* (strange, different) and *Meiacanthus,* the blenniid subgenus to which the new subgenus is most closely related.

Meiacanthus (Allomeiacanthus) ditrema, new species

Specimen Figures 41, 135-136; Tables 4, 15-17.

Description. — Characters of the subgenus. Dorsal fin V-VI, 22-25; total elements 28-30. Anal fin II, 15-18. Pectoral fin 13-15 (rarely 15). Caudal fin: procurrent rays 5-6 + 5-6 = 10-12; segmented rays 11. Vertebrae: precaudal 11-12; caudal 21-24; total 33-35. Pleural ribs on vertebrae 3-10 or 11; epipleural ribs 11-13. Incisor teeth (Fig. 45) in specimens 19-40 mm, n = 30: lower jaw 12-18; upper jaw 12-18.

Sexual dimorphism slight in pelvic fins, well-developed in caudal fin in specimens > 25 mm SL (Fig. 53). Pelvic fin: males 10.8-16.2% SL, females 8.7-11.9% SL; caudal fin: males 23.8-77.5% SL, females 21.2-23.0% SL. Outer lobes of caudal fin elongated in adult males only, interradial membranes of inner rays moderately incised in large males.

Color pattern. — A broad dark stripe, slightly ventral to the mid-line of body, extends from the caudal fin to the pectoral fin. The stripe is confined to the basal part of the caudal fin in females but may extend farther onto the fin in large males. The lateral stripe forks at or just in front of the pectoral fin. The dorsal arm of the fork extends through the middle of the orbit and encircles the snout; the ventral arm extends forward to the gape, where in some specimens it narrowly outlines the mouth. A dorsal stripe, widest anteriorly, extends from the interorbital region along the dorsal contour of the body to the caudal fin. A narrow concentration of dark pigment may also occur on the body near the base of the anal fin. In females this pigment usually extends the entire length of the fin, although best developed anteriorly; in males it usually extends only half the fin length.

In females the dorsal fin is completely pale or has a very narrow dark margin. In males from the Fiji Islands the dorsal fin has a black submarginal stripe slightly wider than the mid-lateral stripe; in males from other localities the entire fin may be heavily pigmented, but no black submarginal stripe is present. The other fins are relatively uniformly pigmented.

Based on a color photograph, taken by Barry C. Russell, of a freshly collected male and female from Fiji, the life colors are as follows: Background coloration of body tan, lateral stripes black. Belly and ventral half of head pale white. Caudal-fin rays dusky; anal and pelvic fins light buff (male) or transparent (female). Dorsal fin light buff except for black submarginal stripe and orange distal margin (male) or transparent except for very narrow dusky margin (female).

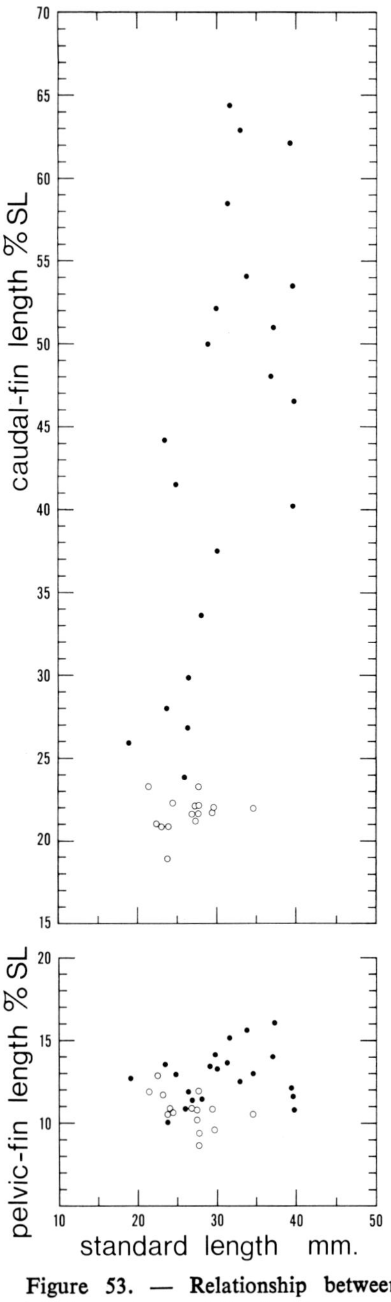

Figure 53. — Relationship between pelvic-fin and caudal-fin lengths versus standard length in *Meiacanthus ditrema.* Closed symbols = males; open symbols = females. (Not plotted was one male, 34.5 mm SL, with caudal fin 77.5 percent SL).

Geographic variation. — A black submarginal stripe is present in the dorsal fin only in males from Fiji.

Etymology. — A combination of the Greek *di* (two) and *trema* (hole), in allusion to the pair of mid-dorsal supratemporal pores. The name is to be treated as a noun in apposition.

Distribution (Fig. 46). — Definitely known only from the Ryukyu, Palau and Fiji Islands. The Banda Sea locality for the Longley "Dutch East Indies" collection is not definite. This collection, made by W. H. Longley during 1926-27, could have come from either Pago Pago, Samoa or from the Banda Sea (Amboina or Banda Island). Springer (1971: 42-43) gives a detailed history of the collection. I follow Springer in tentatively assigning a Banda Sea locality to Longley's Dutch East Indies collection because Samoa was an improbable locality for all four species of *Ecsenius* that he studied from the collection, as well as specimens of *Petroscirtes breviceps* from the same collections cited in this study.

Material examined. — 37 specimens (19.0-39.8 mm SL) from 10 collections.

Holotype. — USNM 205293, male, 34.5 mm SL, Palau Islands, Auluptagel I. in Ngarahelngael Pass, 7°19′13″N, 134°29′28″E; steep slopes of small bay, formed of sand, living and dead coral; depth 0-18 m; 9 Oct. 1957; H. A. Fehlmann *et al.;* original no. GVF 1415.

Paratypes. — RYUKYU IS.: BPBM 7453 (4:35-40), Ishigaki, Taketomi I.; BPBM 18852 (2:40), W. of Sesoko I. PALAU IS.: USNM 205294 (12:24-34), taken with the holotype; BPBM 7456 (2:24-25), Ngargol I.; CAS 28921 (1:27), Urukthapel I., 7°15′57″N, 134°26′56″E; CAS 28922 (2:19-21), Babelthaup I., 7°20′29″N, 134°32′7″E. BANDA SEA: USNM 195706 (2:30-37), "Dutch East Indies", Longley collection — presumably from Banda Sea. FIJI IS.: ANSP 128341 (1:24), ANSP 128490 (3:28-32), ANSP 128491 (2:26-27), Viti Levu, Suva Harbor, Bay of Islands, Labiko I. (Snake I.). E. AUSTRALIA: AMS I.18232-1 (5:22-26), Lizard I., Yonge Reef, 14°40′S, 145°30′E.

Subgenus *Meiacanthus* Norman

Type-species. — *Meiacanthus atrodorsalis ovalauensis* Günther.

Diagnosis. — A subgenus of *Meiacanthus* with major portion of dentary gland positioned dorsolaterally and supported solely by dentary (Fig. 44).

Posttemporal bone well developed, with short ventral arm (Fig. 50a). Infraorbital bones 3. Mid-point of supratemporal canal with a single pore; lateral temporal pores 3; mandibular pores 3. Lateral line present, consisting of a series of simple disconnected tubes that may terminate anywhere from below the base of last spine to eighth segmented ray of dorsal fin. Dorsal-fin spines 3-6 (rarely 6).

Comparisons. — Refer to Table 15 and general discussion of intrageneric relationships.

Meiacanthus (*Meiacanthus*) *lineatus* (De Vis)

Specimen Figures 137-138; Tables 15-17.

Petroscirtes lineatus De Vis, 1884: 698 (type locality: Murray Island, Torres Strait; holotype QMB I.1354).

Description. — Characters of the subgenus. Dorsal fin IV, 25-28; total elements 29-32. Anal fin II, 14-17 (rarely 14). Pectoral fin 14-16. Caudal fin: procurrent rays 4-6 + 5-7 = 9-12; segmented rays 11-13 (usually 11). Vertebrae: precaudal 12-14; caudal 20-23; total 34-36. Pleural ribs on vertebrae 3-12 or 13; epipleural ribs 14-16. Incisor teeth in specimens 22-67 mm, n = 20: lower jaw 16-24; upper jaw 14-24.

Sexual dimorphism well developed in pelvic and caudal fins in specimens > 30 mm SL (Fig. 54). Pelvic fin: males 13.5-33.1% SL, females 10.0-16.6% SL; caudal fin: males 19.9-42.2% SL, females 17.9-22.4% SL. Outer lobes of caudal fin elongated in adult males only, interradial membranes of inner rays slightly incised in large males.

Color pattern. — Head and body with three dark stripes separated by pale interspaces, all of approximately equal width. Dorsalmost stripe extends from top of head along the dorsal contour of body almost to end of dorsal fin. The middle stripe extends from snout to base of caudal fin. Ventralmost stripe extends from lower jaw through pectoral-fin base to lower margin of caudal-fin base. Only the ventralmost stripe extends onto the base of the pectoral fin. Dorsal fin uniformly pale except for narrow, dark submarginal stripe. Other fins unpigmented; some specimens with dusky, submarginal stripe in anal fin.

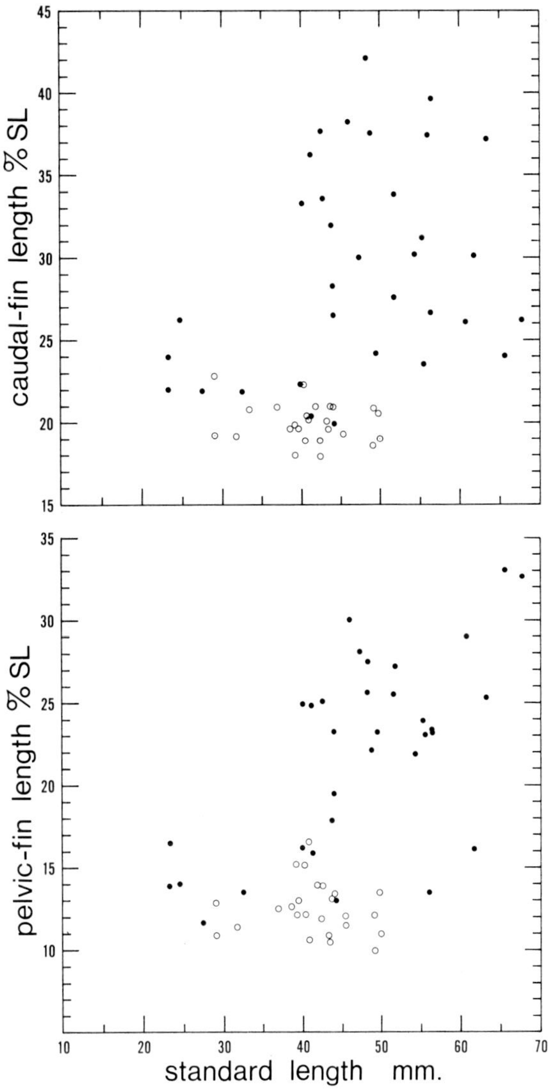

Figure 54. — Relationship between pelvic-fin and caudal-fin lengths versus standard length in *Meiacanthus lineatus*. Closed symbols = males; open symbols = females.

The only adult male available from the Sulu Archipelago (Fig. 137) differs from the above description in having a wider stripe in the dorsal fin, and the two ventralmost stripes partially disrupted by pale blotches (perhaps an artifact of preservation) and extending farther onto the caudal fin.

Based on a color transparency of a freshly dead female from One Tree Island, the life colors are as follows: The dark stripes are black. Proximal area of dorsal fin, caudal fin, all of ventral pale interspace, and anterior third of dorsal interspace yellow. Pelvic fin and remainder of body white. Pectoral and anal fins clear.

Geographic variation. — The only adult male from the disjunct Sulu population differs in color pattern as noted above, and has a longer caudal fin (42.2% SL) than any of the Australian males. Meristic counts for the two populations are not different.

Species comparisons. — The only striped species of *Meiacanthus* with which *lineatus* might be confused is *grammistes*. *M. grammistes* differs in having the lateral stripes broken into rows of spots, especially on the caudal peduncle, and the caudal-fin rays usually spotted. The caudal fin of *grammistes* is not yellow in life, and in large adult males the inner rays of the caudal fin are strongly incised.

Etymology. — The Latin name *lineatus,* meaning streaked or lined, is in reference to the regularly striped appearance of this species.

Mimicry. — Springer and Smith-Vaniz (1972a: 23) documented the strikingly similar appearance of *M. lineatus* and *Petroscirtes fallax,* an eastern Australian endemic, and suggested that the *Petroscirtes* was a Batesian mimic. Allen *et al.* (1975: 55) provided additional observations on this mimetic pair and noted that similarly pigmented juveniles of *Scolopsis bilineatus* (Bloch) also associate with *M. lineatus* at One Tree Island, Great Barrier Reef.

Distribution (Fig. 47). — Known only from the Great Barrier Reef and an apparently disjunct population in the Sulu Archipelago.

Material examined. — 110 specimens (10.9-67.6 mm SL) from 25 collections.

SULU ARCHIPELAGO: USNM 99370 (1:48), Sirum I., 5°35′40″N, 120°47′30″E; USNM 122384 (29:11-30), Jolo I., 5°50′45″N, 120°1′15″E.

E. AUSTRALIA: QMB I.1354 (39.5, holotype of *Petroscirtes lineatus*), Torres Strait, Murray I.; ANSP 109703 (1:25), ANSP 109704 (1:22), Endeavour Reef, 15°45′S, 145°42′E; AMS IA.6116 (1:54), Hayman I., 20°3′S, 148°54′E; AMS IA.6456 (3:40-52), AMS IA.6662 (1: 42), AMS IA.6680 (1:57), AMS IA.7875 (1:56), QMB I.7486 (2:42-44), QMB I.7585 (1:42), Lindeman I.; AMS I.15415-1 (1:44), Swains Reef, Gillet Cay; AMS IB.5400 (1:68), USNM 202483 (2:49-55), USNM 211308 (11:33-66), USNM 211309 (1:41), Great Barrier Reef, Capricorn Group, Heron I.; USNM 201358 (1:55), USNM 201359 (11:15-49), USNM 201360 (4: 20-27), USNM 201361 (8:18-44), USNM 201362 (3:29-43), USNM 201363 (15:22-63), USNM 201364 (8:20-47), USNM 211307 (1:29), Great Barrier Reef, Capricorn Group, One Tree I.

Meiacanthus (*Meiacanthus*) *grammistes* (Valenciennes)

Specimen Figures 42, 139-140; Tables 4, 15-17.

Blennechis grammistes Valenciennes *in* Cuvier and Valenciennes, 1836: 284 (type locality: Java; holotype MNHN A.2083).

Petroskirtes temminckii Bleeker, 1851b: 243 (type locality: Banda; holotype RMNH 26970).

Description. — Characters of the subgenus. Dorsal fin III-V (typically IV), 25-28; total elements 29-31. Anal fin II, 14-16. Pectoral fin 13-16. Caudal fin: procurrent rays 5-8 + 5-7 = 10-15; segmented rays 11-12 (usually 11). Vertebrae: precaudal 13-14; caudal 19-23; total 33-36. Pleural ribs on vertebrae 3-13 or 14; epipleural ribs 13-16. Incisor teeth (Fig. 45) in specimens 22-84 mm, n = 67: lower jaw 18-28; upper jaw 16-26.

Sexual dimorphism well developed in pelvic and caudal fins in specimens > 30 mm SL (Fig. 55). Pelvic fin: males 11.7-35.9% SL, females 7.0-15.6% SL; caudal fin: males 24.9-49.5% SL, females 20.3-29.4% SL. Outer lobes of caudal fin elongated in adult males only, interradial membranes of inner rays strongly incised in large males.

Color pattern. — Head and body with three dark stripes separated by pale interspaces, the width of the stripes variable. Dorsalmost stripe extends from top of head along dorsal contour of body to end of dorsal fin. The middle stripe extends from snout to caudal peduncle. Ventralmost stripe extends from lower jaw almost to end of anal fin. Typically a row of dark spots or blotches, especially noticeable posteriorly, is superimposed on the middle and ventral stripes. In large specimens small spots also frequently occur on the pale interspaces between the stripes. The rays of the caudal fin are usually heavily pigmented and/or spotted. The anal fin may be spotted or plain. The dorsal fin has a black stripe distally that tends to become broken into a series of spots posteriorly. The extent of spotting is highly variable, but seems to be best developed in specimens from Indonesia and the South China Sea. Most populations have the throat immaculate, but 2

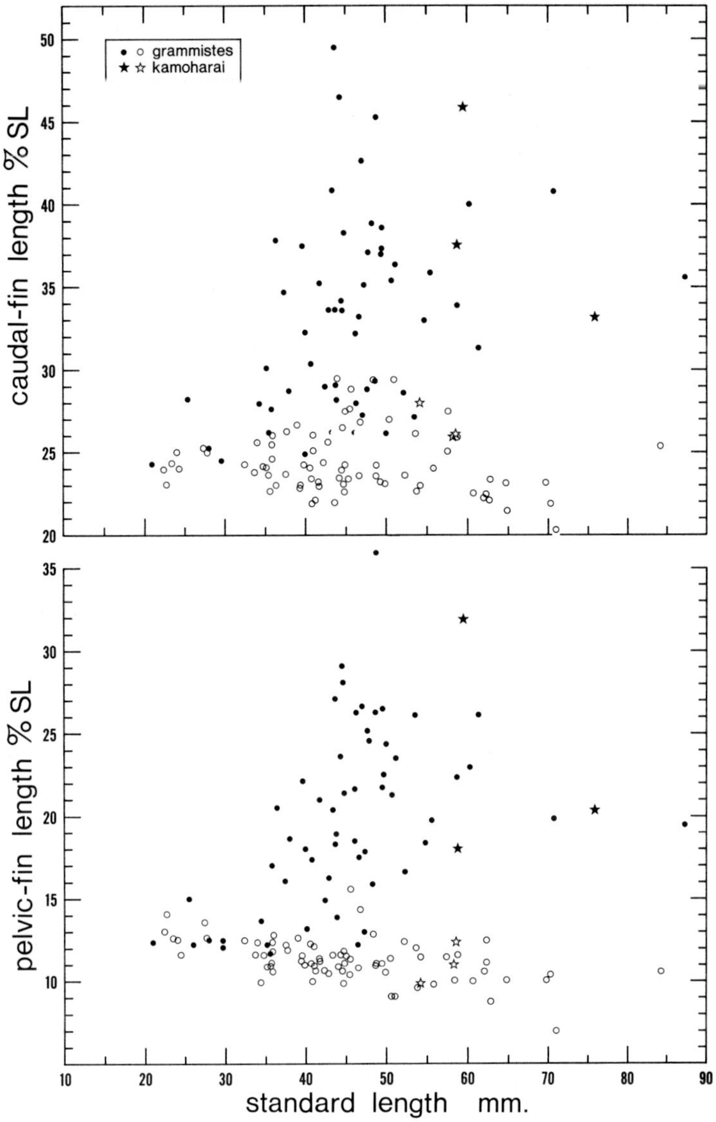

Figure 55. — Relationship between pelvic-fin and caudal-fin lengths versus standard length in *Meiacanthus grammistes* and *kamoharai*. Closed symbols = males; open symbols = females. (Data for 59.9 mm holotype of *M. kamoharai* based on original description.).

or 3 spots are frequently present in specimens from northern Australia, New Guinea, the Solomons and Santa Cruz Islands.

In life, *M. grammistes* is black, white and yellow striped. The yellow areas are confined to the pale anterodorsal regions of the body and dorsal fin. The other pale areas of the body and fins are white, frequently with bluish overtones. Color photographs of *grammistes* are given in Burgess and Axelrod (1973a: 469; 1974: 1334) and Masuda *et al.* (1975: 86).

Geographic variation. — One of two adult specimens of *grammistes* examined from the Ryukyu Islands (ANSP 128828) exhibits possible evidence of intergradation with *kamoharai* (see below). In this specimen the dark stripes are unusually wide, the throat has 6 large blotches and there are faint markings on the belly. The Taiwan population of *grammistes* is typically pigmented, as are all specimens of *kamoharai* from Japan. For now it seems wisest to maintain both taxa as specifically distinct, but their status should be re-evaluated when more specimens from the Ryukyu Islands become available for study.

Species comparisons. — *M. grammistes* is an apparent cognate of *kamoharai,* which is restricted to Japan and the Ryukyu Islands; the pair differ only in color pattern. *M. kamoharai* differs most noticeably in having the ventral surface of the head and belly dark and strongly vermiculated. If dark stripes are present on the head they are approximately twice the width of the pale interspaces (instead of being approximately equal in width), and only a single uninterrupted pale stripe is present on the body. In life, *kamoharai* also apparently differs from *grammistes* in not having any yellow on the body anteriorly. Refer also to species comparisons under account of *M. lineatus.*

Nomenclature. — In the original description, Bleeker did not compare *Petroskirtes temminckii* with any other species. The description agrees well with *M. grammistes,* and my examination of both holotypes confirms that identification.

Etymology. — From the Greek *gramme,* meaning lined or marked, and *-istes,* a suffix indicating a superlative.

Remarks. — Museum collections of *grammistes* have frequently been misidentified as *anema* although the two species are easily distinguished. This confusion probably stems from Günther (1877: pl. 115) and de Beaufort (1951: 356) who gave good illustrations of *anema* labeled as *grammistes.* Munro (1967) greatly added to the confusion by giving illustrations of both species under the name *Meiacanthus grammistes,* indicating that the differences in color pattern were due to sexual dimorphism!

Mimicry. — Springer and Smith-Vaniz (1972a: 24) suggested that the striped color form of *Petroscirtes breviceps* might be a mimic of *grammistes.* Several museum collections (supposedly sorted to species) contained a mixture of the two species, especially those from Palau. The preserved color patterns, especially those of juveniles, were quite similar, but the life colors of the *Petroscirtes* were inadequately known. Field notes taken by J. E. Randall of two juveniles of *breviceps* from Palau record the life colors as "bright yellow on upper part of head and front of body." This agrees very well with the color description of *grammistes* given above.

Distribution (Fig. 47). — Widely distributed in the central Pacific Ocean.

Material examined. — 184 specimens (18.0-87.3 mm SL) from 75 collections.

INDONESIA: MNHN A.2082 (35.5, holotype of *Blennechis grammistes*), Java; RMNH 5050 (1:29), Timor; USNM 211903 (1:54), Celebes, Kabaena I., 5°17′S, 122°4′E; USNM 137835 (1:47), Celebes, Lembeh Strait; USNM 137834 (1:56), Sulu Archipelago, Singaan I.; ZMA 109.337 (1:71), "Moluccas"; USNM 211919 (1:46), Ceram, Kasa I.; RMNH 26970 (65.8, holotype of *Petroskirtes temminckii*), Banda I.; RMNH 4775 (12:28-74), Bleeker Collection, originally in same jar with holotype; RMNH 5030 (2:29-33), USNM 211929 (2:50-56), Banda I.; USNM 195705 (5:35-42), "Dutch East Indies-Longley", presumably Banda Sea; NMV 71775 (1:51), USNM 209759 (1:28), ZMA 109.349 (1:58), Ambon I.

W. AUSTRALIA: WAM uncat. (2:45-51), Dampier Archipelago, Rosemary I., 20°31′S, 116°35′E; AMS IA.4246 (1:55), Cape Leveque.

SOUTH CHINA SEA: SU 35786 (1:54), Singapore, Sultan Shoal; CAS 28946 (1:43), CAS 28947 (2:61-62), South Vietnam, Nha Trang; CAS 28940 (1:60), SU 61236 (1:58), Pratas Reef, 20°45′N, 116°43′E.

PHILIPPINE IS.: USNM 137839 (1:70), Tictauan I., E Zamboanga, 6°55′N, 122°5′E; ANSP 11478 (4:37-48), USNM 113241 (4:37-49), USNM 137836 (3:62-65), USNM 137837 (1:45), USNM 137838 (1:47), USNM 168991 (1:45), USNM 168992 (3:61-65), Cebu Mkt.; USNM 137831 (1:28), USNM 137833 (1:27), Ticao I., San Miguel Harbor; USNM 55977 (1:70), Luzon.

TAIWAN: USNM 137840 (1:71), USNM 207573 (3:44-87).

RYUKYU IS.: HA 329 (1:50), Okinawa; ANSP 128828 (1:45), ZUMT 52873 (1:26), Ishigaki I.

YAP IS.: CAS 28931 (1:48), CAS 28944 (1:44).

PALAU IS.: ANSP 124868 (5:21-48), ANSP 124879 (10:25-50), ANSP 126022 (10:24-47), ANSP 128536 (4:26-46), CAS 28923 (10:24-51), CAS 28924 (5:23-45), CAS 28925 (1:38), CAS 28926 (1:54), CAS 28927 (3:44-48), CAS 28928 (6:28-53), CAS 28929 (2:33-35), CAS 28930 (2:37-47), CAS 28932 (1:42), CAS 28933 (1:37), CAS 28935 (2:45), CAS 28936 (4:43-51), CAS 28937 (9:26-46), CAS 28938 (2:34-41), CAS 28939 (1:19), CAS 28941 (6:18-46), CAS 28942 (2:27-28), CAS 28943 (3:21-40), CAS 28945 (1:24), USNM 114689 (2:44-49).

SANTA CRUZ IS.: ANSP 126026 (1:59), Vanikoro I.
SOLOMON IS.: BPBM 13641 (1:52), Egum Atoll; BPBM 15580 (2:59-62), USNM 212035 (1:47), Guadalcanal I.
NEW BRITAIN: USNM 201482 (5:28-52), Duke of Kirk Group, Keraward I., 4°14'S, 152°26'E.
NEW GUINEA: DASF FO3611 (1:49), Port Moresby, Manubada I.
E. AUSTRALIA: BPBM 15553 (1:50), Lizard I., 14°40'S, 145°30'E; ANSP 109679 (1:24), ANSP 109683 (1:22), Queensland, Big Hope I.; ANSP 109682 (1:23), Queensland, Little Hope I.; ANSP 109677 (4:35-45), ANSP 109678 (1:29), Great Barrier Reef, Endeavour Reef, 15°45'S, 145°42'E.

Meiacanthus (*Meiacanthus*) *kamoharai* Tomiyama
Specimen Figure 141; Tables 15-17.

Meiacanthus kamoharai Tomiyama, 1956b: 1083, Fig. 562-564 (type locality: Kashiwajima, island near Kochi City, Shikoku, Japan; holotype ZITU 18777 [not examined]).

Description. — Characters of the subgenus. Dorsal fin IV, 25-27; total elements 29-31. Anal fin II, 15-16. Pectoral fin 15. Caudal fin: procurrent rays 5-6 + 5-6 = 10-12; segmented rays 11-12. Vertebrae: precaudal 14; caudal 21-22; total 35-36. Pleural ribs on vertebrae 3-13; epipleural ribs 13. Incisor teeth in specimens 26-76 mm, n = 6: lower jaw 17-26; upper jaw 16-22.

Sexual dimorphism well developed in pelvic and caudal fins in specimens > 30 mm SL (Fig. 55). Pelvic fin: males 18.0-31.8% SL, females 9.8-12.4% SL; caudal fin: males 33.2-45.8% SL, females 25.9-28.0% SL. Outer lobes of caudal fin elongated in adult males only, interradial membranes of inner rays strongly incised in large males.

Color pattern. — The color pattern is variable. All specimens agree, however, in having the ventral surface of the head and belly dark and strongly vermiculated. In the original description, the color pattern of the holotype was described as follows:

"After long preservation in formalin, the ground color is dark purplish; a pale narrow longitudinal band across the upper part of the eye is connected with an opposite one on the snout, and extends posteriorly onto the upper part of the caudal peduncle, where the band broadens abruptly below. A similar narrow band across the lower part of the eye runs from the upper lip onto the very dark pectoral base. Below this band, 2 pale longitudinal streaks and a pale median one on the throat make an irregular vermiculation. A pale median streak on the interorbital space reaches to the origin of the dorsal fin. The dorsal fin is darker and crossed by 2 longitudinal series of pale spots; the tips of the rays of the fin are whitish. The anal fin is darker and provided with several pale spots on the basal part. The upper and lower margins of the caudal fin are dark; the proximal part of the fin is dusky; the distal halves of the rays are dark; the membranes between the rays are colorless. The pectoral fins are colorless except for the proximal part which is very dark; the axillae are also dark. The ventral fins are dark proximally and become paler towards the ends."

Several excellent monochrome illustrations of *kamoharai* accompanied the original description. Good color photographs of the species are given in Burgess and Axelrod (1971: 164) (misidentified in the first edition as *M. grammistes*) and Masuda *et al.* (1975: 86). In life, *Meiacanthus kamoharai* is predominantly dark purple-brown with white stripes and vermiculations.

Species comparisons. — Refer to discussion under account of *M. grammistes,* especially comments on geographic variation.

Etymology. — Named after Dr. Toshiji Kamohara, a Japanese ichthyologist at Kochi University.

Distribution (*Fig. 47*). — Endemic to Japan and the Ryukyu Islands.

Material examined. — 10 specimens (12.3-75.9 mm SL) from 5 collections.

RYUKYU IS.: ANSP 88987 (2:15-26), ANSP 88988 (3: 12-21), Aguni Shima; USNM 132705 (2:54-59), Okinawa, W coast Nakadamori; ZUMT 52872 (1:59), Okinawa, Nakagusuku.
JAPAN: BPBM 18962 (2:58-76), Miyake I.

Meiacanthus (Meiacanthus) vittatus, new species
Specimen Figure 142; Tables 15-17.

Meiacanthus grammistes (*non* Valenciennes): Monroe, 1958: 249 (misidentification; Gasmata, New Britain — Fairwind Collection).
Meiacanthus sp. Allen *et al.*, 1975: 55 (illustrated; suggested mimetic model for *Cheilodipterus zonatus* and *Scolopsis margaritifer*).

Description. — Characters of the subgenus. Dorsal fin III-IV (typically IV), 25-27; total elements 29-31. Anal fin II, 14-16. Pectoral fin 13-14. Caudal fin: procurrent rays 4-5 + 5-6 = 9-11; segmented rays 11-13. Vertebrae: precaudal 13-14; caudal 20-22; total 33-35. Pleural ribs on vertebrae 3-12 or 13; epipleural ribs 12-14. Incisor teeth in specimens 25-45.5 mm, n = 10: lower jaw 18-24; upper jaw 16-20.

Sexual dimorphism well developed in pelvic and caudal fins in specimens > 30 mm SL (Fig. 56). Pelvic fin: males 17.4-23.1% SL, females 10.4-13.7% SL; caudal fin: males 39.4-52.3% SL, females 23.9-30.4% SL. Outer lobes of caudal fin elongated in adult males only, interradial membranes of inner rays strongly incised in large males.

Color pattern. — The most distinctive feature of the color pattern of this species is the black mid-lateral stripe extending from snout through eye to caudal fin. The stripe, almost as wide as the eye diameter, is bordered above and below by a narrow pale stripe; the pale border is most obvious on the postorbital area of the head. The stripe envelops the gill-opening and extends ventrally onto the base of the pectoral fin to about the level of the 5th or 6th pectoral-fin ray. Caudal fin pale except for dark, tapering stripe terminating on anterior half of fin. Remainder of body and fins uniformly pigmented.

A color transparency of a freshly dead nuptial male, reveals that the species is essentially gray-white ventrally, with the dorsum grayish-tan. The border of the black stripe is white. The fins are gray except that the modified tips of the posterior rays of the dorsal fin are light tan. Allen *et al.* (1975: 54) give excellent color photographs of *M. vittatus* and the cardinalfish *Cheilodipterus zonatus,* which is a nearly perfect color mimic of the blenny.

Species comparisons. — *M. vittatus* is an obvious cognate of *geminatus,* the only other species of *Meiacanthus* with a single, mid-lateral stripe. These two allopatric species have broadly overlapping characters, but despite their essential similarity there are several differences in color pattern. The most obvious difference is that the stripe is narrower in *geminatus* and does not extend onto the base of the pectoral fin. Also, in *geminatus* the background pigmentation of the body is darker, the dorsal fin has a well-defined pale margin and the stripe is more ventrally positioned on the caudal fin. Although the life colors of *geminatus* are unknown, it almost certainly does not have the predominantly grayish coloration of *vittatus.*

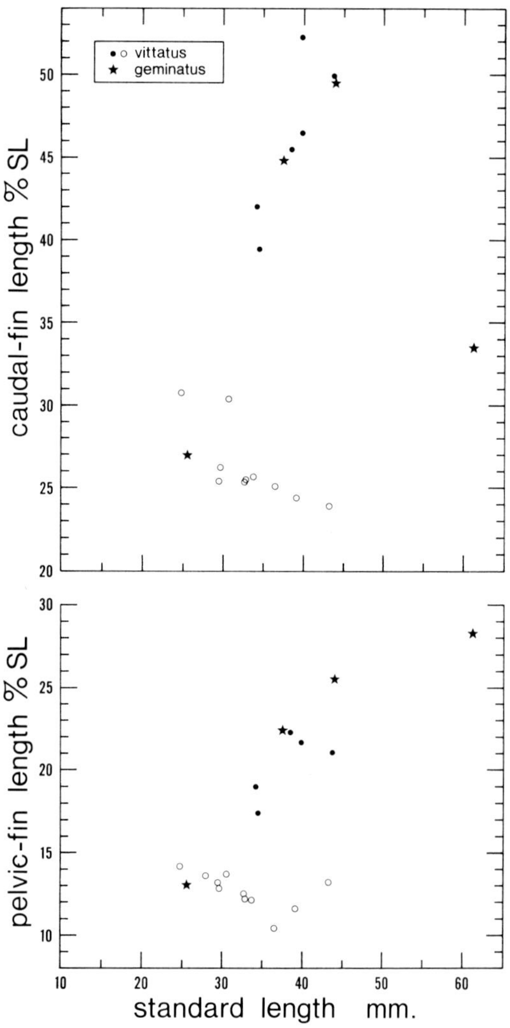

Figure 56. — Relationship between pelvic-fin and caudal-fin lengths versus standard length in *Meiacanthus vittatus* and *geminatus.* Closed symbols = males; open symbols = females.

Etymology. — The Latin name *vittatus* means decorated or bound with a ribbon.

Mimicry. — *Cheilodipterus zonatus* Smith and Radcliffe (Apogonidae) and juveniles of *Scolopsis margaritifer* (Cuvier) (Scolopsidae) were reported to mimic *M. vittatus* (Allen *et. al.,* 1975). These three species are frequently encountered together in shallow lagoon areas. The color patterns and behavior of all three are similar, but that of the *Cheilodipterus* is remarkably like that of the blenny.

Unlike most cardinalfishes, *C. zonatus* is a diurnal species and swims out in the open in quick darting movements exactly duplicating the behavior of a foraging *Meiacanthus*. Inasmuch as juveniles of *Scolopsis* are vulnerable to heavy predation, the mimetic relationship is probably Batesian. The apogonid might also benefit from aggressive mimicry. Similarity to the *Meiacanthus*, which feeds almost exclusively on small invertebrates, might allow *C. zonatus* more easy access to prey items such as small fishes.

Distribution (Fig. 47). — Known only from New Guinea and New Britain.

Material examined. — 19 specimens (18.2-45.5 mm SL) from 9 collections.

Holotype. — USNM 201453, male, 45.5 mm, New Guinea, Madang Harbor; depth 9 m; 5 April 1972; Walter A. Starck.

Paratypes. — NEW GUINEA: AMS I.17148-1 (1:30), Louisiade Archipelago, Rossel I., Pwennegwa Harbor; AMS I.17537-1 (3:25-40), off Port Moresby, Lolorua I.; ANSP 126900 (1:44), D'Entrecasteaux Is., Fergusson I., Seymour Bay; ANSP 128346 (2:34-39), BPBM 15769 (3:33-35), CAS 15794 (1:40), USNM 205677 (4:18-31), Madang, Kranket I.; DASF FO3214 (2:39-43), Port Moresby, local island. NEW BRITAIN: CSIRO B884 (1:28), Gasmata.

Meiacanthus (Meiacanthus) geminatus, new species

Specimen Figure 143; Tables 15-17.

Description. — Characters of the subgenus. Dorsal fin IV, 26-27; total elements 30-31. Anal fin II, 15-18. Pectoral fin 12-16. Caudal fin: procurrent rays 5-6 + 5-7 = 11-13; segmented rays 11. Vertebrae: precaudal 13; caudal 21-23; total 34-36. Pleural ribs on vertebrae 3-12; epipleural ribs 13-14. Incisor teeth in specimens 38-44 mm, n = 2: lower jaw 20-22; upper jaw 19-20.

Sexual dimorphism probably as in *M. vittatus* (Fig. 56), only males have been collected. In specimens > 30 mm SL, pelvic fin 22.4-28.5% SL; caudal fin 33.1-49.5% SL. Outer lobes of caudal fin elongated, interradial membranes of inner rays strongly incised in large males.

Color pattern. — Body uniformly tan except for dark mid-lateral stripe extending from snout to caudal fin. Stripe narrowest on snout and scarcely wider than diameter of pupil at its widest point; stripe envelops gill-opening, but does not extend onto pectoral-fin base. Caudal fin pale except for dark stripe terminating on anterior third of fin. Distal third of dorsal fin pale, proximal part uniformly dark (noticeably darker than body); other fins paler than body. Life colors are unknown.

Geographic variation. — The counts for Australian specimens are invariably higher than those of the Bornean specimens. For example, counts of the pectoral fin are 12-13 vs. 15-16, and for segmented anal-fin rays 15-16 vs. 17-18 in the two populations. The total range for pectoral-fin rays exceeds that of any other species of *Meiacanthus,* although based on only 4 specimens. In view of the extreme range of meristic characters in the two populations, only the Bornean specimens are designated as types.

Species comparisons. — Refer to discussion under account of *M. vittatus.*

Etymology. — The Latin name *geminatus,* meaning to double or pair, alludes to the presumed cognate status of this species and *vittatus.*

Distribution (Fig. 47). — Known only from northern Borneo and Australia.

Material examined. — 4 specimens (25.7-61.3 mm SL) from 3 collections.

Holotype. — USNM 205286, male, 44.0 mm, Borneo, Darvel Bay, Pulau Gaya; 1 Feb. 1965; D. M. Cohen and W. P. Davis; Te Vega Cruise 6, sta. 213.

Paratype. — USNM 208502 (1:26), taken with the holotype.

Other material. — AMS IA.6002 (1:38), Australia, Northern Territory, Darwin; AMS IB.3159/2352 (1:61), Australia, Queensland, Townsville.

Meiacanthus (Meiacanthus) nigrolineatus Smith-Vaniz

Specimen Figures 144-146 and frontispiece; Tables 15-17.

Meiacanthus nigrolineatus Smith-Vaniz, 1969: 351, Fig. 1 (type locality: Red Sea, Straits of Jubal; 27°14′34″N, 33°53′55″E; holotype USNM 200301).

Description. — Characters of the subgenus. Dorsal fin IV-V (usually IV), 23-26; total elements 28-30. Anal fin II, 14-16. Pectoral fin 12-15. Caudal fin: procurrent rays 4-6 + 5-6 = 9-12; segmented rays 11-13 (rarely 13). Vertebrae: precaudal 12-13; caudal 20-22; total 32-35. Pleural ribs on vertebrae 3-12 or 13; epipleural ribs 12-15. Incisor teeth in specimens 19-66 mm, n = 20: lower jaw 14-23; upper jaw 14-21.

Sexual dimorphism slight in pelvic fins, moderate in caudal fin in specimens > 30 mm SL (Fig. 57). Pelvic fins: males 10.8-16.4% SL, females 10.1-13.7% SL; caudal fin: males 31.2-61.7% SL, females 27.4-38.9% SL. Outer lobes of caudal fin elongated in adults of both sexes, interradial membranes of inner rays weakly incised.

Color pattern. — In specimens from the northern Red Sea and tributary Gulfs of Suez and Aqaba, the head is relatively dark with the intensity of the pigment decreasing gradually on the body and becoming pale posteriorly. In adults there is usually a narrow dark stripe or diffuse evidence of one, that extends posteriorly from the eye, darkens considerably on the body, and usually extends to below the posterior third of the dorsal fin. The stripe is frequently broken into a series of small spots posteriorly. Often there is a concentration of pigment on the caudal peduncle near the base of the dorsal procurrent rays of the caudal fin. The dorsal fin has a row of black spots, restricted to the interradial membranes, that extends posteriorly at least half the length of the fin. These spots are best developed anteriorly, where they are restricted to the distal half of the fin. The other fins are pale, except for the outer lobes of the caudal fin which are dusky.

Small juveniles (Fig. 146) have a prominent basicaudal spot, dark mid-lateral stripe, restricted to anterior two-thirds of body, and fainter stripes near the bases of the dorsal and anal fins. The outer rays of the caudal fin are noticeably dark and there is a large black spot anteriorly in the dorsal fin.

In life, adults of *M. nigrolineatus* are slate blue anteriorly grading into pale yellow posteriorly. The dark stripe and the markings in the dorsal fin are black. A color photograph of *nigrolineatus* is given in the frontispiece and in Burgess (1974: 42).

Specimens from the southern Red Sea differ from the above description in having the black markings in the dorsal fin much more extensively developed and the dark body pigment also more extensive and blacker. In some specimens the black stripe on the body is completely obscured by the surrounding dark pigment. The outer lobes of the caudal fin are also more heavily pigmented.

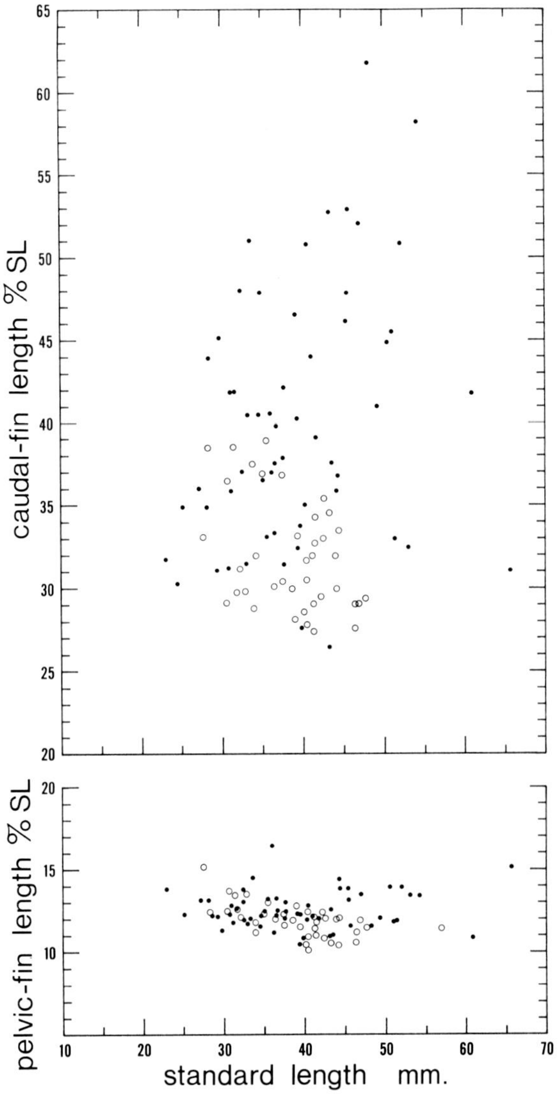

Figure 57. — Relationship between pelvic-fin and caudal-fin lengths versus standard length in *Meiacanthus nigrolineatus*. Closed symbols = males; open symbols = females.

Geographic variation. — The northern and southern Red Sea populations of *nigrolineatus*, separated by a distance of approximately 1500 km, differ in color pattern as noted above. The color pattern of southern specimens is merely an intensification of the dark markings present in northern specimens and there are no apparent meristic or morphometric differences between the two populations.

Species comparisons. — Color pattern will distinguish *nigrolineatus* from the three allopatric Indian Ocean species of *Meiacanthus*. In addition, the outer lobes of the caudal fin are longer in *nigrolineatus* (not elongated in *smithi*). *M. mossambicus* and *fraseri* also differ from *nigrolineatus* in typically having one more dorsal-fin spine.

Etymology. — From the Latin *nigro* (black) and *lineatus* (stripe or line) in reference to the characteristic stripe present in most specimens.

Mimicry. — An apparent mimetic trio (see frontispiece) consisting of *M. nigrolineatus, Ecsenius graveri* (Pellegrin) and *Plagiotremus townsendi* was studied by Springer and Smith-Vaniz (1972a). The coloration and behavior of the *Ecsenius,* a Batesian mimic, is remarkably similar to that of the *Meiacanthus.* Even the dark color pattern of the southern population of *nigrolineatus* is closely duplicated by *E. graveri.*

P. townsendi was also believed to be a mimic of the *Meiacanthus* although the exact nature of the mimetic relationship is unclear (refer to discussion in general section on mimicry). It differs from *nigrolineatus* and *E. graveri* most obviously in having a white stripe passing around the tip of the snout and, in large males (nuptial only?), a small orange spot in the spinous dorsal fin. The only specimen of *townsendi* collected from the southern Red Sea (a recent collection off Port Sudan) is pigmented like northern specimens.

Distribution (Fig. 46). — Known only from the Red Sea.

Material examined. — 137 specimens (11.4-65.7 mm SL) from 33 collections.

GULF OF AQABA: HUJ E-67/29 (3:41-48), HUJ SLR292E (1:32), HUJ 1091E (1:40), HUJ SLR109285 (2:35-36), USNM 204578 (11:18-52), USNM 204579 (7:12-45), USNM 204580 (13:18-44), USNM 204581 (3:25-30), USNM 204582 (9:18-54), USNM 204636 (11:26-43), USNM 204641 (6:11-51), USNM 204642 (15:13-43), USNM 205378 (3:17-61), UTAI 2622 (2: 16-23), UTAI 2632 (1:29), UTAI 3570 (3:37-42), UTAI 3579 (6:39-48), UTAI 3802 (1:31), UTAI 4327 (6:32-44), UTAI 4518 (1:38), UTAI 4786 (1:19).

SINAI PENINSULA: USNM 204586 (1:43), Ras Muhammad; HUJ F-4355 (1:66), HUJ SLR826F (1:46), USNM 204587 (8:31-40), Gulf of Suez, El Tur.

STRAITS OF JUBAL: USNM 200301 (31.7, holotype of *Meiacanthus nigrolineatus*); USNM 200601 (4: 30-34), USNM 201480 (3:36-53), USNM 201485 (1: 41).

SOUTHERN RED SEA: USNM 204485 (1:42), Dahlak Archipelago, Difnein I., 16°36′N, 39°20′E; USNM 204486 (5:38-49), Sheikh el Abu, 16°8′N, 39°26.5′E; USNM 204482 (2:47-51), Isola Delemme, 15°30.5′N, 39°54′E; USNM 204484 (2:44-51), Melita Bay, ca. 15°15′N, 39°49′E; HUJ F4627 (1:33), Zubair Is., 15°0′N, 42°10′E.

Meiacanthus (Meiacanthus) mossambicus Smith

Specimen Figures 147-148; Tables 15-17.

Meiacanthus mossambicus Smith, 1959: 233, pl. 17, Fig. 8 (type locality: Pinda, Mozambique; holotype RUSI uncat. [not examined]).

Description. — Characters of the subgenus. Dorsal fin IV-V (usually V), 24-27; total elements 29-31. Anal fin II, 15-17. Pectoral fin 13-15. Caudal fin: procurrent rays 5-6 + 5-6 = 10-12; segmented rays 11-13 (rarely 13). Vertebrae: precaudal 13-14; caudal 21-23; total 34-36. Pleural ribs on vertebrae 3-12 or 13; epipleural ribs 13-15. Incisor teeth in specimens 29-70 mm, n = 14: lower jaw 17-27; upper jaw 16-23.

Sexual dimorphism absent in pelvic fins, moderate in caudal fin in specimens > 35 mm SL (Fig. 58). Pelvic fin: males 9.6-13.2% SL, females 8.2-11.7% SL; caudal fin: males 36.2-45.6% SL, females 21.1-29.9% SL. Outer lobes of caudal fin elongated in adults of both sexes, interradial membranes of inner rays weakly incised.

Color pattern. — In preserved specimens head, body, dorsal and anal fins heavily and nearly uniformly pigmented, the distal two-thirds of dorsal and anal fins slightly darker than other areas. Posterior third of caudal peduncle and caudal fin abruptly pale. In life the caudal fin is yellow and the body and other fins are chocolate brown, except for the nearly transparent pectoral fins. In nuptial males, at least, the tips of all of the anal-fin rays and those posteriorly in the dorsal fin are pale gray.

Small juveniles (Fig. 147) have a mid-lateral stripe terminating in a prominent basicaudal spot. Stripes are also present on the body near the bases of the dorsal and anal fins, and only the distal third of the dorsal fin is heavily pigmented.

A good color illustration of *mossambicus* accompanied the original description, and Burgess and Axelrod (1973b: 637) present a color photograph of an adult from Mombasa. Burgess and Axelrod (1973b: 469) also present color photographs of a pair of large juveniles, one showing a trace of the basicaudal spot. The body of these juveniles is gray-brown, instead of dark brown as in adults.

Species comparisons. — Refer to discussion under account of *M. fraseri.*

Remarks. — This species, like other members of the genus, can be maintained easily in captivity. Several large individuals of *mossambicus* were observed in public aquaria in Amsterdam and Berlin.

Etymology. — The specific name refers to the African country along whose shores the species was first discovered.

Distribution (Fig. 46). — East African coast from Mombasa to Pinda, Comoro Islands and Madagascar.

Material examined. — 24 specimens (17.5-69.2 mm SL) from 20 collections.

E. AFRICA: RUSI uncat. (2:27-28), RUSI uncat. (1:43), USNM 212028 (1:47), Mombasa; RUSI Z.576 (1:29), RUSI Z.516 (1:18), Zanzibar; BPBM 15949 (1:44), BPBM 15950 (1:55), BPBM 15953 (2:34), Tanzania, Mafia I.; RUSI uncat. (2:67-69), "Mozambique"; ANSP 122784 (1:63), RUSI B.P.319 (1:49), RUSI B.P.1192 (1:62), RUSI B.P.1793 (1:58), Pinda; RUSI uncat. (1:52), probably Pinda; RUSI uncat. (1:52), Wamizi I.

COMORO IS.: USNM 201484 (2:50-53), Pamanzi I., 12°49′0″S, 45°17′29″E; USNM 201483 (1:69), Anjouan I., 12°10′S, 44°23′E.

MADAGASCAR: UMMZ 185821 (1:60), UMMZ 185968 (1:54), Tulear; USNM 201493 (1:38), E. Ambariobe, Nossi Bé, 13°26′10″S, 48°22′54″E.

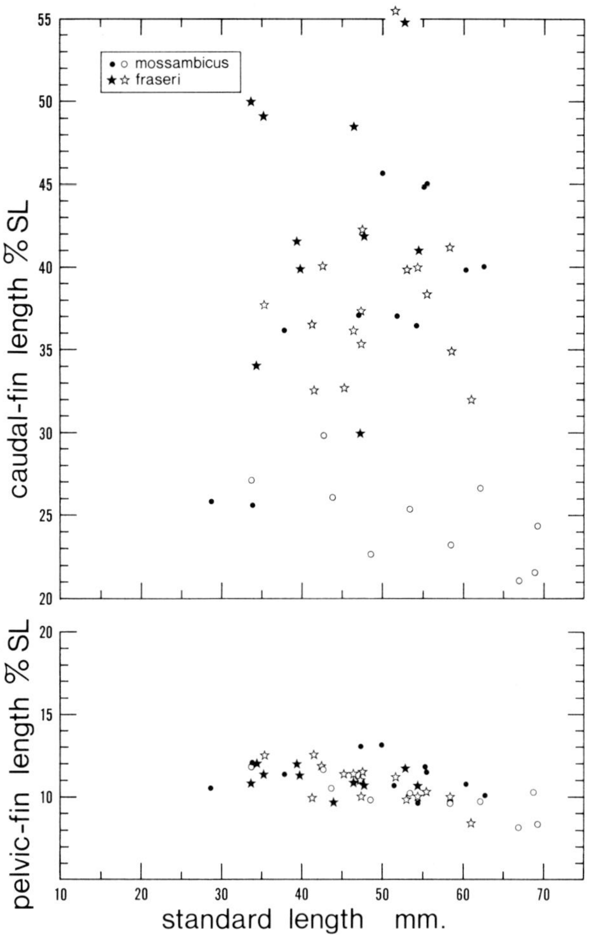

Figure 58. — Relationship between pelvic-fin and caudal-fin lengths versus standard length in *Meiacanthus mossambicus* and *fraseri.* Closed symbols = males; open symbols = females.

Meiacanthus (Meiacanthus) fraseri, new species

Specimen Figure 149; Tables 15-17.

Description. — Characters of the subgenus. Dorsal fin IV-V (usually V), 24 or 25. Anal fin II, 15 or 16. Pectoral fin 14 or 15. Caudal fin: procurrent rays 5-6 + 5-7 = 10-13; segmented rays 11-13 (rarely 13). Vertebrae: precaudal 12 or 13; caudal 21-23; total 34 or 35. Pleural ribs on vertebrae 3-12 or 13; epipleural ribs 13-15. Incisor teeth in specimens 34.3-58.4 mm, n = 18: lower jaw 18-23; upper jaw 18-22.

Sexual dimorphism absent in pelvic fins, slight in caudal fin (Fig. 58) in specimens > 35 mm SL. Pelvic fin: males 9.5-12.1% SL, females 9.8-12.6% SL; caudal fin: males 29.7-54.8% SL, females 32.6-55.5% SL. Outer lobes of caudal fin elongated in adults of both sexes, interradial membranes of inner rays weakly incised.

Color pattern. — In preserved specimens anterior two-thirds of body dusky, darker dorsally, the posterior third gradually becoming pale. Dorsal fin black with narrow, pale fringe. Other fins pale except for dusky to black outer lobes of caudal fin. In life, the dusky areas are slate-gray with bluish violet undertones and the pale areas are very pale yellow, except for the transparent pectoral fins.

Species comparisons. — This species appears to be a cognate of *mossambicus,* the only other species of *Meiacanthus* known from the western Indian Ocean. The only difference between the two species is color pattern and a slightly more filamentous caudal fin (Fig. 58) in *fraseri.*

Etymology. — Named in honor of Dr. Thomas H. Fraser who collected the first specimen of the new *Meiacanthus* and, suspecting that it might be undescribed, informed me of its existence.

Distribution. — Known only from St. Brandon Shoals (Cargados Carajos) in the western Indian Ocean. John E. Randall (pers. comm.) did not observe the species during a recent collecting expedition to Mauritius. Major current systems in the region extend westward around Madagascar to the African coast where, despite a relatively large number of collections, only *M. mossambicus* has been taken.

Material examined. — 27 specimens (33.8-61.0 mm SL) from 7 collections.

Holotype. — USNM 215434, female, 54.3 mm, western Indian Ocean, St. Brandon Shoals (Cargados Carajos), ca. 16°27′S, 59°36′E, about 0.5 mi. S of Raphael Island; depth 4.5-9 m; rocky cliff with incuts, bottom rocky and silty; 12 April 1976; Victor G. Springer, C. Lavett Smith, Thomas H. Fraser, Philip L. Manhard, Paul West and Curtis Marshall; original no. VGS 76-20.

Paratypes (all from St. Brandon Shoals). — USNM 215435 (5:52-58), same data as holotype; USNM 215438 (5:44-54), sta. VGS 76-18; USNM 215436 (5:34-45), sta. VGS 76-12; BPBM 20011 (2:46), sta. VGS 76-12; AMNH 35887 (2:39-48), sta. VGS 76-7; AMS I.18806-001 (2:42-48), sta. VGS 76-8; ANSP 134465 (4:34-47), sta. VGS 76-1; RUSI 74-309 (1:61), Raphael Is., on SW side, 23 March 1971, T. H. Fraser, orig. no. THF-SA-41.

Meiacanthus (*Meiacanthus*) *smithi* Klausewitz

Specimen Figure 150; Tables 15-17.

Meiacanthus smithi Klausewitz, 1961: 17, Fig. 1 (type locality: Maldive Islands, Addu Atoll near Hittadu Island; holotype NFIS 5439).

Description. — Characters of the subgenus. Dorsal fin III-V, 24-26; total elements 28-30. Anal fin II, 15-17. Pectoral fin 13-15. Caudal fin: procurrent rays 5-7 + 5-7 = 10-14; segmented rays 11-13 (rarely 13). Vertebrae: precaudal 12-14; caudal 21-22; total 33-35. Pleural ribs on vertebrae 3-12 or 13; epipleural ribs 13-16. Incisor teeth in specimens 24-58 mm, n = 20: lower jaw 17-27; upper jaw 16-22.

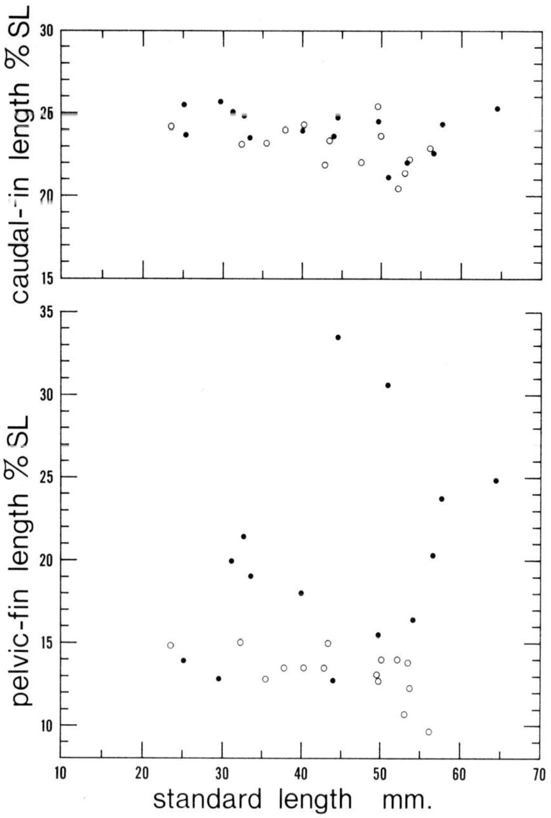

Figure 59. — Relationship between pelvic-fin and caudal-fin lengths versus standard length in *Meiacanthus smithi.* Closed symbols = males; open symbols = females.

Sexual dimorphism well-developed in pelvic fins, absent in caudal fin in specimens > 30 mm SL (Fig. 59). Pelvic fin: males 12.7-33.5% SL, females 9.7-15.0% SL; caudal fin: males 21.2-24.8% SL, females 20.4-25.4% SL. Outer lobes of caudal fin not elongated, interradial membranes of inner rays weakly incised.

Color pattern. — Head and body uniformly pigmented except for dark bar, half pupil diameter in width, extending from posterodorsal margin of orbits to origin of dorsal fin. In some specimens this bar has a narrow pale border. Dorsal fin with broad dark stripe through center of fin, tapering posteriorly. Caudal fin with dorsal and ventral margins of fin and posterior half of segmented rays dark.

In life the dark markings are brown, those of the dorsal fin nearly black. The general body coloration and the proximal area of the dorsal fin may be pale olive-brown or pale gray. The other fins and the pale distal border of the dorsal fin is essentially colorless.

Species comparisons. — *M. smithi* does not seem to be closely related to any other *Meiacanthus.* It superficially resembles some specimens of *atrodorsalis* that have postorbital bars and a well-developed dorsal-fin stripe; *smithi* differs from *atrodorsalis* most obviously in not having a dark spot in the axil of the pectoral fin or an unmarked caudal fin with the outer rays elongated in adults.

Etymology. — Named after the late Dr. J. L. B. Smith, a famous South African ichthyologist.

Mimicry. — Springer and Smith-Vaniz (1972a: 23) first suggested, on the basis of circumstantial evidence, that *Meiacanthus smithi* and *Plagiotremus phenax* were involved in a mimetic relationship. The exact nature of the mimetic relationship was unclear (refer to discussion in general section on mimicry). They also noted that one of three color forms of *Ecsenius bicolor* might be a mimic of *smithi.* At that time all three species were known from Sri Lanka but had not been collected together, and the life colors of the *Meiacanthus* were unknown. In January 1973, General Alfred L. Wolf and Rodney Jonklass sent me several live *smithi* and *P. phenax* that they had collected in the Maldive Islands, where both species occur in close proximity. Except for an orange spot in the spinous dorsal fin of some males of *phenax* (a sexually dimorphic character probably restricted to nuptial males), life colors are virtually identical in the pair. The *Plagiotremus* also lacks the postorbital bar of *smithi,* a minor difference compared to the remarkable similarity of their more conspicuous markings. *Ecsenius bicolor* does not closely resemble *smithi* in its life coloration, thus it seems unlikely that it is a mimic of *smithi.*

Wolf and Jonklass also noted (*in litt.*) that juveniles of an unidentified species of *Scolopsis* from the Maldive Islands had the identical markings of *smithi.* Juveniles of *Scolopsis* also mimic three other species of *Meiacanthus* (see Fig. 9).

Distribution (Fig. 48). — Widely distributed in the eastern Indian Ocean where suitable reef habitats occur. Apparently absent from the Cocos-Keeling Islands where it was not observed during six weeks of intensive collecting.

Material examined. — 36 specimens (11.7-64.4 mm SL) from 15 collections.

MALDIVE IS.: NFIS 5439 (53.7, holotype of *Meiacanthus smithi*), Addu Atoll, near Hittadu I., 0°38′S, 73°10′E; ANSP 122787 (5:25-50), Bandos I., ca. 5 mi. N of Male; FMNH 71349 (1:50), S Male Atoll, Dunidu I.; FMNH 71355 (4:40-53), S Nilandu Atoll, Kuda Huvadu I.; FMNH 71356 (5:25-58), Addu Atoll, Wilingili I.

SRI LANKA (CEYLON): All from Trincomalee, USNM 205284 (1:54), USNM uncat. (1:50), sta. PCH 69-277, USNM 214394 (1:15).

THAILAND: USNM 201489 (1:38), Phuket, Patong, 7°52′N, 98°22′E.

MALACCA STRAIT: USNM 199422 (6:32-50), Jarak I., 3°59′N, 100°6′E.

SUMATRA: USNM 201488 (1:64), NE of Simeulue, Penju I., 2°50′58″N, 95°56′24″E; USNM 199483 (1:56), Mentawei Is., Siburu I., 2°0′0″S, 99°35′40″E.

JAVA SEA: USNM 211999 (1:24), Seribu Is., ea. 5°51′S, 106°34′E; USNM 211982 (1:12), USNM 211987 (4: 31-36), Karimundjawa I., 5°52′S, 110°25′E; USNM 212004 (2:16-21), either Seribu I. or Karimundjawa I.

Meiacanthus (Meiacanthus) atrodorsalis (Günther)

Specimen Figures 151-155; Tables 4, 15-18.

(For synonymy, see both subspecies)

Description. — Characters of the subgenus. Dorsal fin III-VI (typically IV), 25-28; total elements 29-32. Anal fin II, 15-18. Pectoral fin 13-15. Caudal fin: procurrent rays 4-7 + 5-8 = 9-15; segmented rays 11-13 (rarely 13). Vertebrae: precaudal 12-14; caudal 20-25; total 33-37. Pleural ribs on vertebrae 3-12 or 13; epipleural ribs 13-16. Incisor teeth in specimens 19-67 mm, n = 42: lower jaw 16-26; upper jaw 16-24.

Sexual dimorphism absent in pelvic fins, moderate in caudal fin in specimens > 30 mm SL (Fig. 60). Pelvic fins: males 8.4-16.2% SL, females 8.4-14.3% SL; caudal fin: males 29.8-98.1% SL, females 26.3-47.0% SL (n = 64), except two females 39.7 and 51.8 mm SL with caudal fins 58.4 and 63.4% SL respectively. Outer lobes of caudal fin elongated in adults of both sexes, interradial membranes of inner rays weakly incised.

Color pattern. — The combination of an unstriped body, uniformly pale caudal fin and the axil (underside) of the pectoral fin with a conspicuous, dark spot (slightly smaller than pupil diameter) distinguishes *atrodorsalis* from all other species of *Meiacanthus*.

Geographic variation. — Populational variation in frequencies of selected meristic characters is shown in Table 18.

Species comparisons. — Color pattern (see above) will distinguish *atrodorsalis* from all other species of *Meiacanthus*. See also discussion under account of *M. procne*.

Meiacanthus atrodorsalis atrodorsalis (Günther)

Petroscirtes atrodorsalis Günther, 1877: 198, pl. 115, Fig. B (type locality: Samoa; holotype BMNH 1876.5.1.37).
Petroscirtes herlihyi Fowler, 1946: 183, Fig. 47 (type locality: Aguni Shima, Ryukyu Islands; holotype ANSP 72053).

Color pattern. — The color pattern of *M. a. atrodorsalis,* except for the diagnostic characters given previously, is quite variable. A dark postorbital bar (sometimes with pale border) extends to the dorsal-fin origin but the bar is often absent or only partially developed. Similarly, a dark submarginal stripe in the dorsal fin may be variously developed, although usually present at least anteriorly. Specimens with complete and incomplete postorbital bars and dorsal-fin stripes frequently occur in the same collection. In preserved specimens the pigment is darkest on the head, decreases gradually as it extends on the body and becomes pale posteriorly. The anal fin is pale.

Ten specimens from Kabaena Island, Celebes, all agree in being relatively uniformly pigmented, devoid of any dark markings except for the axillary spot. Unlike *M. a. ovalauensis,* which is also uniformly pigmented (except for the axillary spot), these specimens were not bright yellow in life, but appeared whitish or pale blue-gray (V. G. Springer, pers. comm.).

In life, the dark markings on the body and dorsal fin are black or dark brown. The pale border of the postorbital bar, if present, is white. The head and anterior half to two-thirds of the body is blue-gray, changing to yellow posteriorly. The outer lobes of the caudal fin are yellow, the central area of the fin is hyaline. The anal fin may be partially blue-gray, yellow or nearly colorless. Losey (1972: 131) ob-

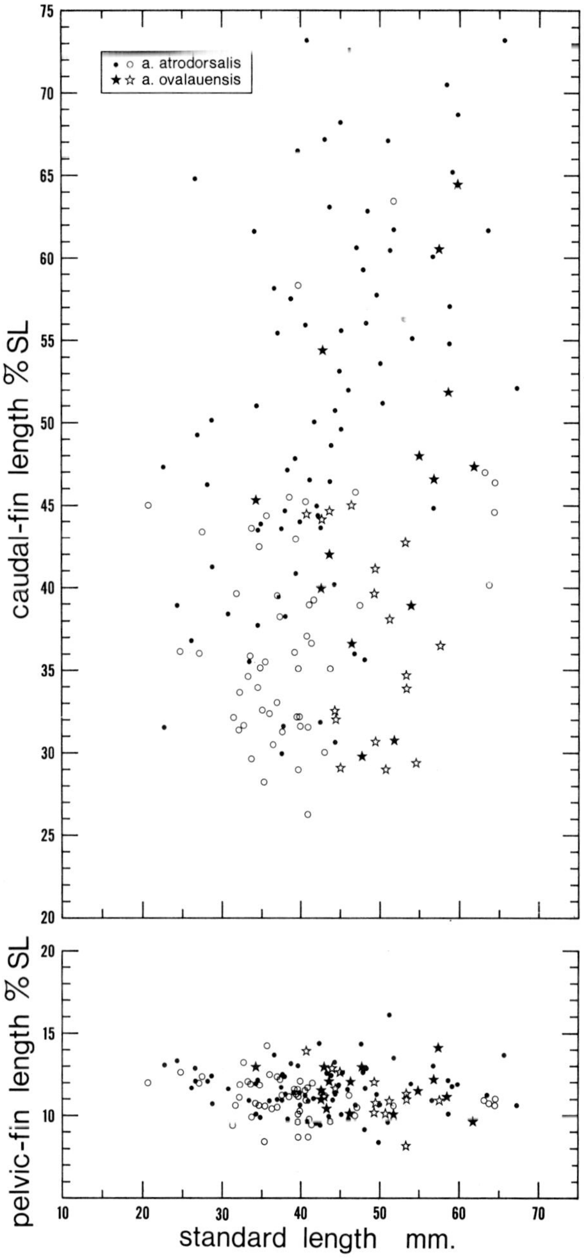

Figure 60. — Relationship between pelvic-fin and caudal-fin lengths versus standard length in *Meiacanthus atrodorsalis* (subspecies plotted separately). Closed symbols = males; open symbols = females. (Not plotted were two males of *M. a. atrodorsalis* 47.7 and 48.6 mm SL with caudal fins 93.9 and 98.1 percent SL.).

TABLE 18. — Geographic variation in frequencies of certain meristic characters in *Meiacanthus atrodorsalis* as indicated by specimens from selected localities.

Localities	Total vertebrae							Total dorsal-fin elements					
	33	34	35	36	37	N	x̄	29	30	31	32	N	x̄
Celebes	3	6	1			10	33.8	5	5			10	29.5
Banda Sea		6	8			14	34.6	2	8	5		15	30.2
Ryukyu Is.			2	4	1	7	35.9		3	3	1	7	30.7
Mariana Is.	1	1	10	1		13	34.8		7	6	1	14	30.6
Palau Is.		2	4	2		8	35.0		6	2		8	30.2
Ifaluk Atoll		1	4	3	1	9	35.4		3	4	2	9	30.9
Kapingamarangi		3	12			15	34.8		13	3		16	30.2
Ponape Is.			3			3	35.0			3		3	31.0
Marshall Is.			10	3		13	35.2		3	10		13	30.8
Gilbert Is.			6			6	35.0			6		6	31.0
Samoa		1	–	1		2	35.0		1	1		2	30.5
Tonga			1	6		7	35.9			4	3	7	31.4
Fiji		2	21	15	2	40	35.4		5	29	8	42	31.1
New Hebrides		1				1	34.0		1			1	30.0
Santa Cruz Is.		2	1			3	34.3	1	2			3	29.7
Solomon Is.		3	2			5	34.4	1	2	2		5	30.2
New Guinea	1	6	3			10	34.2	4	6			10	29.6
E. Australia		2	2			4	34.5		3	2		5	30.4
New Caledonia			1			1	35.0		2	–	1	3	30.7

Localities	Segmented anal-fin rays						Total pectoral-fin rays						
	15	16	17	18	N	x̄	26	27	28	29	30	N	x̄
Celebes	4	5	1		10	15.7			10			10	28.0
Banda Sea	2	7	6		15	16.3		1	7	–	7	15	28.9
Ryukyu Is.		3	3	1	7	16.7			1	–	6	7	29.7
Mariana Is.		8	6		14	16.4			5	2	7	14	29.1
Palau Is.		5	3		8	16.4			7	1		8	28.1
Ifaluk Atoll		1	7	1	9	17.0			4	3	2	9	28.8
Kapingamarangi		8	7		15	16.5		2	10	2	2	16	28.2
Ponape Is.			3		3	17.0			2	–	1	3	28.7
Marshall Is.		3	8	2	13	16.9		2	9	–	2	13	28.2
Gilbert Is.			6		6	17.0			2	2	2	6	29.0
Samoa		2			2	16.0			1			1	28.0
Tonga			3	4	7	17.6			4	2	1	7	28.6
Fiji		4	28	10	42	17.1	1	3	32	1	1	38	27.9
New Hebrides	1				1	15.0			1			1	28.0
Santa Cruz Is.		2	1		3	16.3			2	–	1	3	28.7
Solomon Is.		3	2		5	16.4	1	–	3	1		5	27.8
New Guinea	3	6	1		10	15.8	1	–	5	3	1	10	28.3
E. Australia		5			5	16.0			5			5	28.0
New Caledonia		1	1	1	3	17.0			2	1		3	28.3

served that *M. a. atrodorsalis,* when hiding, adopts a fright coloration with numerous white spots, but its background coloration remains the same. Burgess and Axelrod (1972: 163; 1973a: 470) and Masuda *et al.* (1975: 86) present color photographs of this subspecies and Losey (1972: 132) gives a monochrome illustration of the typical color pattern of specimens from Eniwetok Atoll.

Nomenclature. — *Petroscirtes herlihyi,* clearly a junior synonym of *P. atrodorsalis,* was not compared with any other species in the original description. The Latin name, a combination of *atro* (black) and *dorsalis* (of the back), presumably refers to the dark submarginal stripe usually present in the dorsal fin of this subspecies.

Mimicry. — Springer and Smith-Vaniz (1972a: 22) first suggested that *Ecsenius bicolor* (Day) and *Plagiotremus laudandus* might be mimics of *M. atrodorsalis.* Losey (1972) also published an account of this mimetic trio based on predation experiments and field observations at Eniwetok Atoll. Losey's study indicated that *E. bicolor* is a Batesian mimic of the *Meiacanthus,* but the exact nature of the relationship involving the *Plagiotremus* was unclear (refer to general section on mimicry). See also comments on mimicry under account of *M. atrodorsalis ovalauensis.*

Distribution (Fig. 48). — Widely distributed in the central and western Pacific Ocean, with the notable exception of the Fiji Islands where it is replaced by *M. a. ovalauensis.*

Material examined. — 158 specimens (12.5-67.9 mm SL) from 84 collections.

INDONESIA: USNM 211902 (1:38), USNM 211912 (9: 21-36), Celebes, Kabaena I.; USNM 209710 (1:39), USNM 209843 (1:29), Ceram, Piru Bay; USNM 210141 (1:23), Saparua I.; USNM 211952 (2:27-31), USNM 195707 (8:41-58), "Dutch East Indies — Longley", presumably Banda Sea; USNM 211931 (1:44), USNM 211938 (1:35), USNM 211942 (1:38), USNM 211943 (1:32), Banda I. and vicinity.

RYUKYU IS.: ANSP 72053 (23.8, holotype of *Petroscirtes herlihyi*), ANSP 72055-6 (2:19-21), taken with holotype, ANSP 72054 (1:23), Aguni Shima; ANSP 128831 (1:35), BPBM 7458 (1:55), BPBM 7461 (2:48-57), Ishigaki I.

MARIANA IS.: UG 5106 (3:42-59), Saipan I.; UG 5175 (2:37-41), Tinian I.; CAS 28920 (1:19), UG 4243 (1: 43), UG 4322 (7:40-45), Guam I.

PALAU IS.: 124873 (1:42), BPBM 9635 (1:59), CAS 28918 (2:31-34), CAS 28919 (3:36-45).

SOROL ATOLL: ANSP 126029 (1:51), 8°9′N, 140°21′E.

SAMOA IS.: BMNH 1876.5.1.37 (56.7, holotype of *Petroscirtes atrodorsalis*), "Museum Godeffroy"; SU 8996 (1: 43), Apia; USNM 52242 (1:35), Pago Pago.

TONGA IS.: USNM 203765 (7:48-67), Vava'u Group, Neiafu I.

NEW HEBRIDES IS.: USNM 201594 (1:20), Espiritu Santo I.

SANTA CRUZ IS.: USNM 201561 (3:24-28), Vanikoro I.

SOLOMON IS.: USNM 212032 (1:41), Guadalcanal I.; USNM 201481 (2:28-35), Tautsina I., E of Kieta Peninsula, Bougainville I.; BPBM 15648 (2:38-50), Florida I.

NEW BRITAIN: CSIRO B871 (1:28), Möwe Harbor; USNM 201479 (2:34-35), Rabaul, Dawapia Rocks (Beehives).

NEW GUINEA: USNM 205682 (1:33), USNM 205683 (1:42), Madang Harbor; USNM 205680 (4:17-38), USNM 205681 (1:40), Trobriand Is.; DASF FO3179 (2: 37-42), Port Moresby.

E. AUSTRALIA: ANSP 109680 (1:41), ANSP 109681 (3:15-48), ANSP 109725 (3:13-14), ANSP 109726 (1: 17), ANSP 109728 (1:13), Queensland, Endeavour Reef, 15°45′S, 145°42′E; AMS I.17671-1 (1:47), USNM 201593 (1:50), Great Barrier Reef, Capricorn Group, One Tree I.

NEW CALEDONIA: AMS IB.3835 (1:63), AMS IB.3848 (1:37), AMS IB.4508 (1:48), BPBM 11463 (1:43), CAS 15799 (1:38), CAS 15800 (1:44).

IFALUK ATOLL: ANSP 124872 (1:59), CAS 24698 (2: 41-49), CAS 28902 (3:40-60), CAS 28903 (1:47), CAS 28904 (2:41-52).

TRUK IS.: BPBM 7448 (2:25-40), UG 4628 (2:46-51).

KAPINGAMARANGI ATOLL: CAS 28905 (1:34), CAS 28906 (1:40), CAS 28907 (1:27), CAS 28908 (1:37), CAS 28909 (3:34-41), CAS 28910 (1:38), CAS 28911 (1:35), CAS 28912 (2:14-18), CAS 28913 (2:32-37), CAS 28914 (3:29-39), CAS 28915 (1:37).

PONAPE IS.: CAS 28916 (2:37-40), CAS 28917 (1:48).

MARSHALL IS.: BPBM 6258 (1:45), Eniwetok Atoll; USNM 142222 (1:36), USNM 142223 (1:66), USNM 142224 (3:46-54), Rongelap Atoll; USNM 201490 (6: 23-44), USNM 201492 (2:27-44), Taka Atoll.

GILBERT IS.: AMS uncat. (3:27-50), sta. DFH 73-36, AMS uncat. (2:41-45), sta. DFH 73-37, Apaiang Atoll; USNM 167327 (1:40), Onotoa Atoll.

Meiacanthus atrodorsalis ovalauensis (Günther)

Petroscirtes oualanensis[1] Günther, 1880: 35 (type locality: Ovalau, Fiji Islands; lectotype BMNH 1879.5.14.582).

Petroscirtes auratus Seale, 1936: 85 (type locality: Suva, Fiji Islands; holotype CAS 5527).

---

[1] The original specific name was misspelled and is amended as per discussion of etymology.

Color pattern. — In preservation the body and fins are uniformly pale, except for the dark spot in the axil of the pectoral fins. This subspecies is most appropriately referred to in the aquarium literature as "canary blenny." In life *M. atrodorsalis ovalauensis* is essentially brilliant yellow; the central area of the caudal fin is hyaline and there is a brown axillary spot (usually hidden from view by the pectoral fin). In large aquarium specimens, at least, the top of the head is frequently olive.

Springer (1970: 54) and Burgess and Axelrod (1973a: 470) present the same color photograph of this subspecies. An excellent color photograph is also given in Ryan (1973).

Nomenclature. — *Petroscirtes auratus* was not compared with any other species in the original description.

Etymology. — In the original description the type locality was given as *"ovalau"*, although the specific name (obviously in reference to the capture locality) was spelled *"oualanensis."* Manuscripts were all handwritten in Günther's day and it would have been easy for the printer to misread u's, v's, and n's. Were only the initial "v" altered one might suspect that Günther had merely latinized Ovalau but the two errors argue strongly for typographical error. Article 32c of the International Code of Zoological Nomenclature states that an original spelling that is an inadvertent error (such as a copyist's or printer's error) is to be corrected wherever it is found. In accordance with this recommendation, the new amended spelling for *Petroscirtes oualanensis* Günther adopted in this study is *P. ovalauensis.*

Taxonomic status. — The occurrence of a discrete spot in the axil of the pectoral fin in both *M. atrodorsalis* and *ovalauensis* is indicative of close kinship. This spot, which is normally hidden from view by the pectoral fin, may serve an important role in courtship or species recognition since it is highly visible when exposed (Figs. 154-155) and occurs in no other species of *Meiacanthus.*

M. a. atrodorsalis is broadly distributed in the central Pacific Ocean (including Tonga, Samoa and the New Hebrides Islands), but does not occur in the Fiji Islands where the endemic *ovalauensis* is common and has been known for almost a century. This distributional pattern does not appear to be an artifact of uneven collecting effort, and there are no known geographical barriers that would prevent typical *atrodorsalis* from becoming established at Fiji. One possible explanation for the allopatric distributions of the two forms is that competition from the closely related and better adapted Fijian *Meiacanthus* provides the barrier to colonization. An alternative and (I believe) more plausible explanation is given below.

Individuals of *M. atrodorsalis ovalauensis* are indistinguishable from the nominal subspecies except for their all-yellow color pattern. This coloration is apparently genetically controlled because individuals of *ovalauensis* maintained in aquaria for many months retain their normal coloration. Xanthochroism has been documented in several diverse families of fishes (Dawson, 1964; see also discussion of mimicry below), and xanthic individuals occur with considerable frequency in several species of American groupers (Smith, 1971: 75). Conceivably, a gene for xanthic coloration was present in a small hypothetical founder population of *atrodorsalis* that colonized Fiji. Due to the small initial size of the founder population a relatively large number of the initial progeny might have possessed the gene. Selection pressure from predation acting in consort with the unique defense mechanism provided by the dentary glands of *Meiacanthus* might have favored the survival of xanthic individuals. (Under identical conditions the conspicuous xanthic phenotype might provide a more effective aposematic function that the bicolored blue and yellow pattern of the nominal subspecies.) Once established the two allopatric populations of *Meiacanthus* would persist, given a low rate of outside recruitment, because resident predators would be more likely to associate a negative experience with the locally dominant color form.

Species of *Meiacanthus* are relatively easy to rear and maintain in aquaria (Fishelson, 1975). It might prove rewarding to test the above hypothesis by attempting to cross-breed *atrodorsalis* and *ovalauensis* in the laboratory. Because of the allopatric distributions of the two forms, only by laboratory experiments will it be possible to determine their true taxonomic status. For the present I prefer to emphasize their similarities by recognizing *ovalauensis* as a subspecies of *atrodorsalis.*

Aside from *Plagiotremus laudandus* (see below),

I am aware of only one other situation involving marine tropical shorefishes where two essentially allopatric populations are maintained with the only reliable distinguishing character being the entirely yellow coloration of one of the forms. Randall (1955) documented such a situation involving *Zebrasoma flavescens* (Bennett) and *Z. scopas* (Cuvier). Unlike the distributional pattern of *atrodorsalis-ovalauensis,* the range of one of the forms of *Zebrasoma* does not completely encircle that of the other. Randall elected to treat both forms as different species, but he also acknowledged that the available data were insufficient to determine with assurance if his action was correct.

Mimicry. — Springer and Smith-Vaniz (1972a: 23) noted that an undescribed subspecies of *Plagiotremus laudandus* (distinguished from the nominal subspecies by its entirely brillant yellow coloration) endemic to the Fiji Islands was probably a mimic of *M. a. ovalauensis.* Refer to species account of *P. laudandus* for discussion of taxonomic status.

Allen *et. al.* (1975) reported that in the Fiji Islands, juveniles of the coral bream, *Scolopsis bilineatus,* also mimic *M. atrodorsalis ovalauensis.* They stated that "the scolopsid, which is very different in shape when compared to the model, enhances its disguise by keeping the dorsal fin folded most of the time and swims in quick *Meiacanthus*-like darting motions interspersed with stationary hovering a short distance above the bottom. Only juvenile coral bream under 2-3 inches are effective mimics. Larger fish, which are less vulnerable to attack from predators and are therefore not dependent on mimicry for protection, gradually assume the adult pattern typified by the loss of yellow coloration and the development of an oblique dark-edged pale line extending from the mouth to the base of the last dorsal spines." Only in the Fiji Islands are the juveniles of *S. bilineatus* entirely bright yellow.

Distribution. — Endemic to the Fiji Islands.

Material examined. — 42 specimens (34.2-67.9 mm SL) from 13 collections.

FIJI IS.: BMNH 1879.5.14.582 (61.8, here designated lectotype of *Petroscirtes oualanensis*), BMNH 1890.2. 26.115 (53, paralectotype of *Petroscirtes oualanensis*), Ovalau I., "Challenger"; CAS 5527 (42.6, holotype of *Petroscirtes auratus*), CAS 5528 (1:39), CAS 7490 (3: 42-46), CAS 24699 (8:43-59); ANSP 128238 (2:43-44), Viti Levu, Mbengga I.; BPBM 11387 (2:49), Viti Levu, Rat-tail Pass, near Suva Harbor; USNM 201455 (5:41-59), Great Astrolabe, reef N of Vuro I., 18°52′S, 178°20.5′E; USNM 201677 (1:60), Wailangilala I., 16°45′S, 179°7′W; USNM 207574 (1:51), Mbulia I., 18°50′S, 178°32′E; NMV 71773 (1:51), "Viti Inselin"; ANSP 124267 (7:34-46), LACM 9431-1 (1:55), LACM 9633-14 (7:53-68), aquarium specimens, "Fiji Is."

Meiacanthus (Meiacanthus) procne, new species

Specimen Figure 156; Tables 15-17.

Description. — Characters of the subgenus. Dorsal fin IV-V (typically V), 25-26; total elements 30-31. Anal fin II, 15-16. Pectoral fin 13-14. Caudal fin: procurrent rays 5-6 + 5-6 = 10-12; segmented rays 11-12. Vertebrae: precaudal 13-14; caudal 21-22; total 35-36. Pleural ribs on vertebrae 3-13; epipleural ribs 13-14. Incisor teeth in specimens 28-56 mm, n = 10; lower jaw 17-21; upper jaw 16-20.

Sexual dimorphism absent in pelvic fins, moderate in caudal fin in specimens > 30 mm SL (Fig. 61). Pelvic fin: males 9.7-13.0% SL, females 8.5-10.5% SL; caudal fin: males 47.9-70.9% SL, females 33.8-54.0% SL. Outer lobes of caudal fin elongated in adults of both sexes, interradial membranes of inner rays weakly incised.

Color pattern. — Head and body dark, especially dorsally. Large specimens with spotted or speckled pattern on central area of caudal fin and on posterior half of body ventrally. Venter, and pelvic and anal fins pale. Dorsal fin dark with broad pale stripe through center of fin. Outer lobes of caudal fin black with pale edge proximally. Life colors are unknown.

Species comparisons. — *M. procne* is perhaps most closely related to *atrodorsalis,* which was taken with the new species in the same station at Tonga. It differs from *atrodorsalis* most obviously in lacking a spot in the axil of the pectoral fin, having the outer lobes of the caudal fin strongly pigmented and in having typically 5 dorsal-fin spines. *M. procne* is the only central Pacific species of *Meiacanthus* with heavily pigmented caudal-fin lobes that has a pale anal fin.

Etymology. — The Greek name *procne,* meaning a kind of swallow, is in allusion to the produced caudal-fin lobes of the species. The name is to be treated as a noun in apposition.

Distribution (*Fig. 48*). — Known definitely from the Tonga Islands and questionably from Tahiti. The record of *procne* from Tahiti constitutes the easternmost locality for any species of *Meiacanthus.* I question the validity of this record because the collection data is imprecise for the single specimen and John E. Randall (pers. comm.) did not encounter the species while making extensive collections in the Society Islands.

Material examined. — 15 specimens (15.9-44.0 mm SL) from 2 collections.

Holotype. — USNM 203275, male, 44.0 mm, Tonga Islands, Vava'u Group, Neiafu I., 18°39'0"S, 174°0'30"W; depth 0-16 m; 28 June 1965; R. L. Bolin *et al.;* Te Vega Cruise 8, sta. 308.

Paratypes. — USNM 205291 (13:16-43), taken with the holotype. USNM 195810 (1:38), Tahiti; Jan.-April 1931; Zane Grey.

Meiacanthus (Meiacanthus) reticulatus, new species

Specimen Figure 157; Tables 15-17.

Description. — Characters of the subgenus. Dorsal fin V, 26. Anal fin II, 17. Pectoral fin 14. Caudal fin: procurrent rays 7 + 7 = 14; segmented rays 11. Vertebrae: precaudal 14; caudal 22. Pleural ribs on vertebrae 3-12; epipleural ribs 14. Incisor teeth: lower jaw 24; upper jaw 20.

Sexual dimorphism probably as in *M. procne* (Fig. 61), female holotype only known specimen; pelvic fin 10.2% SL, caudal fin 53.9% SL. Outer lobes of caudal fin elongated, interradial membranes of inner rays weakly incised.

Color pattern. — Dorsal fin, belly, anal fin (including small area above fin anteriorly) and outer lobes of caudal fin heavily pigmented. A somewhat lighter dark area on dorsum anteriorly extending obliquely from snout through eye to origin of last spine of dorsal fin. Central area of caudal fin and remainder of head and body consisting of a dark reticulated or speckled pattern superimposed on a pale background. Life colors are unknown.

Species comparisons. — The unique reticulated color pattern readily distinguishes this species from all other members of the genus.

Etymology. — The Latin name *reticulatus,* meaning reticulated, refers to the dominant feature of the color pattern.

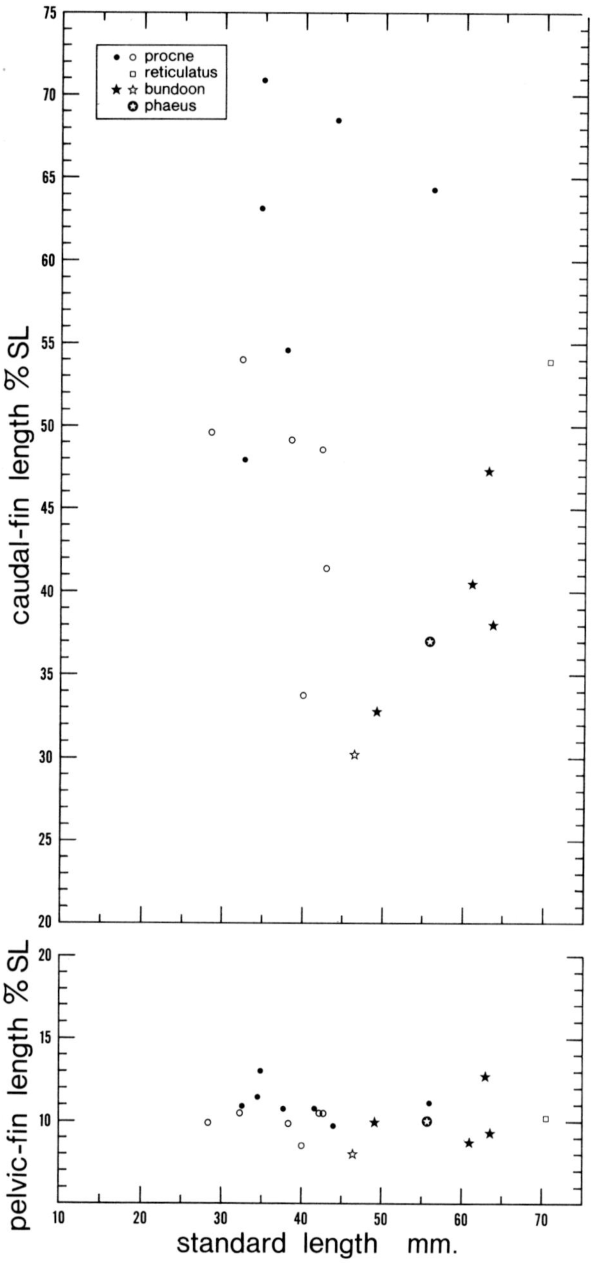

Figure 61. — Relationship between pelvic-fin and caudal-fin lengths versus standard length in four presumably closely related species of *Meiacanthus.* Closed symbols = males; open symbols = females.

Distribution (*Fig. 48*). — Known from a single collection in the Solomon Sea.

Material examined. — Holotype, only known specimen.

Holotype. — BPBM 13648, female, 70.5 mm, Solomon Sea, Egum Atoll in lagoon near Egum Island, 9°20'S, 151°57'E; depth 6 m; 11 June 1972; G. R. Allen.

Meiacanthus (Meiacanthus) bundoon, new species
Specimen Figure 158; Tables 15-17.

Description. — Characters of the subgenus. Dorsal fin IV-V (usually V), 26-27; total elements 31-32. Anal fin II, 16-17. Pectoral fin 15-16. Caudal fin: procurrent rays 5-6 + 6-7 = 11-13; segmented rays 11-13. Vertebrae: precaudal 13; caudal 23-25; total 36-38. Pleural ribs on vertebrae 3-13; epipleural ribs 13-14. Incisor teeth in specimens 46-64 mm, n = 5; lower jaw 21-24; upper jaw 20-22.

Sexual dimorphism absent in pelvic fins, probably moderate in caudal fin (Fig. 61) in specimens > 30 mm SL. Pelvic fin: males 8.7-12.7% SL, female 8.0% SL; caudal fin: males 32.7-47.3% SL, female 30.2% SL. Outer lobes of caudal fin elongated in adults of both sexes, interradial membranes of inner rays weakly incised.

Color pattern. — Body rather uniformly dark except for pale, wedge-shaped area beginning on postorbital region of head (from just above gill opening to lower base of pectoral fin) extending mid-laterally approximately two-thirds length of anal fin. Caudal-fin base and outer rays heavily pigmented, central area of fin pale. Pectoral fins essentially transparent except for narrow concentration of melanophores outlining rays.

Life colors, based on a series of color slides taken by B. C. Carlson underwater of living specimens, are: head and anterior dorsum dark olive; cheeks and opercle rich green; pale wedge-shaped area on sides yellow-orange. Remainder of body, dorsal and anal fins brown. In some individuals the dorsal and anal fins are pale orange or yellow proximally. Outer lobes of caudal fin dark brown (nearly black); central area of caudal fin transparent or pale blue.

Species comparisons. — *M. bundoon* is perhaps most closely related to *phaeus,* a new species known only from New Caledonia; structurally the two are nearly identical. The more slender appearance of *bundoon* is at least partly a reflection of the higher vertebral count of the illustrated specimen and is probably not a reliable distinguishing character. The life colors of *phaeus* are unknown, but in the single available specimen the body is uniformly pigmented and noticeably paler than the dark body coloration in preserved specimens of *bundoon.* In view of the differences in coloration of preserved specimens, and the spectacular life colors of *bundoon,* it seems unlikely that these two apparently allopatric forms are conspecific.

Etymology. — The specific name is in honor of the yacht BUNDOON, which was used for transportation during the collection of the new species.

Distribution (Fig. 48). — Known only from the Fiji Islands. In addition to the type locality, the species has been observed (Carlson, *in litt.*) at Frigate Passage on the leeward barrier reef of Mbengga (Bega) Island.

Material examined. — 5 specimens (46.3-63.7 mm SL) from one collection.

Holotype. — USNM 212257, male, 63.0 mm, Fiji Islands, Lau Group, Moala I., 18°34'S, 179°56'E, patch reef ca. ½ mi. off N side of island; depth to 15 m; much living coral with channels and caves in reef; 3 July 1974; B. Carlson, B. Goldman and B. Russell.

Paratypes. — AMS I.17917 (1:49), ANSP 128523 (1:64), BPBM 17873 (1:46), USNM 212258 (1:61), taken with the holotype.

Meiacanthus (Meiacanthus) phaeus, new species
Specimen Figure 159; Tables 15-17.

Description. — Characters of the subgenus. Dorsal fin IV, 27. Anal fin II, 17. Pectoral fin 16. Caudal fin: procurrent rays 6 + 7 = 13; segmented rays 11. Vertebrae: precaudal 14; caudal 22. Pleural ribs on vertebrae 3-13; epipleural ribs 13. Incisor teeth: lower jaw 22; upper jaw 20.

Sexual dimorphism probably as in *M. bundoon* (Fig. 61), female holotype only known specimen; pelvic fin 10.0% SL, caudal fin 37.0% SL. Outer lobes of caudal fin elongated, interradial membranes of inner rays weakly incised.

Color pattern. — Head and body uniformly pale brown. Anal fin somewhat darker brown and dorsal fin with distal third noticeably darker than remainder of fin. Outer lobes of caudal fin black, central part of fin pale, and base of fin brown like rest of body. Pectoral fins essentially transparent except for narrow concentration of melanophores outlining rays. Life colors are unknown.

Species comparisons. — Refer to discussion under account of *M. bundoon.*

Etymology. — The Greek name *phaeus,* meaning of the hue or color of twilight, dusky, brown or grayish brown, refers to the dominant coloration of preserved specimens.

Distribution (Fig. 48). — Known only from New Caledonia where it occurs sympatrically with *M. atrodorsalis.*

Material examined. — Holotype, only known specimen.

Holotype. — AMS I.17466-4, female, 55.8 mm, New Caledonia, Puetege Reef, 22°22'S, 167°8'E; depth 12 m; 15 June 1973; G. R. Allen; original number GRA-11.

PLAGIOTREMUS Gill

Plagiotremus Gill, 1865: 138 (type-species: *P. spilistius* Gill, 1865, by monotypy).

Runula Jordan and Bollman, 1890: 171 (type-species: *R. azalea* Jordan and Bollman, 1890, by original designation).

Atopoclinus Vaillant, 1894: 73 (type-species: *A. ringens* Vaillant, 1894 [= *Runula azalea* Jordan and Bollman, 1890], by monotypy).

Macrurrhynchus Ogilby, 1896: 136 (type-species: *M. maroubrae* Ogilby, 1896 [= *Petroskirtes tapeinosoma* Bleeker, 1857], by subsequent designation, in parentheses[1], of Ogilby, 1910: 55).

Lembeichthys Herre, 1936a: 283 (type-species: *L. celebesensis* Herre, 1936, by original designation).

Musgravius Whitley, 1961: 177 (as a subgenus of *Pescadorichthys* Tomiyama [= *Ecsenius* McCulloch]; type-species: *Pescadorichthys (Musgravius) laudandus* Whitley, 1961, by original designation).

Diagnosis. — Major characters useful in distinguishing the genera of Nemophini are summarized in Tables 1, 3 and 4. The nature of the dentary incisors (Figs. 72-73) and the absence of premaxillary canines and a lateral line easily distinguish *Plagiotremus* from the other genera of Nemophini. The general physiognomy of the head of representative species of *Plagiotremus* is shown in Figures 62-67.

Nomenclature. — None of the junior synonyms of *Plagiotremus* was compared with the type-species in the original descriptions. *Runula* was compared only with *Petroscirtes,* and *Runula* and *Atopoclinus* were based on the same type-species. *Macrurrhynchus,* based on a juvenile *P. tapeinosoma,* was not contrasted with any other genus. The type-species of *Plagiotremus* and Herre's *Lembeichthys* are very closely related or possibly conspecific (refer to account of *P. celebesensis*). Whitley erected *Musgravius* as a subgenus of *Pescadorichthys* (a junior synonym of *Ecsenius* — tribe Salariini), therefore his comparison was irrelevant.

The species of *Plagiotremus,* as here defined, share so many unique characters that it is readily apparent that they comprise a monophyletic group. They can be separated, however, into four well-defined species groups (refer to discussion of intrageneric relationships).

Etymology. — The generic name is a combination of the Greek *plagio* (oblique) and *trema* (hole); gender masculine.

Biology. — Anyone who has observed a species of *Plagiotremus* in the field would have been surprised initially at seeing these small blennies attacking other fishes many times their own size. During the day *Plagiotremus* can usually be found hovering a meter or so above the reef from which position they maintain a continual watch for unwary prey. *Plagiotremus* typically attacks its prey from behind, although head-on approaches have been observed. Attacks are executed with an amazing burst of speed but can be aborted just as rapidly if the intended victim turns toward the blenny. In addition to "ambushing" prey from behind, some species also apparently benefit from aggressive mimicry (refer to general section on mimicry for discussion). Several species of *Plagiotremus* regularly occur among aggregations of similarly colored fishes which they only superficially resemble. As noted by Russell (1976: 328), the similarity is sufficient to deceive passing prey fishes, thereby increasing attack success. The color patterns of the prominently striped species of *Plagiotremus* may also affect the success of attacks. In a fish moving rapidly forward, the stripe should make it difficult for prey to fix on a particular reference point. The brilliant blue stripes of *rhinorhynchos* would seem to be particularly well suited for such a function.

The bite of *Plagiotremus* probably causes little, if any, permanent damage to prey but it is obviously an unpleasant experience. Hobson (1968: 42) stated

[1] For a discussion of type-species designations in parentheses, see Hubbs *et al.* (1963).

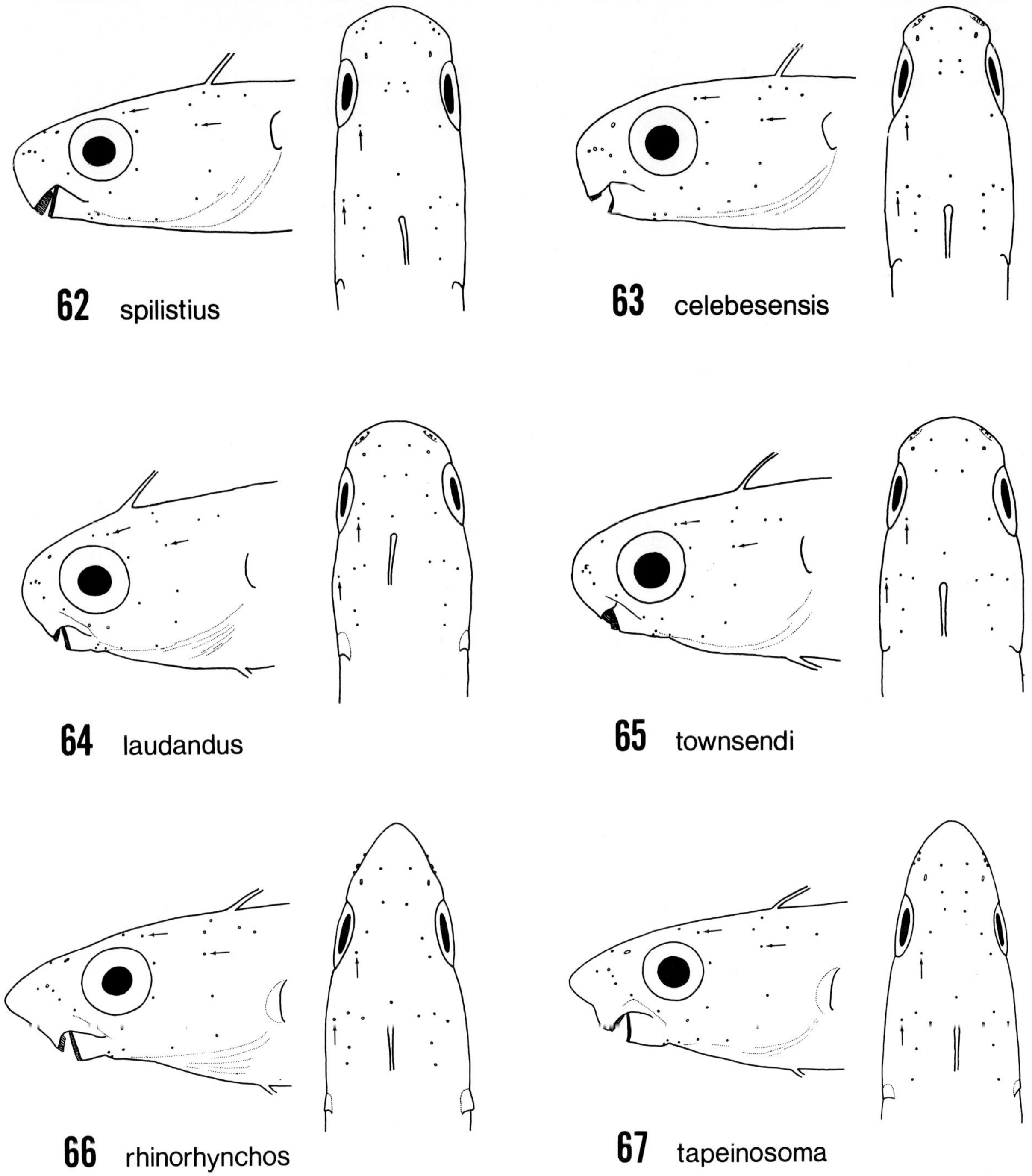

Figures 62-67.—General physiognomy and cephalic sensory pore distributions in selected species of *Plagiotremus:* Fig. 62, *spilistius,* MNHN 1964-628, male, 164.3 mm, South Vietnam; Fig. 63, *celebesensis,* SU 67356, juvenile, 54.9 mm, Celebes, Lembeh Strait; Fig. 64, *laudandus,* USNM 205270, male, 52.0 mm, Marshall Islands, Eniwetok; Fig. 65, *townsendi,* USNM 204577, male, 35.4 mm, Gulf of Aqaba; Fig. 66, *rhinorhynchos,* ANSP 124856, male, 64.3 mm, Yap Islands; Fig. 67, *tapeinosoma,* ANSP 124880, male, 63.9 mm, Mariana Islands. Small arrows identify ventralmost supraorbital and dorsalmost preopercular pores in both lateral and dorsal views.

that the actions of only four to six tiny *azaleus* were sufficient to keep a large school of *Haemulon sexfasciatus* Gill, numbering many hundreds of individuals, over sandy bottom away from protective rocks. Quoting G. W. Barlow, Hobson also reported that, in the Galapagos Islands, spawning of *Abudefduf troschelii* (Gill) was sharply curtailed by attacks of *azaleus* in areas where *azaleus* was common. In the Gulf of Aqaba I observed a single individual of *townsendi* restrict four or five individuals of *Holocentrus* to cover underneath a small ledge. Every time one of the *Holocentrus* would expose itself the blenny would strike, forcing the larger fish to retreat back to its original cover. I watched this performance repeated numerous times. One of the most vivid descriptions of the aggressive nature of *Plagiotremus* (Read, 1970: 31) appeared in a popular article. The following account records the behavior of *P. azaleus:*

> ". . . small rays were common and lay on the bottom covered with sand. Flitting around nearby rocks were schools of slender two-inch-long blennies, *Runula.* When I idly touched the tail of one of the rays it oozed into slow motion; immediately the blennies spotted the moving ray and sped at it like a salvo of little rockets, biting unmercifully. The ray shot off over the sand pursued by the tiny harpies. It then stopped and frantically shrugged sand over itself; the attackers desisted at once. I repeated the performance several times with the same results. I later noticed that it was fairly common to see individual *Runula* biting other fish . . ."

As suggested above, *Plagiotremus* preys upon a relatively large variety of species. However, piscivorous species are generally avoided (see also Springer and Smith-Vaniz, 1972a: 6). Preferred prey seem to be fishes of the families Acanthuridae, Chaetodontidae, Pomacentridae, Mullidae, Pomadasyidae, and especially Holocentridae. During the course of extensive underwater observations in the Gulf of California, Hobson (1968: 34) reported being struck on his legs hundreds of times by *azaleus* that responded to him exactly as they do to larger fishes. *Plagiotremus* exhibit little fear of divers initially and, if not approached too rapidly, can usually be fooled into attacking a fine meshed dip net used for collecting.

The food of *Plagiotremus* is as unusual as its feeding habits. The stomachs of the relatively large number of specimens that I have examined, when not empty, invariably contained a large amount of mucus and usually a few fish scales. Mucus and dermal tissue taken with the scales is undoubtedly the primary source of nourishment. Losey (1972: 135) reported that the gut contents of *P. laudandus* consisted "almost entirely of fish scales and a few fish eggs." Wickler (1962) sated that when *P. rhinorhynchos* is kept in an aquarium without other fishes it rejects all the aquarist's standard food supply, but reacts in typical aggressive fashion when other fish are introduced. He also stated (p. 19) that it chooses the least scaly portions of its prey. The last statement, however, conflicts with field observations. A careful study of interspecific competition and trophic niche overlap among sympatric species of *Plagiotremus* should be made. Three species of *Plagiotremus* are common at many localities, but only one member of any species group occurs at any given locality. Although fundamentally similar, the shapes of the dentary incisors of the four species groups (Fig. 72) are surprisingly different, and there also is little overlap in number of incisors (Fig. 74) (see also Table 4).

Plagiotremus is apparently unique among the Blenniidae in its scale-eating habit, but at least one other blenniid ingests mucus. Hobson (1974: 1000) reported that *Exallias brevis* (Kner) (tribe Salariini) may be exceptional in feeding mostly on the tissue and mucus of scleractinian corals. Scale ingestion has also been reported in a number of other unrelated fish groups: Characoids (Roberts, 1970); Trichomycteridae (Roberts, 1972: 138); Cichlidae (Trewavas, 1947; Fryer *et al.,* 1955; Liem and Stewart, 1976); Ariidae (Springer and Woodburn, 1960; Hoese, 1966); Carangidae (Major, 1973; Smith-Vaniz and Staiger, 1973: 233-237); and possibly the Triacanthodidae (Tyler, 1968: 46).

The jaws and dentition of *Plagiotremus* are highly specialized to accommodate their bizarre mode of feeding. The tips of the dentary incisors are thin and wedge-shaped, ideally suited to snag and tear out scales. The attachment of the functional dentary incisors is remarkable. In the other genera of Nemophini these incisors are essentially immobile, with their broad bases firmly ankylosed to the dentary bone. In *Plagiotremus* the dentary incisors have narrow bases that are not basally attached to the

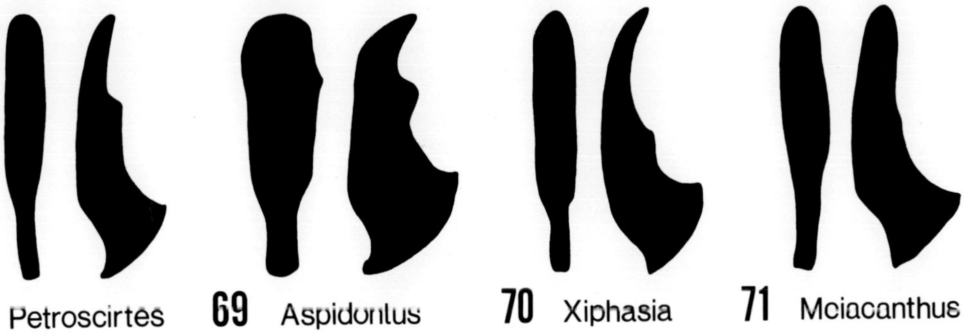

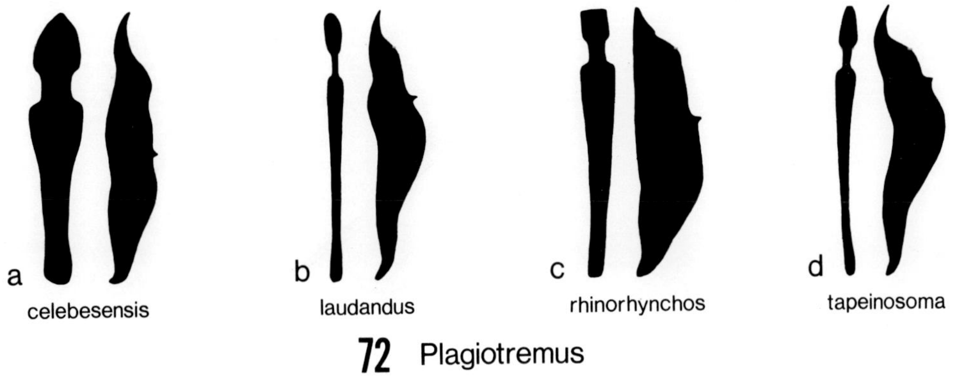

Figures 68-72. — Anterior and lateral views of dentary incisors of selected species of Nemophini (not all drawn to same scale): Fig. 68, *Petroscirtes variabilis,* USNM 138250, 49.9 mm; Fig. 69, *Aspidontus taeniatus,* ANSP 122374, 48.2 mm; Fig. 70, *Xiphasia setifer,* ANSP 128699, 282 mm; Fig. 71, *Meiacanthus atrodorsalis,* ANSP 124872, 59.1 mm; Fig. 72, *Plagiotremus:* a, celebesensis, SU 67356, 54.9 mm; b, *laudandus,* USNM 201413, 31.4 mm; c, *rhinorhynchos,* ANSP 124876, 67.8 mm; d, *tapeinosoma,* ANSP 124878, 38.9 mm.

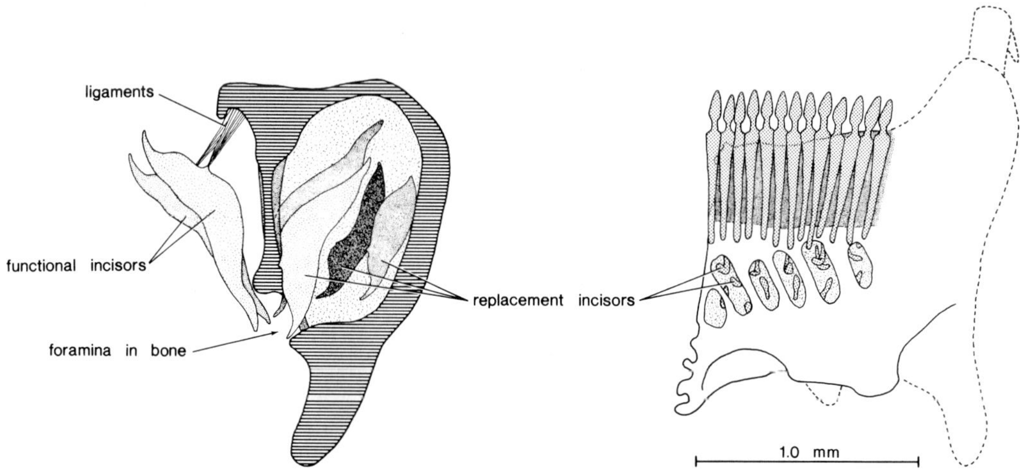

Figure 73. — Semidiagrammatic illustration of left dentary of *Plagiotremus tapeinosoma* showing functional and replacement incisors. Left, cross section of dentary; right, frontal view of dentary.

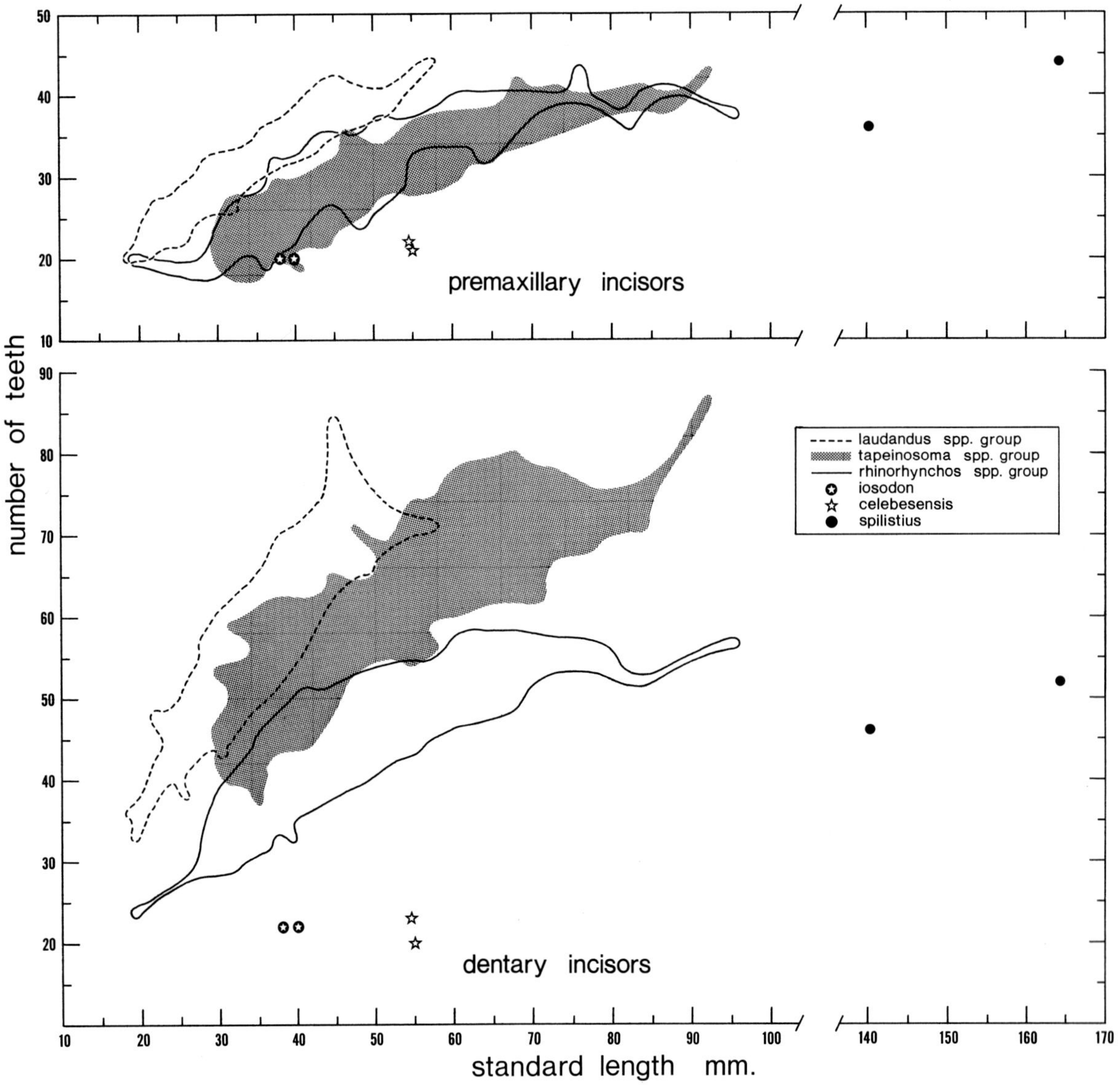

Figure 74. — Relationship between number of teeth versus standard length in species of *Plagiotremus.* (Envelopes for *laudandus, rhinorhynchos* and *tapeinosoma* species groups based on Figures 75-77).

dentary bone. Each dentary incisor has a unique bony spur posteromesially (Fig. 72). A short ligament extends from this spur to the underside of a flange on the anterodorsal margin of the dentary (Fig. 73). The skin covering of the incisors tends to hold the teeth in position, while the ligamentous attachment allows a certain amount of movement, both laterally and anteroposteriorly. This can be demonstrated readily by moving the tips of the dentary incisors with a small probe. I believe that this unique arrangement enables the incisors to make slight frictional adjustments as they drag across the body surfaces of the prey. The arrangement of the teeth would, of course, greatly facilitate scraping and snagging of scales.

The distal ends of the ascending processes of the premaxillae are broadly rounded and overlapped by the nasals, median and lateral ethmoids (Figs. 92 and

94j). The latter bones are also broadly expanded and firmly ankylosed to each other and to the frontals. (The external joints of these bones are evident since the synostosis has taken place internally only.) The frontal bones are fused together, with no external joint evident. These specializations serve to strengthen the skull and allow the ball-joint arrangement of the movable premaxillae (immovable in all other genera of Blennidae). The underslung lower jaw is capable of rotating through a wide arc (see Springer, 1968: pl. 7). The gut, posterior caudal vertebrae and pterygiophores of the dorsal-fin spines of *Plagiotremus* also exhibit specializations that appear to be closely correlated with feeding behavior (refer to general section on evolutionary trends).

Plagiotremus has a formidable pair of dentary canines, but they are used solely for protection or territorial defense (Wickler, 1962; Hobson, 1968: 43). Because most species of *Plagiotremus* are rather small fishes and highly visible during the daylight hours, it is surprising that they are not subjected to heavy predation. All species of *Plagiotremus* possibly have a noxious taste that makes them unacceptable as prey to most predators. This has been partially demonstrated for *P. townsendi* (Springer and Smith-Vaniz, 1972a: 21) and unpublished data (Springer, pers. comm.). Each species (and subspecies) in the *laudandus* species group strongly resembles a sympatric species of *Meiacanthus*. As a result these species probably are less vulnerable to predation, either through Batesian or Müllerian mimicry. If pursued by a diver, *Plagiotremus* will seek refuge in abandoned mollusk or worm tubes. These retreats are also utilized as shelter at night.

Strasburg *et al.* (1968: 426) reported sighting *P. ewaensis* at a depth of 55 meters, while making observations from a small submarine. Most collections of *Plagiotremus* have come from much shallower depths. Small juveniles of all species of the subgenus *Runula* are strongly banded, unlike the adults. At this stage they seem to be primarily pelagic, surface dwellers (see also "remarks" under account of *P. tapeinosoma*). No information is available on the reproductive biology of *Plagiotremus*.

Zoogeography. — The distribution of *Plagiotremus* (Figs. 78-80) extends from the northern

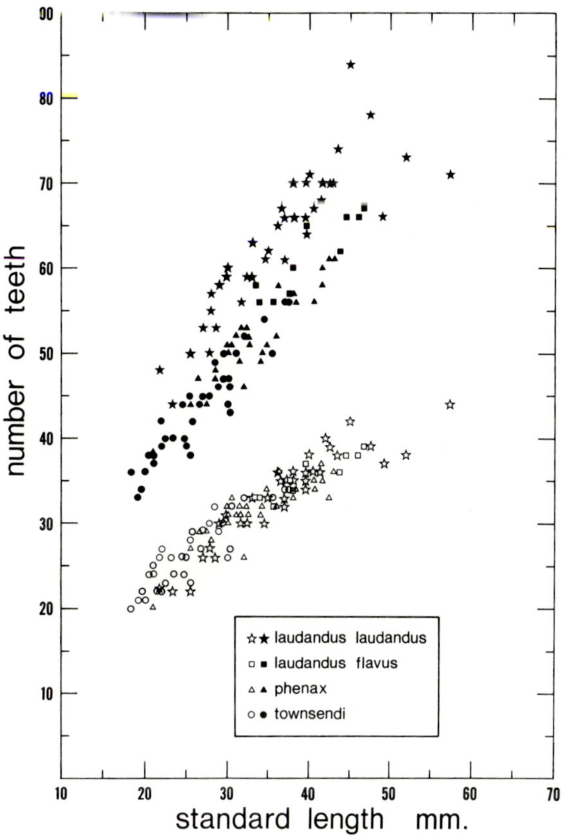

Figure 75. — Relationship between number of teeth versus standard length in members of the *Plagiotremus laudandus* species group. Closed symbols = dentary incisors; open symbols = premaxillary incisors.

Red Sea and southern Africa to the Hawaiian Islands and the eastern Pacific Ocean. The absence of collections of the otherwise broadly distributed *P. rhinorhynchos* and *tapeinosoma* from the Persian Gulf Arabian Sea and most of western Australia probably is a reflection of inadequate collecting effort.

As discussed under the account of *P. laudandus flavus*, there is strong circumstantial evidence that the distribution of species of *Meiacanthus* has influenced (or controlled) the distribution and/or speciation of the *laudandus* species group. The absence of a member of the *laudandus* species group in the western Indian Ocean is enigmatic. Considering the number of collections that have been made in the region (and the numerous records of *P. rhinorhynchos* and *tapeinosoma*) it is difficult to believe

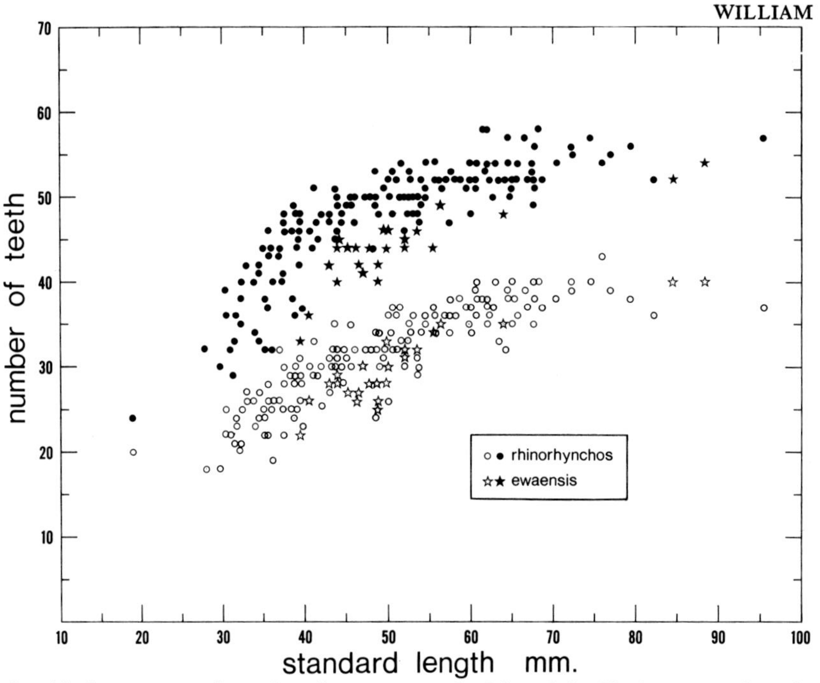

Figure 76. — Relationship between number of teeth versus standard length in *Plagiotremus rhinorhynchos* and *ewaensis* (see also Fig. 84). Closed symbols = dentary incisors; open symbols = premaxillary incisors.

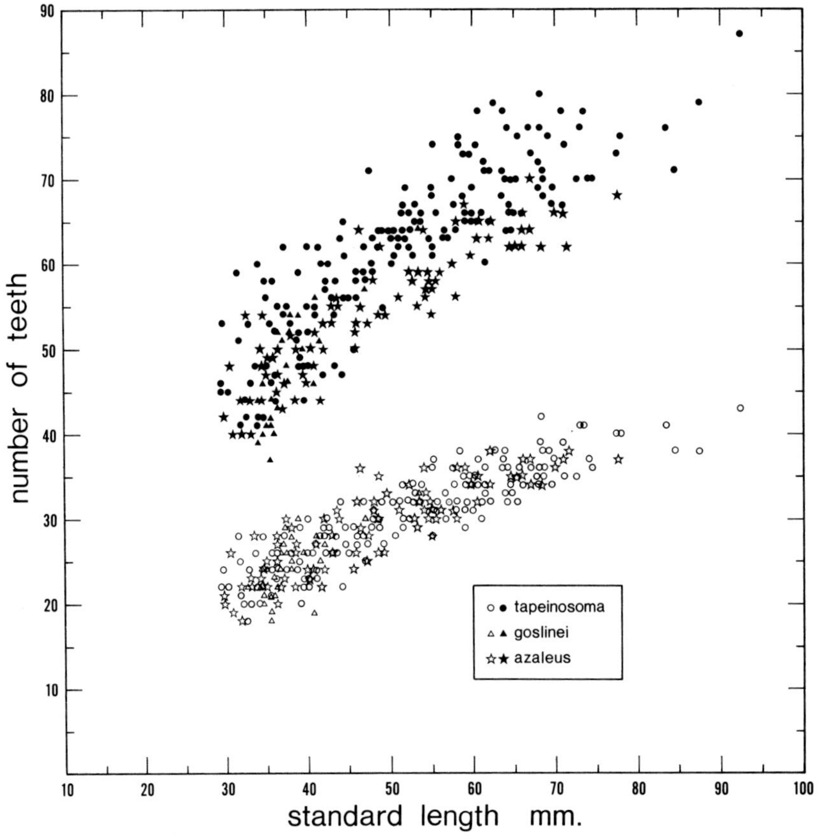

Figure 77. — Relationship between number of teeth versus standard length in members of the *Plagiotremus tapeinosoma* species group. Closed symbols = dentary incisors; open symbols = premaxillary incisors.

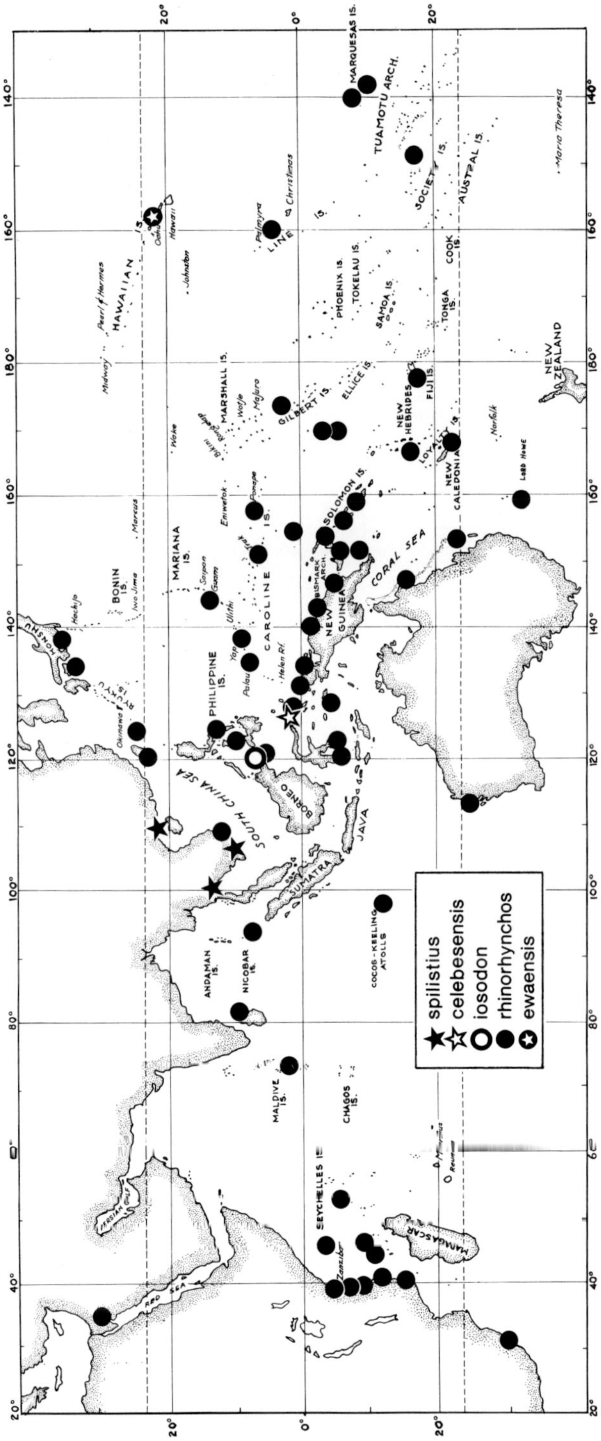

Figure 78. —Distributional records for selected species of *Plagiotremus*.

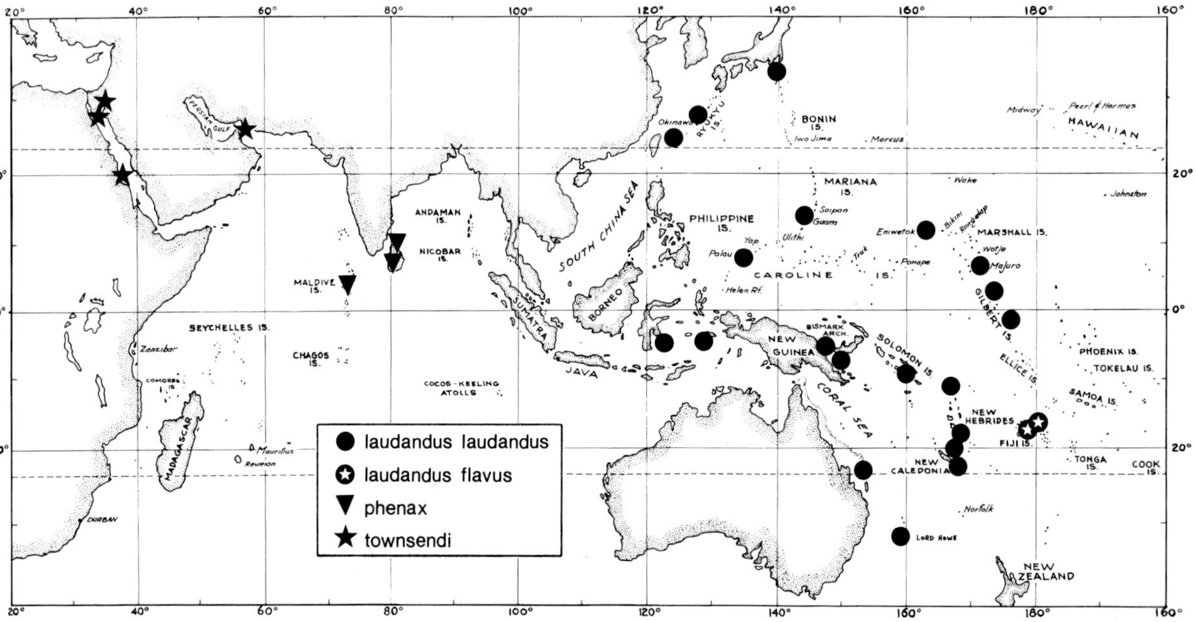

Figure 79. — Distributional records for members of the *Plagiotremus laudandus* species group.

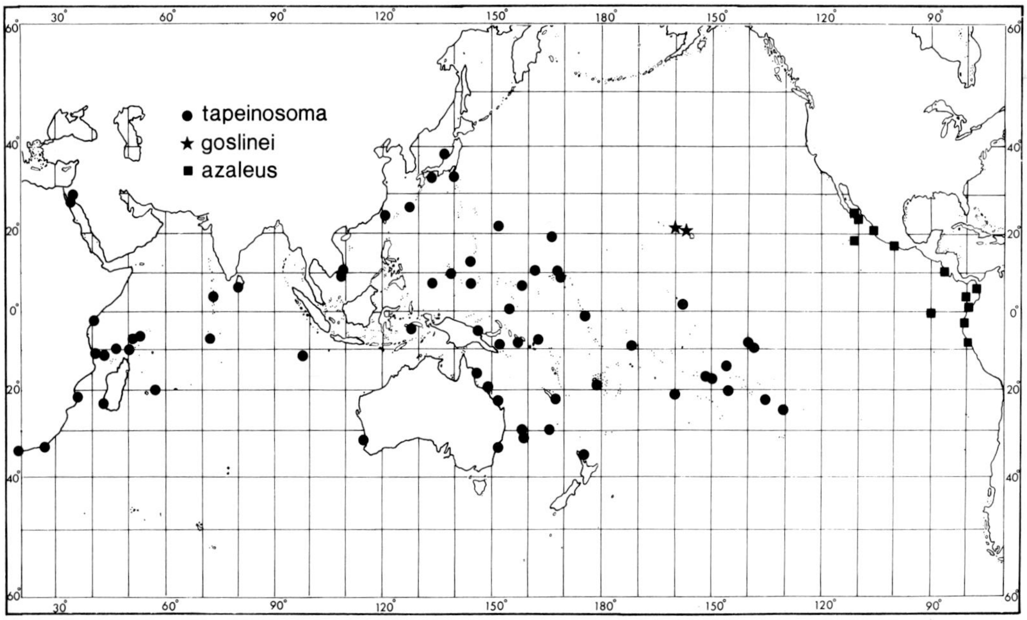

Figure 80. — Distributional records for members of the *Plagiotremus tapeinosoma* species group.

that a member of this species group might actually occur in the region but has escaped collection. Habitat and food certainly could not be limiting factors, and two species of *Meiacanthus* occur in the region. Considering the distributions of the other members of the *laudandus* species group, one would expect to find an endemic *Plagiotremus* in the western Indian Ocean.

I consider the subgenus *Plagiotremus* to contain the most generalized species. It is of interest to note that this is the only subgroup of *Plagiotremus* whose members are found exclusively in the western Pacific Ocean.

Intrageneric relationships. — The four species groups of *Plagiotremus* are compared in Table 19. The following character analysis establishes the monophyletic nature of these groupings. I tentatively recognize the *spilistius* and *laudandus* species groups as distinct subgenera because generic-group names are available for each and I am unable to identify any apomorphic characters that unite them. Rigid application of the principles of phylogenetic (cladistic) systematics dictates that the *laudandus* and *tapeinosoma* species groups be accorded a sister-group relationship because they share the greatest number of apomorphic characters (characters B, I, J and K). However, all of these shared specializations represent minor evolutionary changes involving labile characters of a reductive or additive (increase in number of dentary incisors) nature, and I believe their occurrence in both groups is a case of parallel evolution. Snout morphology is weighted heavily in the hypothetical phylogeny suggested in Figure 81. If my interpretation of a fleshy, conical snout as an apomorphic character state is correct, then on that basis the *rhinorhynchos* and *tapeinosoma* groups constitute a monophyletic grouping for which *Runula* is available as a subgeneric name. The virtually identical head strip pattern (Fig. 82c-d) of the adults of both groups also seems to support the proposed classification.

Numbers given below correspond to those shown on Figure 81. For each character (or character complex) the plesiomorphic condition is given a lower case letter and the apomorphic condition a capital letter.

1. Snout of adults not fleshy, blunt (a). Snout of adults fleshy, conical (A). A fleshy conical snout has evolved independently in several unrelated groups of fishes, and in each case seems to be correlated with specialized feeding habits (refer to general section on evolutionary trends). It seems unlikely that such a character would have evolved twice in a small group of closely related fishes or have been secondarily lost once it had evolved.

2. Ratio between number of upper and lower incisors (see Table 4). Most genera of Nemophini have approximately equal numbers of incisors in each jaw (*Aspidontus* is an exception in usually having more premaxillary incisors). The lower incisor teeth ratio found in the *spilistius* and *rhinorhynchos* species groups is considered to be the plesiomorphic condition (b). The considerably higher ratio found in the *laudandus* and *tapeinosoma* groups is the apomorphic condition (B).

3. Shape of dentary incisors (Fig. 72). When total number of teeth attained by the adults (Fig. 74) is taken into consideration, it becomes apparent that the dentary incisors of the *rhinorhynchos* group are the most dissimilar in shape. The high number of dentary incisors in the *laudandus* and *tapeinosoma* groups almost certainly accounts for the slender, essentially identical shape of the teeth in each group. Similarly, the robust nature of the dentary incisors of the *spilistius* group is due to the lower number of these teeth attained by the adults (a plesiomorphic condition, see above). Thus, it would be relatively easy to derive the dentary tooth-shape characteristics of the *laudandus* and *tapeinosoma* groups from a *spilistius*-like progenitor, while the uniqueness of the shape of the *rhinorhynchos* dentary incisors is assumed to represent a specialization of a different nature. Dentary incisors with tips arrowhead-shaped, ovoid or bluntly lanceolate and slightly recurved (c). Dentary incisors with tips broadly chisel-shaped (Fig. 72c) (C).

4. Innermost pair of segmented caudal-fin rays without well-developed bony spurs proximally (d). Innermost pair of segmented caudal-fin rays with well-developed bony spurs proximally (Fig. 96c) (D). Only *P. rhinorhynchos* and *ewaensis* have these bony spurs, and both species have relatively specialized caudal skeletons. Presumably these structures provide increased surface for attachment of the caudal-fin musculature.

5. Pigmentation of innermost caudal-fin ray of

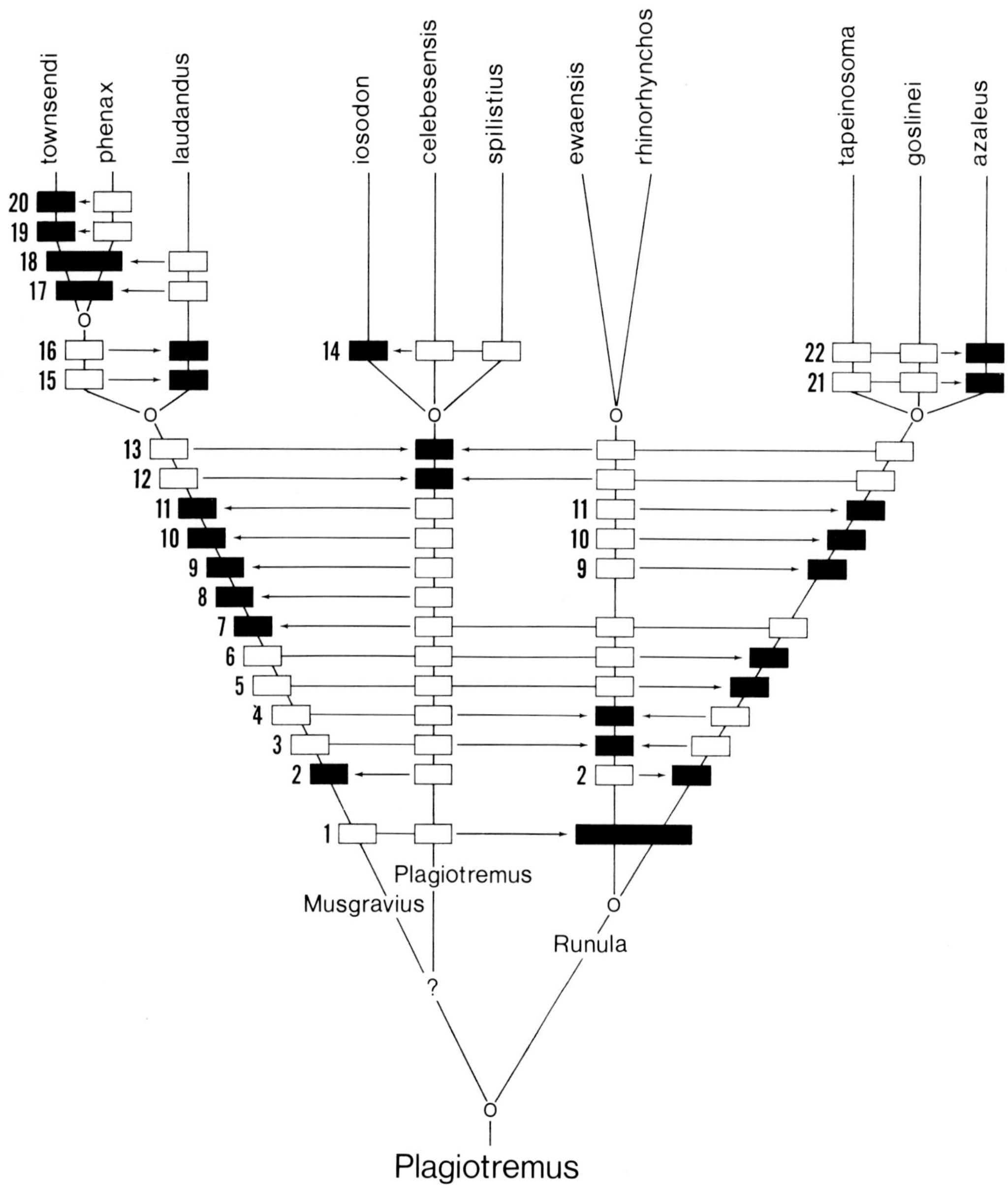

Figure 81. — A phylogenetic analysis of the eleven species of *Plagiotremus*. Open squares are primitive (plesiomorph) character expressions and shaded squares are derived (apomorph) character expressions. Numbers correspond to characters discussed in text. Arrows are inferred directions of evolutionary change and do not imply ancester-descendent relationships.

adults like adjacent rays (e). Pigmentation of innermost caudal-fin ray of adults unlike adjacent rays (E).

6. Pelvic-fin rays, if present, not sexually dimorphic (f). Pelvic-fin rays sexually dimorphic in adults (Fig. 85) (F).

7. Adults not involved in mimetic relationships with species of *Meiacanthus* (g). Adults apparently involved in mimetic relationships (as Müllerian and/or aggressive mimics) with species of *Meiacanthus;* refer to discussion under account of *P. laudandus flavus* (G). It seems reasonable to hypothesize that the occurrence of this type of specialized interaction exclusively and consistently in a small group of species has a genetic or behavioral basis indicative of a monophyletic origin.

8. Snout uniformly pigmented (h). Snout with a conspicuous, broad stripe (Figs. 82a-b); stripe secondarily lost in *P. laudandus flavus* (H).

9. Dorsal-fin spines 10-12 (i). Dorsal-fin spines 6-9 (rarely 10) (see Table 20) (I).

10. Mandibular pores usually 3 (see Table 23) (j). Mandibular pores 2 (J).

11. Interorbital pores 4 (k). Interorbital pores 2 (K).

12. Supraorbital pores 2; supraorbital pores secondarily reduced in *P. townsendi* (l). Typically a single supraorbital pore (each side) (L).

13. Total vertebrae fewer than 54 (m). Total vertebrae 59-76 (M). I believe that the extremely high number of vertebrae in the *spilistius* group (Table 22) is a specialization because, excluding *Xiphasia* (for which a sister-group relationship with *Plagiotremus* seems improbable), the other genera of Nemophini and blenniid tribes have 48 or fewer total vertebrae. The progenitor of *Plagiotremus* probably had an intermediate number of vertebrae; the low number (38-41) found in *P. townsendi* is an obvious secondary specialization.

The tendency in *Plagiotremus* for the neural and haemal spines of the last few caudal vertebrae to become expanded seems to be an obvious specialization associated with caudal locomotion. Although this condition is least well developed in the *spilistius* group (compare Figs. 95 f and 95 g), this does not necessarily refute the preceding character analysis because fishes with high numbers of vertebrae tend to employ a more undulating, anguilliform mode of swimming with less emphasis on the caudal fin as a means of propulsion.

14. Total vertebrae 69-76 (n). Total vertebrae 59 (N). Hennig (1966) has shown that in determining phylogenetic relationships, characters treated as apomorphic at one level of a classification must be treated as plesiomorphic at the next higher level (refer to discussion of preceding character state).

15. Incisor teeth ratio lower, mean 1.64 (see Table 4) (o). Incisor teeth ratio higher, mean 1.84 (O).

16. Outer lobes of caudal fin not elongate in adults (Fig. 83) (p). Outer lobes of caudal fin elongate in adults (P).

17. Dorsal-fin not sexually dichromic (q). Dorsal-fin sexually dichromic; large males, presumably nuptial, differ from females in having a conspicuous orange spot in the fin anteriorly (Q).

18. Dorsal-fin spines typically 9 (see Table 20) (r). Dorsal-fin spines 6-8 (R).

19. Dorsal-fin spines typically 8 (s). Dorsal-fin spines typically 7 (S).

20. Supraorbital pores typically 2 (see Table 23) (t). A single supraorbital pore usually present (T).

21. Innermost segmented caudal-fin ray not noticeably longer than adjacent rays in large adults (u). Innermost segmented caudal-fin ray noticeably elongate in large adults (U).

22. Caudal fin of adults rarely exceeds 20 percent SL (v). Caudal fin of adults frequently exceeds 20 percent SL (Fig. 86) (V).

KEY TO SPECIES OF *PLAGIOTREMUS*

1. Preanal length 22.8-28.2 percent SL; total dorsal-fin elements 66-71; segmented anal-fin rays 54-58 .... **2**

Preanal length 34.0-52.8 percent SL; total dorsal-fin elements 32-54; segmented anal-fin rays 19-40 .... **3**

2. Pelvic fins absent; (South China Sea) .... *spilistius*
Minute pelvic fins (1.5-1.6 percent SL) present; (Celebes Sea) .... *celebesensis*

3. Pelvic fins absent; segmented anal-fin rays 40; total dorsal-fin elements 53-54; dentary incisors with tips arrowhead-shaped (Fig. 72a); (Sulu Archipelago) .... *iosodon*

Pelvic fins present; segmented anal-fin rays 19-33; total dorsal-fin elements 32-48; dentary incisors with tips chisel-shaped, ovoid or bluntly lanceolate (Figs. 72b-d) **4**

4. Segmented anal-fin rays 19-24; snout of adults not fleshy, blunt (Figs. 64-65); body of adults without stripes **5**

Segmented anal-fin rays 27-33; snout of adults fleshy, conical (Figs. 66-67); body of adults striped **7**

5. Longest caudal-fin ray 25.5-51.6 percent SL (Fig. 83); dorsal-fin spines 7-10 (usually 9); (Pacific Ocean) .... *laudandus*

Longest caudal-fin ray 18.0-25.5 percent SL; dorsal-fin spines 6-8 .... **6**

6. Dorsal fin with dark stripe entire length of fin; total dorsal-fin elements 35-37; (eastern Indian Ocean) .... *phenax*

Pigmentation of dorsal fin not as above; total dorsal-fin elements 32-34; (Red Sea and Gulf of Oman) .... *townsendi*

7. Two pairs of interorbital pores; dorsal-fin spines 10-12 (usually 11); innermost caudal-fin ray of adults pigmented like adjacent rays; dentary with tips chisel-shaped (Fig. 72c) **8**

Single pair of interorbital pores; dorsal-fin spines 7-9 (usually 8); innermost caudal-fin ray of adults distinctly darker than adjacent rays, at least proximally; dentary incisors with tips ovoid or bluntly lanceolate (Fig. 72d) **9**

8. Background coloration of adults usually brown or brownish-black; dentary incisors of adults usually more numerous (Figs. 76 and 83; Table 24); (widespread Indo-Pacific) .... *rhinorhynchos*

Background coloration of adults yellow-orange; dentary incisors of adults usually fewer; (Hawaiian Islands) .... *ewaensis*

9. Anal fin immaculate; posterior two-thirds of dorsal fin with intense dark stripe; caudal fin of adults frequently exceeds 20 percent SL (Fig. 86); (eastern Pacific Ocean) .... *azaleus*

Anal fin dusky, at least distally; pigmentation of dorsal fin not as above; caudal fin of adults rarely exceeds 20 percent SL **10**

10. Dark mid-lateral stripe with very irregular margin, especially dorsally, the stripe largely broken into a series of vertically elongate blotches; (widespread Indo-Pacific) .... *tapeinosoma*

Dark mid-lateral stripe of nearly uniform intensity, if spots are present they are always connected by stripe; (Hawaiian Islands) .... *goslinei*

TABLE 19. — Comparison of the four species groups of *Plagiotremus*.

Characters	(Plagiotremus) *spilistius* group	(Musgravius) *laudandus* group	(Runula) *rhinorhynchos* group	(Runula) *tapeinosoma* group
Snout of adults	blunt	blunt	fleshy, conical	fleshy, conical
Color pattern of adults	unstriped	unstriped	striped	striped
Interorbital stripes (Fig. 82)	absent	joined anteriorly	not joined anteriorly	not joined anteriorly
Tips of dentary incisors (Fig. 72)	arrowhead-shaped	ovoid or bluntly lanceolate	chisel-shaped	ovoid or bluntly lanceolate
Dentary incisor teeth ratio (upper/lower)[1]	1.0-1.2	1.4-2.1	1.3-1.8	1.6-2.4
Pelvic fins	minute or absent	short not sexually dimorphic	short not sexually dimorphic	elongate sexually dimorphic (Fig. [illegible]5)
Pigmentation of innermost caudal-fin ray of adults	like adjacent rays	like adjacent rays	like adjacent rays	unlike adjacent rays
Segmented caudal-fin rays	innermost pair of rays without well-developed bony spurs proximally	innermost pair of rays without well-developed bony spurs proximally	innermost pair of rays with well-developed bony spurs proximally (Fig. 96c)	innermost pair of rays without well-developed bony spurs proximally
Interorbital pores	4	2	4	2
Mandibular pores[2]	3	2 or 3 (usually 2)	3	2
Dorsal-fin spines	10	6-10 (rarely 10)	10-12	7-9
Segmented dorsal-fin rays	43-61	25-30	31-37	32-39
Total dorsal-fin elements	53-71	32-39	43-48	40-47
Segmented anal-fin rays	40-58	19-24	29-33	27-33
Total vertebrae	59-76	38-45	48-53	47-53

[1] See Table 4 for frequency distributions; see also Figure 74.

[2] See Table 23 for frequency distributions.

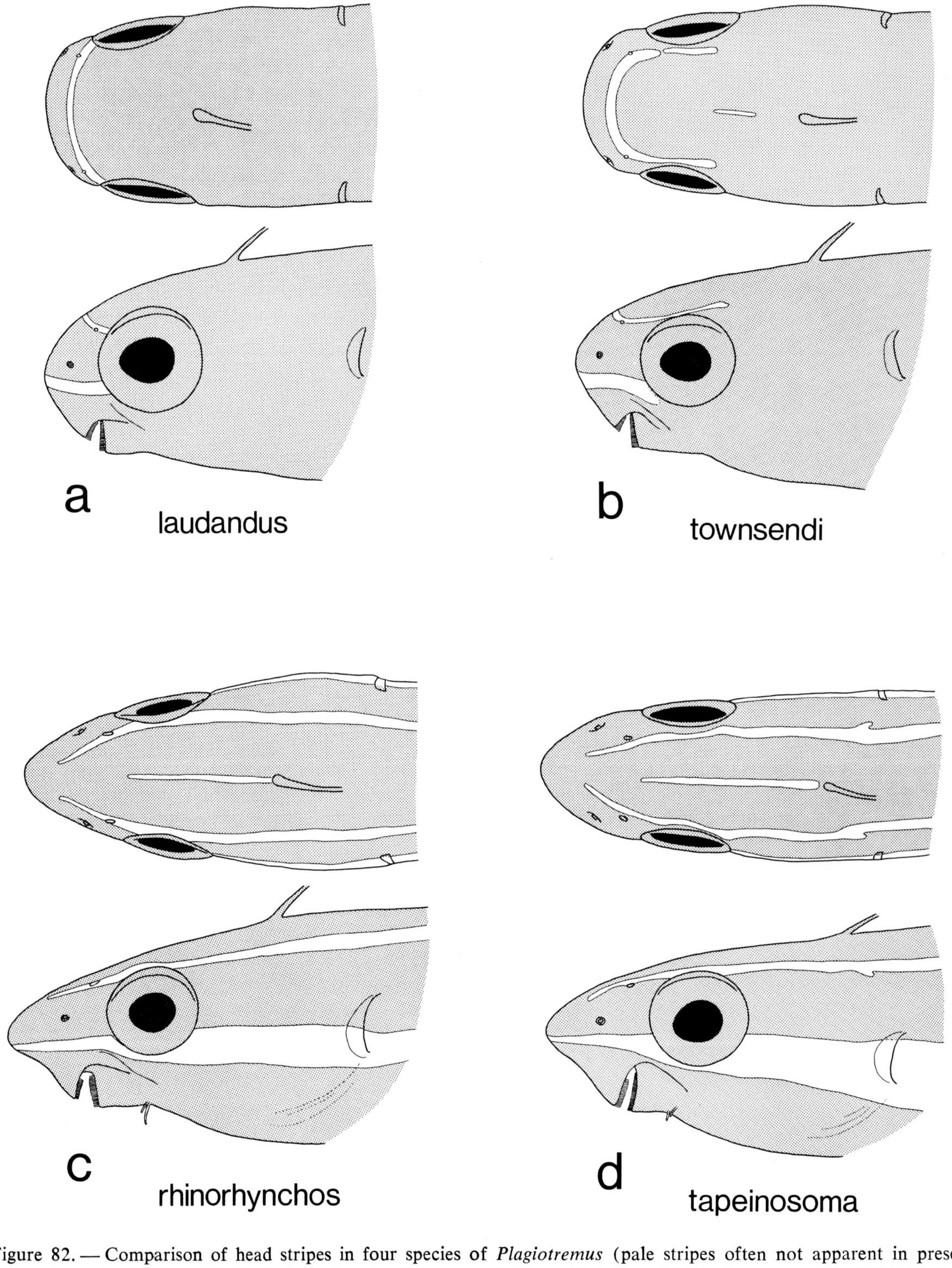

Figure 82. — Comparison of head stripes in four species of *Plagiotremus* (pale stripes often not apparent in preserved specimens): a, *laudandus,* USNM 211901, female, 38.1 mm, Celebes, Kabaena Island; b, *townsendi,* USNM 205379, female, 24.7 mm, Gulf of Aqaba; c, *rhinorhynchos,* USNM 209845, male, 57.4 mm, Moluccas, Ceram; d, *tapeinosoma,* USNM 211117, male, 47.7 mm, Society Islands, Huahine Is.

TABLE 20. — Frequency distributions of dorsal-fin rays in species of *Plagiotremus*.[1]

Species	Dorsal-fin spines 6	7	8	9	10	11	12	N	x̄	Segmented dorsal-fin rays 25	26	27	28	29	30	31	32	33	34
spilistius					3			3	10.0										
celebesensis					2			2	10.0										
iosodon					2			2	10.0										
l. laudandus		1	2	53	2			58	9.0			14	32	11	1				
l. flavus			2	8*				10	8.8				5*	3	2				
phenax		4	27*					31	7.9			10	18*	3					
townsendi	2	65	4					71	7.0	11	53	7							
rhinorhynchos					16	209	6	231	11.0							1	22	125	50
ewaensis					1	25	1	27	11.0										5
tapeinosoma		15	291	11				317	8.0										8
goslinei		3	34					37	7.9										6
azaleus		27	208	8				243	7.9								2	51	146

Species	Segmented dorsal-fin rays 35	36	37	38	39	40	41	42	43	44	—	56	57	58	59	60	61	N	x̄
spilistius														2	–	1		3	58.7
celebesensis												1	–	–	–	–	1	2	58.5
iosodon									1	1*								2	43.5
l. laudandus																		58	28.0
l. flavus																		10	28.7
phenax																		31	27.8
townsendi																		71	25.9
rhinorhynchos	21	10	2															231	33.5
ewaensis	15	7																27	35.1
tapeinosoma	75	89	77	57	11													317	36.4
goslinei	24	6	1															37	35.1
azaleus	38	6																243	34.0

[1] When counts vary, those for holotypes of new taxa are indicated by an asterisk.

Plagiotremus (*Plagiotremus*) *spilistius* Gill
Specimen Figures 62, 160; Tables 4, 19-23.

Plagiotremus spilistius Gill, 1865: 140, pl. 3, Figs. 1-2 (type locality: Chinese seas [= South China Sea]; holotype USNM 21722).

Lembeichthys furcocaudalis Sze-Chung, 1957: 229, pl. 1, Figs. 1-4 (type locality: Chich-shih [= Kitchioh, Kwang tung Prov., China]; holotype ZIAS 29936 [not examined]).

Description. — Dorsal fin X, 58-60; total elements 68-70. Anal fin II, 55-57. Pectoral fin 12. Caudal fin: procurrent rays 5 + 5 = 10; segmented rays 6 + 5. Vertebrae: precaudal 16; caudal 57-59; total 73-75. Pleural ribs on vertebrae 3-16; epipleural ribs 14-16. Incisor teeth (Fig. 74), in specimens 140-164 mm, n = 2: lower jaw 46-52; upper jaw 42-44; ratio (lower/upper) 1.1-1.2, mean 1.14.

Dentary incisors with tips arrowhead-shaped (Fig. 72a). Snout blunt, broadly rounded. Interorbital pores 4; supraorbital pores 1; mandibular pores 3. Preanal length 22.8-25.2% SL. Pelvic fins absent. Caudal fin sexually dimorphic: male (164 mm) 49.8% SL, female (140 mm) 19.4% SL; caudal fin damaged in other available female. Sze-Chung (1957) gives proportional measurements for the caudal fin of *Lembeichthys furcocaudalis* (see below) as follows: males (6, 133-174 mm) 2.6-4.2 times head length, females (5, 122-153 mm) 1.2-1.7 times HL. Outer lobes of caudal fin filamentous (both sexes); innermost ray not noticeably longer than adjacent rays.

TABLE 20. — Continued.

Species	Total dorsal-fin elements																			
	32	33	34	35	36	37	38	39	40	41	42	43	44	45	46	47	48	49	50	51
spilistius																				
celebesensis																				
iosodon																				
l. laudandus					16	31	9	2												
l, flavus						6*	3	1												
phenax				12	18*	1														
townsendi	8	56	7																	
rhinorhynchos												29	121	50	20	10	1			
ewaensis														5	15	7				
tapeinosoma											10	73	91	74	59	10				
goslinei											8	23	5	1						
azaleus									3	55	150	34	1							

Species	Total dorsal-fin elements																					
	52	53	54	55	56	57	58	59	60	61	62	63	64	65	66	67	68	69	70	71	N	x̄
spilistius																	2	–	1		3	68.7
celebesensis															1	–	–	–	–	1	2	68.5
isodon		1	1*																		2	53.5
l. laudandus																					58	36.9
l. flavus																					10	37.5
phenax																					31	35.6
townsendi																					71	33.0
rhinorhynchos																					231	44.1
ewaensis																					27	46.1
tapeinosoma																					317	44.4
goslinei																					37	43.0
azaleus																					243	41.9

Color pattern. — Body with a row of 30-38 dark spots, slightly smaller than pupil diameter, extending dorsally from just above gill-opening to caudal-fin base; another row of 25-42 considerably smaller spots extends from in front of anus just above base of anal fin to end of fin. Dorsal and anal fins heavily pigmented, the dorsal fin with numerous black spots smaller than pupil diameter; caudal fin also spotted. Pectoral fin pale except for conspicuous dark spot on base of upper two rays. A good illustration accompanied the original description of *furcocaudalis.*

According to Sze-Chung (1957), the life colors are: "Head and body yellowish-brown, spots on body bluish-brown. Dorsal and anal fins grayish-brown, membrane of dorsal fin with numerous minute black spots. Pectoral fin pale yellow, except for black dorso-basal spot. Caudal fin yellow, in the central anterior half with about 3-5 small darkish black spots (females) and with a larger spot above anterocentral end and many small black spots on the posterior half, their filiform rays black, upper margin also black or blackish-brown (males)."

Species comparisons. — As noted in the account of *celebesensis,* the only reliable character that distinguishes it from *spilistius* is the presence of pelvic fins (*spilistius* is known only from adults, *celebesensis* only from juveniles). Both are easily distinguished by their high meristic values, *i.e.* 54-58 segmented anal-fin rays versus 19-40 in the other species of *Plagiotremus.* In view of their allopatric distributions it is conceivable that they are conspecific and that the presence or absence of pelvic fins is only a populational character. Breder (1926: 230) observed that the pelvic fins are those that fishes can best dispense with, not infrequently being entirely

TABLE 21. — Frequency distributions of certain meristic characters in species of *Plagiotremus*.[1]

Species	Segmented anal-fin rays 19	20	21	22	23	24	25	26	27	28	29	30	31	32	33	34	35	36	37	38	39	40	—	54	55	56	57	58	N	x̄
spilistius																									1	1	1		3	56.0
celebesensis																								1	–	–	–	1	2	56.0
iosodon																						2							2	40.0
l. laudandus				23	27	8																							58	22.7
l. flavus				2	7*	1																							10	22.9
phenax		1	12	17*	1																								31	21.6
townsendi	15	50	6																										71	19.9
rhinorhynchos											7	83	93	25	19														227	30.9
ewaensis													1	13	13														27	32.4
tapeinosoma										4	57	98	85	64	8														316	30.5
goslinei										8	17	8	2																35	29.1
azaleus									7	18	134	11																	230	28.6

Species	Total pectoral-fin rays							Procurrent caudal-fin rays: Dorsal					Ventral					Total										
	22	23	24	25	26	N	x̄	5	6	7	8	9	5	6	7	8	9	10	11	12	13	14	15	16	17	18	N	x̄
spilistius			3			3	24.0	3					3					3									3	10.0
celebesensis			2			2	24.0		2					2						2							2	12.0
iosodon			2			2	24.0	1*	1				1*	1				1*	–	1							2	11.0
l. laudandus		1	53	3		57	24.0		6	38	9			8	36	9				5	4	32	6	6			53	14.1
l. flavus			10			10	24.0		1	8*	1			1	7	2				1	–	7	1*	1			10	14.1
phenax		2	29*			31	23.9			11	16*			2	12	13*					2	9	4	12*			27	15.0
townsendi	1	2	62	2	1	68	24.0		2	34	15			2	29	20					4	25	9	13			51	14.6
rhinorhynchos		7	156	9	4	176	24.1		4	115	84	4	2	66	129	10			1	2	63	55	76	7	3		207	14.1
ewaensis			25	1		26	24.0		2	18	5			16	9					1	16	3	5				25	13.5
tapeinosoma	3	2	138	4	2	149	24.0		14	228	43			46	210	29				13	33	184	38	17			285	14.0
goslinei			24			24	24.0		1	16	4			4	16	1				1	3	13	3	1			21	14.0
azaleus	1	2	109	6	1	119	24.0		29	166	27	3		17	167	39	2			12	21	143	27	19	2	1	225	14.1

[1] When counts vary, those for holotypes of new taxa are indicated by an asterisk.

TABLE 22. — Frequency distributions of vertebrae in species of *Plagiotremus*.[1]

Species	Precaudal vertebrae						Caudal vertebrae											
	13	14	15	16	N	x̄	24	25	26	27	28	29	30	31	32	33	34	35
spilistius				3	3	16.0												
celebesensis			1	1	2	15.5												
iosodon			1	1*	2	15.5												
l. laudandus		23	32	1	56	14.6				17	27	12						
l. flavus			9*	1	10	15.1				1	8*	–	1					
phenax		14*	15		29	14.5			12	14*	3							
townsendi	6	64	1		71	13.9	1	35	31	4								
rhinorhynchos		126	100		226	14.4											17	58
ewaensis		5	22		27	14.8												
tapeinosoma			168	141	309	15.5									3	25	82	84
goslinei			19	17	36	15.5										17	18	1
azaleus		34	190	10	234	14.9									31	134	62	7

Species	Caudal vertebrae																	N	x̄
	36	37	38	39	40	41	42	43	44	—	54	55	56	57	58	59	60		
spilistius														1	1	1		3	58.0
celebesensis											1	–	–	–	–	–	1	2	57.0
iosodon								1*	1									2	43.5
l. laudandus																		56	27.9
l. flavus																		10	28.1
phenax																		29	26.7
townsendi																		71	25.5
rhinorhynchos	99	37	13	2														226	35.9
ewaensis	5	16	5	1														27	37.1
tapeinosoma	76	38	1															309	35.0
goslinei																		36	33.6
azaleus																		234	33.2

[1] When counts vary, those for holotypes of new taxa are indicated by an asterisk.

missing. This would seem to be especially true of the reduced pelvic fins of *P. celebesensis* (1.5-1.6% SL) which obviously serve no function in swimming or as a support when resting on the substrate. The pelvic bones are very small in those species of *Plagiostremus* that have pelvic fins and are absent in *spilistius* and *iosodon* which lack these fins.

Nomenclature. — In the original description, *Lembeichthys furcocaudalis* was compared only with *celebesensis.* Except for the reported presence of vomerine teeth in *furcocaudalis,* the description and illustrations agree well with *spilistius.* Vomerine teeth are not present in any of the three specimens of *spilistius* available to me (I have been unable to borrow paratypes of *furcocaudalis*) or in any other species of *Plagiotremus;* therefore, I conclude that the original observation was erroneous.

Etymology. — A combination of the Greek *spilos* (spot, blemish) and *histion* (sail), in reference to the spotted fins.

Biology. — Specimens of *furcocaudalis* were reported "always with sandy mud in their mouths." They were also stated to occur over the sand bottom of shallow seas. Presumably the specimens were caught with trawls, and the contents of their mouths could have resulted from the method of capture. The stomach of the holotype of *spilistius* contained 13 fish scales, from at least two species, and a large quantity of unidentified material that appeared to be mucus. As discussed in the general

TABLE 22. — Continued.

Species	Total vertebrae																			
	38	39	40	41	42	43	44	45	46	47	48	49	50	51	52	53	54	55	56	57
spilistius																				
celebesensis																				
iosodon																				
l. laudandus				3	26	23*	3	1												
l. flavus						9	–	1												
phenax			2	19*	8															
townsendi	2	37	29	3																
rhinorhynchos											1	37	106	50	24	8				
ewaensis														7	16	4				
tapeinosoma											11	54	98	73	62	11				
goslinei											5	25	6							
azaleus										35	143	54	2							

Species	Total vertebrae																			N	x̄
	58	59	60	61	62	63	64	65	66	67	68	69	70	71	72	73	74	75	76		
spilistius																1	1	1		3	74.0
celebesensis												1	–	–	–	–	–	–	1	2	72.5
iosodon		2																		2	59.0
l. laudandus																				56	42.5
l. flavus																				10	43.2
phenax																				29	41.2
townsendi																				71	39.5
rhinorhynchos																				226	50.4
ewaensis																				27	51.9
tapeinosoma																				309	50.5
goslinei																				36	49.0
azaleus																				234	48.1

TABLE 23. — Frequency distributions for selected cephalic sensory pores in species of *Plagiotremus*.[1]

Species	interorbital			supraorbital			mandibular	
	2	3	4	1-1	1-2	2-2	2	3
spilistius	–	–	3	3	–	–	–	6
celebesensis	–	–	1	1	–	–	–	2
iosodon	–		2	2	–	–	–	4
laudandus	22	–	–	–	2	20	30	12
phenax	12	–	–	–	2	10	24	–
townsendi	30	–	–	25	2	3	60	–
rhinorhynchos	–	2	55	–	3	54	–	114
ewaensis	–	–	10	–	–	10	–	20
tapeinosoma	50	–	–	–	–	50	100	–
goslinei	10	–	–	–	–	10	20	–
azaleus	41	1	–	–	2	40	42	–

[1] Refer to Figure 1 and text for methodology of pore counts.

section on biology, scale predation is a characteristic feeding habit of all species of *Plagiotremus.*

Distribution (Fig. 78). — Known only from the South China Sea.

Material examined. — 3 specimens (119.7-164.3 mm SL) from 3 collections.

SOUTH CHINA SEA: USNM 21722 (140.3, holotype of *Plagiotremus spilistius*), "China Seas"; collected by Dr. William Stimpson during the Voyage of the North Pacific Exploring Expedition. MNHN 1964-628 (1:164), South Vietnam; P. Fourmanoir. NFIS 12257 (1:120), Gulf of Thailand, Bangkok Fish Mkt.; 27 March 1973; T. Wongratana.

Plagiotremus (Plagiotremus) celebesensis (Herre)

Specimen Figures 63, 161; Tables 4, 19-23.

Lembeichthys celebesensis Herre, 1936a: 283, pl. 1, Fig. 3 (type locality: Lembeh Strait, Celebes; holotype SU 29081).

Description. — Dorsal fin X, 56-61; total elements 66-71. Anal fin II, 54-58. Pectoral fin 12. Caudal fin: procurrent rays 6 + 6 = 12; segmented rays 6 + 5. Vertebrae: precaudal 15-16; caudal 54-60; total 69-76. Pleural ribs on vertebrae 3-15; epipleural ribs 14. Incisor teeth (Fig. 74), in specimens 54-59 mm, n = 2: lower jaw 20-23; upper jaw 21-22; ratio (lower/upper) 1.0.

Dentary incisors with tips arrowhead-shaped (Fig. 72a). Snout bunt, broadly rounded. Interorbital pores 4; supraorbital pores 1; mandibular pores 3. Preanal length 26.2-28.2% SL. Pelvic fins minute 1.5-1.6% SL. Sexual dimorphism of caudal fin unknown (only juveniles available) 10.2% SL in holotype. Outer lobes of fin not elongate, but longer than inner rays (the caudal fin is damaged in the illustrated specimen).

Color pattern. — The available specimens are both faded, but the illustration of *celebesensis* given in the original description is here reproduced as Figure 161. Herre gave the color in alcohol as follows:

> "Yellow with large brown spots between the pectoral and dorsal, and similar but much smaller spots covering the entire head and pectoral base; the dorsal has 10 blackish brown spots or vertical bands; the pectoral is brown spotted, with black tips; the other fins are colorless."

Species comparisons. — As noted in the original description, the outermost premaxillary teeth are larger and more pointed than the inner teeth, indicating that the available specimens of *celebesensis* are juveniles (refer to discussion in methods section). *P. spilistius* and *celebesensis* have identical meristic counts but only the latter has pelvic fins. It is possible that the presence or absence of pelvic fins is a populational rather than specific difference (refer to account of *spilistius* for discussion). Other differences could be related to state of maturity (only adults of *spilistius* are known). Until more specimens become available for study it seems best to maintain the taxonomic status quo.

Etymology. — Named *celebesensis,* in reference to the type locality.

Distribution (Fig. 78). — Known only from the Celebes Sea.

Material examined. — 2 specimens (54.5-54.9 mm SL) from one collection.

CELEBES: SU 29081 (54.5, holotype of *Lembeichthys celebesensis*), Lembeh Strait, 1°25′N, 125°17′E; 13 June 1929; A. W. C. T. Herre. SU 67356 (1:55), taken with holotype.

Plagiotremus (Plagiotremus) iosodon, new species

Specimen Figure 162; Tables 4, 19-23.

Description. — Dorsal fin X, 43-44; total elements 53-54. Anal fin II, 40. Pectoral fin 12. Caudal fin: procurrent rays 5-6 + 5-6 = 10-12; segmented rays 6 + 5. Vertebrae: precaudal 15-16; caudal 43-44; total 59. Pleural ribs on vertebrae 3-15 or 16; epipleural ribs 15. Incisor teeth (Fig. 74), in specimens 38-40 mm, n = 2: lower jaw 22; upper jaw 20; ratio (lower/upper) 1.1.

Dentary incisors with tips arrowhead-shaped (Fig. 72a). Snout blunt, broadly rounded. Interorbital pores 4; supraorbital pores 1; mandibular pores 3. Preanal length 34.0-36.1% SL. Pelvic fins absent. Sexual dimorphism of caudal fin unknown (only juveniles available) 11.5-12.0% SL.

Color pattern. — There is a fine peppering of brownish chromatophores on the body and the caudal fin has a wedge-shaped area of pigment proximally. The dorsal and anal fins are speckled and heavily pigmented. The color pattern of *iosodon* is very similar to that of *celebesensis,* the only difference being that there are no vertical bands in the dorsal fin as reported for *celebesensis* and the dorsal fin has a narrow, somewhat indistinct, dark border. Life colors are unknown.

Species comparisons. — Only *P. iosodon* and *spilistius* lack pelvic fins, but these species are easily distinguished by several meristic characters. *Plagiotremus iosodon* has 40 segmented anal-fin rays (55-57 in *spilistius*), 43 or 44 segmented dorsal-fin rays (58-60 in *spilistius*) and 59 total vertebrae (73-75 in *spilistius*).

Etymology. — A combination of the Greek *ios* (arrow) and *odon* (tooth). The name is to be treated as a noun in apposition.

Distribution (*Fig. 78*). — Known only from the Sulu Archipelago.

Material examined. — 2 specimens (38.0-39.9 mm SL) from one collection.

Holotype. — USNM 99392, male, 38.0 mm, Sulu Sea, Jolo Island, Tutu Bay, 5°50′45″N, 121°1′15″E; depth 3-9 m; dynamite collection; 19 Sept. 1909; M/V ALBATROSS.

Paratype. — USNM 205372 (1:40), taken with holotype.

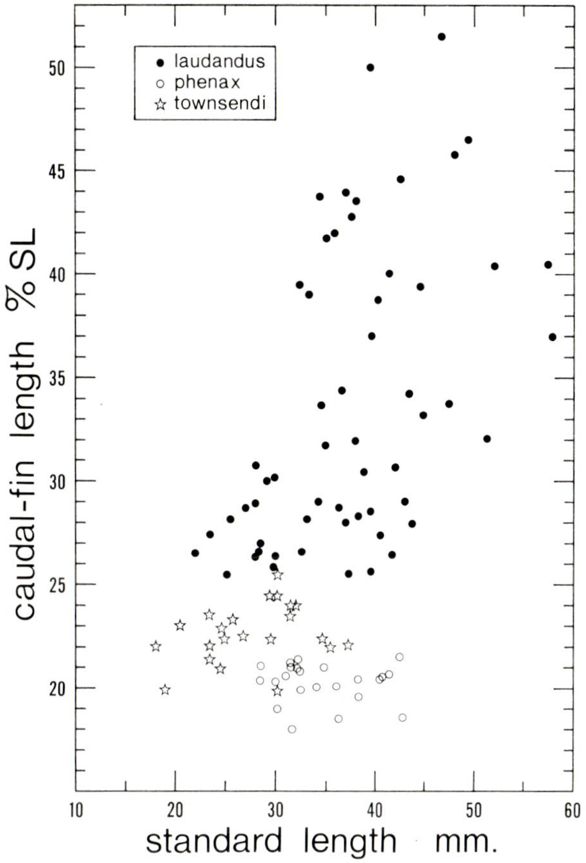

Figure 83. — Relationship between caudal-fin length versus standard length in members of the *Plagiotremus laudandus* species group.

Plagiotremus (*Musgravius*) *laudandus* (Whitley)

Specimen Figures 64, 82a, 163-165; Tables 4, 19-23.

(For synonymy, see both subspecies)

Description. — Dorsal fin VII-X, 27-30; total elements 36-39. Anal fin II, 22-24. Pectoral fin 11-13 (typically 12). Caudal fin: procurrent rays 6-8 + 6-8 = 12-16; segmented rays 6 + 5. Vertebrae: precaudal 14-16 (rarely 16); caudal 27-30; total 41-43. Pleural ribs on vertebrae 3-14 or 15; epipleural ribs 14-16. Incisor teeth (Figs. 74-75), in specimens 22-57 mm, n = 47: lower jaw 48-84; upper jaw 22-44; ratio (lower/upper) 1.6-2.0, mean 1.87.

Dentary incisors with tips ovoid or bluntly lanceolate (Fig. 72b). Snout blunt, broadly rounded. Interorbital pores 2; supraorbital pores 1-2, typically 2; mandibular pores 2-3, usually 2. Preanal length 41.3-50.5% SL. Pelvic fins not sexually dimorphic 3.7-8.4% SL. Caudal fin not sexually dimorphic 25.5-51.6% SL. Outer lobes of fin elongate (Fig. 83); innermost ray not noticeably longer than adjacent rays.

Color pattern. — Radically dissimilar overall coloration and allopatry are the bases for my recognition of subspecies in *laudandus.* In both subspecies, however, the outer lobes of the caudal fin have a very narrow black border, otherwise the fin is immaculate. In life the central area of the caudal fin is hyaline and the base and outer lobes are bright yellow. Unlike *P. phenax* and *townsendi, laudandus* apparently is not sexually dichromic; nuptial males lack an orange spot in the spinous portion of the dorsal fin.

Species comparisons. — *P. laudandus, phenax* and *townsendi* comprise a trio of closely related allopatric species (and subspecies). In addition to a large number of characters that distinguish them collectively as a group (Table 19), each species (or subspecies) appears to mimic a different species (or subspecies) of *Meiacanthus*. No other species of *Plagiotremus* are known or suspected to mimic a *Meiacanthus*.

The color pattern (see above) and elongate outer lobes of the caudal fin easily distinguish *laudandus* from its nearest relatives. The dark markings in the caudal fin also distinguish *phenax* from the other two species. All three species exhibit strong modal differences in vertebral and dorsal- and anal-fin element counts.

Plagiotremus laudandus laudandus (Whitley)

Pescadorichthys (*Musgravius*) *laudandus* Whitley, 1961: 63, Fig. 2 (type locality: barrier reef of Noumea, about 10 miles from Baie des Citrons, New Caledonia; holotype AMS IB.4509).

Color pattern. — The head is relatively dark, the pigment gradually decreasing on the body and becoming pale posteriorly. A dark submarginal stripe in the dorsal fin is variously developed; usually it is restricted to the spinous part of the fin but may extend the entire length of the fin. A narrow stripe unites the orbits anterodorsally; a slightly wider stripe encircles the snout. Both stripes may be pale or dusky in preserved specimens.

In life the submarginal stripe in the dorsal fin is black, bordered below by a narrow white stripe (often indistinct). The snout is orange and the interorbital and snout stripes are white; otherwise the head and anterior half to two-thirds of the body is blue-gray, gradually changing to yellow posteriorly. The coloration of the dorsal and anal fins coincides with the adjacent body coloration, except that posteriorly the fins are often transparent distally. The caudal fin was described above.

Etymology. — The derivation of the name was not given in the original description. Apparently it is from the Greek *laudo* (to praise) and *danos* (a gift). The construction appears to be improper but does not require emendation.

Mimicry. — An apparent mimic of *Meiacanthus a. atrodorsalis* as suggested by Springer and Smith-Vaniz (1972a: 22) and Losey (1972).

Distribution (*Fig. 79*). — Widely distributed in the central and western Pacific oceans, with the notable exception of the Fiji Islands where it is replaced by *P. laudandus flavus*. The following acceptable literature record for *laudandus* (misidentified in the first edition as *P. townsendi*), for which no specimens were examined, is plotted on the distributional map: Sesoko I., Ryukyu Is. (Masuda *et. al.*, 1975: 265).

Material examined. — 58 specimens (22.0-57.9 mm SL) from 39 collections.

INDONESIA: USNM 211901 (1:38), USNM 211911 (4: 30-33), Celebes, Kabaena I.; USNM 209744 (3:22-41), USNM 211948 (1:37), Ambon I.; USNM 211933 (1: 38), USNM 211937 (1:42), Banda I.
RYUKYU IS.: ANSP 128830 (1:39), ANSP 128832 (1: 40), BPBM 7459 (1:43), Ishigaki I.
JAPAN: ANSP 128823 (1:44), Miyake I.
MARIANA IS.: BPBM 7463 (1:35), UG 4238 (1:45), Guam I.
PALAU IS.: BPBM 9634 (2:42-47), BPBM 9936 (1:23).
MARSHALL IS.: USNM 205270 (3:40-52), USNM 205271 (1:37), USNM 205272 (1:32), Eniwetok Atoll; BPBM 9896 (1:42), Majuro Atoll.
GILBERT IS.: AMS uncat. (1:25), sta. DFH 73-32, AMS uncat. (1:43), sta. DFH 73-37, AMS uncat. (1:34), sta. DFH 73-42, AMS uncat. (6:25-39), sta. DFH 73-49, Apaiang Atoll; BPBM 12466 (1:41), Onotoa Atoll.
NEW HEBRIDES IS.: AMS uncat. (1:38), sta. GRA-19.
SANTA CRUZ IS.: USNM 201413 (3:28-32), Vanikoro I.
SOLOMON IS.: BPBM 15557 (1:28), USNM 212036 (1: 29), Guadalcanal I.
NEW GUINEA: USNM 205685 (3:36-37), Trobriand I.; AMS I.16671-61 (2:26-34), CAS 15795 (1:26), USNM 205684 (2:29-35), USNM 205686 (3:27-40), Madang.
E. AUSTRALIA: AMS I.17671-2 (1:51), Queensland, Capricorn Group, One Tree I.; AMNH 37017 (1:35), Queensland, Northern Escape Reef [Not plotted on distribution map].
NEW CALEDONIA: ANSP 126230 (2:40-49), BPBM 9052 (1:57), AMS IB.4509 (57.9, holotype of *Pescadorichthys laudandus*).
LOYALTY IS.: AMS uncat. (1:48), sta. GRA-13, Uvea Atoll.
LORD HOWE IS.: BPBM 14915 (1:28).

Plagiotremus laudandus flavus, new subspecies

Plagiotremus laudandus subsp. Springer and Smith-Vaniz, 1972a: 23, pl. 3 (illustration; suggested mimic of *Meiacanthus atrodorsalis ovalauensis*).

Color pattern. — In preservation essentially immaculate except typically with very narrow black border on outer lobes of caudal fin (often difficult to see without magnification), and rarely with the tips of the dorsal- and anal-fin rays dark (see Fig. 165). In life brilliant yellow, the central area of the caudal fin hyaline. In some specimens the top of head and snout is olive-yellow, and the tips of the posterior rays of the dorsal and anal fins may be black. Unlike the nominal subspecies, the dorsal fin never has a black submarginal stripe, the pale interorbital and snout stripes are absent and there is no blue-gray pigmentation anteriorly.

Etymology. — The Latin name *flavus* for yellow, refers to the life coloration.

Taxonomic status. — Only coloration will distinguish *flavus* from the allopatric nominal subspecies, and their respective distributions are almost certainly influenced by the distributions of the two subspecies of *Meiacanthus atrodorsalis* which they mimic. If the coloration of *flavus* is only an ecophenotypic manifestation, subspecific recognition is unjustified. However, *Plagiotremus laudandus, phenax* and *townsendi* (distinct but closely related species) each have color patterns remarkably similar to a sympatric species of *Meiacanthus,* and each of these *Plagiotremus-Meiacanthus* pairs is allopatric. This suggests that speciation in this closely related species group has been influenced (or controlled) by the distribution of a species of *Meiacanthus.*

Mimicry. — Refer to discussion under *Meiacanthus atrodorsalis ovalauensis.*

Distribution. — Endemic to Fiji Islands.

Material examined. — 10 specimens (33.3-46.8 mm SL) from 5 collections.

Holotype. — USNM 201676, female, 46.8 mm, Fiji Islands, Wailangilala Island, 16°45′S, 179°7′W; 26 May 1965; Te Vega Cruise 7, sta. 295.

Paratypes (all from Fiji Is.). — USNM 207577 (1:34), taken with holotype; USNM 201452 (1:44), Great Astrolabe, reef N of Vuro I., 18°52′S, 178°20.5′E; BPBM 11343 (1:40), Viti Levu, pass on SE side of Makaluva I.; BPBM 11393 (1:46), Viti Levu, Nukulau Pass SE side Makaluva Is.; ANSP 128024 (5:33-44), Viti Levu, Suva, Lauthala Bay, Nukumbutho Pass, reef S of University of South Pacific.

Plagiotremus (Musgravius) phenax, new species

Specimen Figure 166; Tables 4, 19-23.

Plagiotremus sp. Springer and Smith-Vaniz, 1972a: 23, pl. 33 (illustration; suggested mimic of *Meiacanthus smithi*); Burgess and Axelrod, 1973b: 636, Fig. 97 (color photograph of specimen from Maldive Islands).

Description. — Dorsal fin VII-VIII, 27-29; total elements 35-37 (rarely 37). Anal fin II, 20-23. Pectoral fin 11-12 (typically 12). Caudal fin: procurrent rays 7-8 + 6-8 = 13-16; segmented rays 6 + 5. Vertebrae: precaudal 14-15; caudal 26-28; total 40-42. Pleural ribs on vertebrae 3-13 or 14; epipleural ribs 14-16. Incisor teeth (Figs. 74-75), in specimens 21-43 mm, n = 31: lower jaw 38-61; upper jaw 20-37; ratio (lower/upper) 1.5-1.9, mean 1.64.

Dentary incisors with tips ovoid or bluntly lanceolate (Fig. 72b). Snout blunt, broadly rounded. Interorbital pores 2; supraorbital pores 1-2, typically 2; mandibular pores 2. Preanal length 41.1-51.3% SL. Pelvic fins not sexually dimorphic 5.6-8.1% SL. Caudal fin not sexually dimorphic 18.0-21.5% SL. Outer lobes of fin not elongate (Fig. 83); innermost ray not noticeably longer than adjacent rays.

Color pattern. — The typical color pattern is shown in Figure 166. The most diagnostic features are the dark markings on the caudal fin and the wide submarginal stripe in the dorsal fin. In life the body, anal fin and proximal area of the dorsal fin is light olive-brown. The dark markings in the caudal fin and caudal peduncle are brown, those of the dorsal fin nearly black. The snout is orange to yellow-amber. Two narrow, pale turquoise stripes unite the orbits; one crosses the interorbital region anteriorly, the other encircles the snout (these stripes often indistinct in preserved specimens). In some males (presumably nuptial) the spinous dorsal fin has an orange stripe, slightly narrower than the dark submarginal stripe. Both the dorsal and anal fins have a narrow white or straw-colored margin.

Burgess and Axelrod (1973b: 636) present a color photograph of a specimen of *phenax* from the Maldives.

Species Comparisons. — Refer to discussion under account of *P. laudandus.*

Etymology. — The Greek name *phenax,* meaning imposter, alludes to the striking similarity of this species and *Meiacanthus smithi.* The name is to be treated as a noun in apposition.

Mimicry. — As suggested by the specific name, *phenax* apparently is a mimic of *Meiacanthus smithi* (refer to discussion under account of *M. smithi*).

Distribution (*Fig. 79*). — Known only from the Maldive Islands and Sri Lanka.

Material examined. — 31 specimens (21.0-42.9 mm SL) from 9 collections.

Holotype. — USNM 204699, male, 34.1 mm, Sri Lanka (Ceylon), Trincomalee, cove N of harbor, 8°34′N, 81°13′E; depth 1-3 m; large rock outcrop devoid of coral; 26 June 1969; W. F. Smith-Vaniz; sta. S-V 69-119.

Paratypes. — SRI LANKA: USNM 204700 (4:28-32), taken with holotype; MCZ 46931 (1:21), Trincomalee Bay; CAS 28724 (3:25-30), Galle, SE end of Bay; BPBM 12459 (1:29), Weligama, near resthouse. MALDIVE IS.: ANSP 122786 (7:26-42), ANSP 128561 (2:36-38), Bandos Is., ca. 5 mi. N of Male; FMNH 71350 (1:41), S Male Atoll, Dunidu I.; FMNH 71351 (5:31-43), S Nilandu Atoll, Kuda Huvadu I.; FMNH 71352 (5:32-43), Addu Atoll, Wilingili I.

Other material. — USNM 212029 (1:41), "Mombasa, Kenya; 1973; H. R. Axelrod, T. F. H. Publications." This collection is believed to have come from the Maldive Islands despite the original locality data that accompanied the specimen. Burgess and Axelrod (1973b: 636, Fig. 97) present a color photograph of an individual of *Plagiotremus phenax* stated to have come from the Maldive Islands that appears identical to the USNM specimen.

Plagiotremus (*Musgravius*) *townsendi* (Regan)

Specimen Figures 65, 82b, 167 and Frontispiece; Tables 4, 19-23.

Petroscirtes townsendi Regan, 1905: 328, pl. 3, Fig. 7 (type locality: Gulf of Oman, Mekran Coast, Jask; holotype BMNH 1899.5.8.100).

Description. — Dorsal fin VI-VIII, 25-27; total elements 32-34. Anal fin II, 19-21. Pectoral fin 11-13 (typically 12). Caudal fin: procurrent rays 6-8 + 6-8 = 12-16; segmented rays 6 + 5. Vertebrae: precaudal 13-15 (rarely 15); caudal 24-27; total 38-41. Pleural ribs on vertebrae 3-13 or 14; epipleural ribs 14-15. Incisor teeth (Figs. 74-75), in specimens 19-38 mm, n = 56: lower jaw 33-56; upper jaw 19-34; ratio (lower/upper) 1.4-1.9, mean 1.64.

Dentary incisors with tips ovoid or bluntly lanceolate (Fig. 72b). Snout blunt, broadly rounded. Interorbital pores 2; supraorbital pores 1-2, usually 2; mandibular pores 2. Preanal length 44.9-53.0% SL. Pelvic fins not sexually dimorphic 5.6-9.4% SL. Caudal fin not sexually dimorphic 19.7-25.5% SL. Outer lobes of fin not elongate (Fig. 83); innermost ray not noticeably longer than adjacent rays.

Color pattern. — The head and anterior half of the body is dark, gradually becoming pale posteriorly. A narrow white stripe encircles the snout, and a more indistinct stripe connects the orbits anterodorsally. The spinous part of the dorsal fin has a black submarginal stripe, approximately a third the width of the fin. The stripe is bordered above by a very narrow white stripe and below by a slightly wider white stripe; otherwise the pigmentation generally conforms to that of the body. Both the dorsal and anal fins are nearly transparent posteriorly. The caudal fin is also hyaline except for the outermost rays of each lobe which are dusky.

In life (see frontispiece) the head, proximal part of spinous dorsal fin and anterior half of the body are slate blue, becoming pale yellow posteriorly. The snout, except for the narrow white stripe, is dark orange-amber (especially in males). Adult males differ from females and juveniles in having a brilliant orange stripe in approximately the anterior third of the dorsal fin. The orange pigment is confined to the proximal area of the fin.

Species comparisons. — Refer to account of *P. laudandus*.

Etymology. — Named for F. W. Townsend, who collected the holotype.

Mimicry. — Springer and Smith-Vaniz (1972a) gave a detailed treatment of an apparent mimetic trio (see frontispiece) consisting of *Plagiotremus townsendi, Meiacanthus nigrolineatus* and *Ecsenius graveri. P. townsendi* was believed to be a mimic of the *Meiacanthus* but the exact nature of the mimetic relationship is unclear (refer to discussion in general section on mimicry).

Distribution (*Fig. 79*). — Locally abundant in the Gulf of Aqaba; elsewhere known from single specimens from the Red Sea off Port Sudan and from the type locality in the Gulf of Oman.

Material examined. — 79 specimens (18.1-37.5 mm SL) from 21 collections.

GULF OF OMAN: BMNH 1899.5.8.100 (30.2, holotype of *Petroscirtes townsendi*), Merkran Coast, Jask.

GULF OF AQABA: ANSP 134463 (2:21-25), ANSP 134464 (5:19-22), USNM 204575 (2:20-21), USNM 204576 (2:29), USNM 204577 (10:18-37), USNM 205379 (30:20-38), USNM 205093 (1:30), USNM uncat. (1:20), sta. VGS 69-1, USNM uncat. (2:19-23), sta. VGS

69-3, USNM uncat. (1:24), sta. VGS 69-8, USNM uncat. (2:21-26), sta. VGS 69-21, USNM uncat. (1:26), sta. VGS 69-24, UTAI 3922 (3:18-30), UTAI 4192 (2:22-29), UTAI 4539 (1:18), UTAI 5055 (1:32), UTAI 5059 (1:30).

RED SEA: USNM 204574 (1:20), USNM 204640 (9:21-28), Strait of Jubal; BPBM 17897 (1:22), Port Sudan.

Plagiotremus (*Runula*) *rhinorhynchos* (Bleeker)

Specimen Figures 66, 82c, 168-171; Tables 4, 19-25.

Petroskirtes rhinorhijnchos Bleeker, 1852a: 273 (type locality: Wahia, Ceram; lectotype RMNH 26470).

Petroskirtes amblyrhynchos Bleeker, 1857a: 64 type locality: Amboina; holotype RMNH 4464).

Runula fasciata Fourmanoir, 1971: 496, Fig. 4 (type locality: R/V CYCLONE Cruise 3, sta. 11, 2°51′S, 169°55′E; holotype MNHN 1973-37).

Description. — Dorsal fin X-XII, 31-37; total elements 29-48. Anal fin II, 29-33. Pectoral fin 11-13 (typically 12). Caudal fin: procurrent rays 6-9 + 5-8 = 11-17; segmented rays 6 + 5. Vertebrae: precaudal 14-15; caudal 34-39; total 48-53. Pleural ribs on vertebrae 3-14 or 15; epipleural ribs 15-17. Incisor teeth (Figs. 74 and 76), in specimens 28-79 mm, n = 161: lower jaw 29-58; upper jaw 18-40; ratio (lower/upper) 1.3-1.8, mean 1.51.

Dentary incisors with tips chisel-shaped (Fig. 72c). Snout fleshy, conical (in aduts). Interorbital pores 4; supraorbital pores 2; mandibular pores 3. Preanal length 36.6-44.8% SL. Pelvic fins not sexually dimorphic 5.7-8.2% SL. Caudal fin not sexually dimorphic 17.2-22.8% SL. Outer lobes of fin not elongate; innermost ray not noticeably longer than adjacent rays.

Color pattern. — Adults of *P. rhinorhynchos* are characterized by two pale stripes (brilliant blue in life) on the sides of the body. The dorsal stripe, which touches the dorsum of the eye, extends from near the tip of the snout to the dorsum of the caudal peduncle. The ventral stripe encircles the snout, touches the ventral margin of the eye and extends through the pectoral-fin base to the ventral half of the base of the caudal fin. The pale stripes have diffuse dark borders that are most apparent on the head. A narrow, pale predorsal stripe (Fig. 82c) is also present. In most, but not all, specimens the interspace between the pale stripes becomes darker posteriorly and ends as a basicaudal spot, often covering the entire base of the caudal fin. The uniformly pigmented dorsal and anal fins have a very narrow dark margin. The caudal fin is pale, the rays sometimes faintly outlined. The background coloration is highly variable and may be pale or very dark. Juveniles (usually smaller than 34 mm SL) are banded (Fig. 168), have a distinct dark blotch in the anterior part of the dorsal fin and have the pale lateral stripes poorly developed or absent.

The only consistent feature of the life coloration is the pair of brilliant blue stripes on the sides, which usually have diffuse dark brown or black borders. The background color may be yellow, orange-red, brown or black. The venter and fins are usually paler than the sides of the head and body. The pectoral fins are essentially transparent, but the rays may be faintly outlined in black. See also discussion of mimicry below.

Tomiyama (1956c) gives a good monochrome illustration of an adult from Japan. Color illustrations of adult *rhinorhynchos* are given in Smith (1949: pl. 17), Burgess and Axelrod (1974: 160-161) and Masuda *et al.* (1975: 86).

Geographic variation. — Populational variation in frequencies of selected meristic characters is shown in Table 24.

Species comparisons. — *P. rhinorhynchos* and *ewaensis* comprise a closely related species pair; refer to account of *ewaensis* for comparison. Both are easily distinguished from the other species of *Plagiotremus* by the combination of 10-12 dorsal-fin spines and 31-37 segmented rays. They are also unique in having dentary incisors with chisel-shaped tips (Fig. 72c).

Nomenclature. — The holotype of *P. amblyrhynchos* is a 35-mm juvenile *rhinorhynchos* with a characteristically blunt snout and banded color pattern. Bleeker was unaware of the changes in color pattern and snout shape that occur with growth in this species. Fourmanoir repeated Bleeker's mistake when he described *fasciata* based on four juveniles (18-35 mm) of *rhinorhynchos*. In the original description Fourmanoir gave the count for the dorsal fin as 34-35, but failed to indicate that this did not include the dorsal-fin spines.

Etymology. — The specific name, a combination of the Greek *rhinos* (beak, bill) and *rhynchos* (snout), refers to the fleshy conical snout of the adults.

TABLE 24.—Geographic variation in frequencies of certain meristic characters in the *Plagiotremus rhinorhynchos* species group as indicated by specimens from selected localities.

	Total vertebrae								Total dorsal-fin elements								Segmented anal fin rays						
Species	48	49	50	51	52	53	N	x̄	43	44	45	46	47	48	N	x̄	29	30	31	32	33	N	x̄
rhinorhynchos																							
Gulf of Aqaba			11	11	1		23	50.6	3	9	10	1			23	44.4		8	12	3		23	30.8
Seychelles		1	–	2	11	3	17	51.9	1	–	3	9	5		18	45.9		1	–	9	8	18	32.3
Africa			2	7	3		12	51.1	4	4	5				13	44.1			1	4	8	13	32.5
Maldive Is.			3	1			4	50.2		2	2				4	44.5		2	2			4	30.5
Sri Lanka		8	5	2			15	49.6	3	9	3				15	44.0		7	8			15	30.5
Celebes		2	5				7	49.7	3	3	1				7	43.7		6	1			7	30.1
Banda Sea		6	20	3			29	49.9	4	19	5	1			29	44.1	2	13	12	2		29	30.5
Philippines		1	2				3	49.7	1	2					3	43.7		1	2			3	30.7
Ryukyu Is.		1	1	2			4	50.2		3	1				4	44.2		1	3			4	30.7
Palau-Yap Is.		2	16	4			22	50.1	2	19	1				22	43.9	1	13	8			22	30.3
Truk Is.		1	2				3	49.7	2	1					3	43.3		3				3	30.0
Ponape Is.		1	1				2	49.5		2					2	44.0		1	1			2	30.5
Gilbert Is.	1	4	3				8	49.2	2	5	1				8	43.9		3	4	1		8	30.7
Marquesas Is.					2	1	3	52.3				1	1	1	3	47.0					3	3	33.0
Fiji		1	–	–	1		2	50.5		1	1				2	44.5		1	–	1		2	31.0
New Hebrides			1	1			2	50.5			2				2	45.0			2			2	31.0
Solomon Is.			10	5			15	50.3	2	12	1				15	43.9		8	7			15	30.5
New Britain		4	7	1			12	49.7	2	8	2				12	44.0	2	7	3			12	30.1
New Guinea		1	10	4			15	50.2	2	9	3	1			15	44.2	1	5	9			15	30.9
E. Australia			3	1			4	50.2		3	1				4	44.2		1	1	2		4	31.2
Lord Howe Is.			1	2			3	50.7		3					3	44.0			3			3	31.0
ewaensis																							
Hawaiian Is.				7	16	4	27	51.9			5	15	7		27	46.1			1	13	13	27	32.5

In the original description the specific name was spelled *"rhinorhijnchos"*, undoubtedly due to a printers error. In a handwritten manuscript "y" could easily have been mistaken for "ij". In Bleeker's subsequent papers the name is *correctly* spelled *"rhinorhynchos."*

Mimicry. — Barry C. Russell (*in litt.*) noted that at Rabaul, New Britain, *Plagiotremus rhinorhynchos* apparently mimics the yellow female phase of *Anthias mortoni* Macleay. Individuals of *P. rhinorhynchos* observed mingling with schools of the anthiine were nearly uniform yellow except for the characteristic blue stripes, which were unusually narrow. Museum specimens of apparently yellow *rhinorhynchos,* as well as individuals with more "typical" color patterns, are also available from New Guinea and the Solomon Islands.

Juveniles of the cleaner wrasse, *Labroides dimidiatus,* have a black body with a brilliant blue stripe along each side of the back. Young *P. rhinorhynchos* are virtually identically colored and perhaps benefit from the resemblance in a similar fashion to *Aspidontus taeniatus.* Russell (1976: 328) believes that a loose form of aggressive mimicry persists even in the adults:

> "As they grow older, however, the colour pattern of *P. rhynorhynchos* changes and a second blue stripe develops along the lower side of the body. Although the blue and black colouration generally resembles that of *Labroides* there the similarity ends; the blenny is much more elongate in form and swims with an undisguised sinuous body movement. At this stage *P. rhynorhynchos* usually is solitary and does not aggregate with other fishes. Its superficial resemblance to the cleanerfish, however, seems to give some mimetic advantage. Although small territorial reef fishes recognise the predatory blenny and avoid it, larger roving species such as snappers, parrotfishes and siganids often approach unaware. At a distance these fishes may well mistake the blenny for a cleanerfish — a common and harmless species. The deception is sufficient for the blenny to benefit. By employing a sneak hit-and-run type of attack and concentrating on roving fishes — prey which in effect never learn to recognise their attacker — a loose form of aggressive mimicry can be maintained."

Distribution (*Fig. 78*). — Widely distributed throughout the Indo-Pacific, but rare east of longitude 180° despite a relatively large number of collections from the region.

Whitley (1968: 81) listed *rhinorhynchos* from the Kermadec Islands based on Waite (1910: 380) who did not make a positive identification. Waite stated: "(?) *Petroscirtes rhinorhynchos,* Bleeker. A small example, 48 mm in length, is provisionally referred to this species." I have been unable to locate Waite's specimen. The following acceptable literature records for *rhinorhynchos,* for which no specimens were available, are plotted on the distributional map: Susaki, Kochi Prefecture, Japan (Tomiyama, 1956c: 1129); Moroiso, Kanagawa, Japan (Burgess and Axelrod, 1972: 161, illustrated specimen); and Izu Peninsula, Japan (Masuda *et. al.,* 1975: 264).

Material examined. — 287 specimens (18.9-95.5 mm SL) from 161 collections.

GULF OF AQABA: USNM 204635 (5:32-37), USNM 204637 (4:33-37), USNM uncat. (1:39), sta. VGS 69-4, USNM uncat. (1:40), sta. VGS 69-5, USNM uncat. (2: 36-39), sta. VGS 69-6, USNM uncat. (1:38), sta. VGS 69-8, USNM uncat. (1:42), sta. VGS 69-21, UTAI 4218 (1:44), UTAI 4764 (1:37).

SINAI PENINSULA: USNM 204573 (6:39-67), Ras Muhammad.

SEYCHELLES: ANSP 109805 (1:62), ANSP 110999 (4: 32-52), ANSP 111009 (1:58), ANSP 111013 (1:70), ANSP 111015 (2:53-57), ANSP 111038 (2:70-79), ANSP 114818 (1:35), ANSP 123716 (2:39-64), ANSP 123718 (1:72), ANSP 123721 (2:60-68), USNM uncat. (1:63), Amirante Is.; ANSP 111012 (1:77), Curieuse I.

ALDABRA IS.: RUSI 2202 (1:59), USNM 213477 (1: 67), USNM uncat. (1:61), sta. HA 67-60, USNM uncat. (1:69), sta. HA 67-71.

AGALEGA IS.: USNM 215443 (2:49-52).

GRANDE COMORE IS.: CAS 34381 (5:31-41), CAS 34385 (3:34-40).

E. AFRICA: LACM 31299 (1:36), RUSI M.200 (1:63), RUSI M.651 (1:58), Malindi; RUSI uncat. (1:65), Mombasa; RUSI Z.1410 (1:60), Zanzibar; BPBM 15952 (1:41), Tanzania, Mafia I.; RUSI uncat. (1:50), "Mozambique"; RUSI uncat. (1:66), Ibo; BMNH 1917.5.1.8 (1:96), RUSI uncat. (3:36-82), Durban.

MALDIVE IS.: FMNH 71353 (1:61), S Male Atoll; NFIS 9761 (2:40-53), Ari Atoll; NFIS 9762 (1:55), Fadiffulu Atoll.

SRI LANKA (CEYLON): ANSP 123717 (1:39), ANSP 123720 (1:46), BMNH 1903.5.12.5 (1:70), BMNH 1961.8.17.3-6 (3:32-39), USNM 204698 (7:33-41), USNM uncat. (2:58-62), sta. PCH 69-279, USNM uncat. (1:35), sta. CCK 69-131B, USNM uncat. (2:39-44), sta. CCK 69-136, USNM uncat. (1:32), sta. TI 70-346.

NICOBAR IS.: NFIS 9768 (2:39-53).

COCOS-KEELING IS.: ANSP 128050 (1:76).

INDONESIA: USNM 211909 (4:54-59), Celebes, Kabaena I.; USNM 211971 (2:43-62), Celebes, Buton I.; SU 28842 (1:30), Celebes, Lembeh Strait, 1°25′N, 125°17′E;

ZMA 109.353 (1:32), Celebes, Salajar I. (Saleyer); FMNH 51793 (1:56), Borneo, Sandakan District; SU 32282 (2:62-68), Sulu Archipelago, Sitankai; USNM 111971 (1:49), Sulu Arch., Jolo I.; USNM 137859 (1: 65), Sulu Arch., Tulayan I., vicinity Jolo I.; SU 29141 (1:33), Moluccas, Ternate I.; RMNH 26470 (49.5, lectotype by present designation of *Petroskirtes rhinorhynchos*), Ceram, Wahai; RMNH 4770 (13:42-68) originally in same jar with lectotype; RMNH 4464 (35.2, holotype of *Petroskirtes amblyrhynchos*), Amboina (Ambon I.); RMNH 5051 (1:44), Ceram; USNM 209845 (1:57), USNM 210204 (2:36-52), Ceram, Piru Bay; USNM 209867 (1:38), Nusa Laut I., 3°40′S, 128°47′E; ISZZ 5365 (4:39-48), USNM 209765 (5:30-49), Ambon I.; USNM 195703 (7:28-65), "Dutch East Indies-Longley", presumably Banda Sea; ZMA 109.352 (7:30-34), Banda I.

W. AUSTRALIA: WAM P8774-5 (2:49-54), "Blow holes" N Carnarvon, 24°51′S, 113°45′E.

SOUTH CHINA SEA: CAS 29045 (1:44), South Vietnam, Nha trang.

PHILIPPINE IS.: SU 28451 (1:43), Negros I., Dumaguete; USNM 99371 (1:54), Luzon, Ragay Gulf.

TAIWAN: USNM uncat. (1:36), sta. VGS 68-2.

RYUKYU IS.: ANSP 128829 (1:67), BPBM 7391 (2:42-45), BPBM 7457 (1:73), Ishigaki I.

MARIANA IS.: BPBM 7281 (1:53), Guam I.

YAP IS.: ANSP 124856 (2:64-65), CAS 29042 (2:51-62), CAS 29043 (1:68).

PALAU IS.: ANSP 124876 (1:68), ANSP 126027 (1:67), BPBM 7454 (1:47), BPBM 7786 (2:56-61), BPBM 9636 (2:48-50), CAS 29033 (2:50-62), CAS 29034 (1:36), CAS 29035 (1:54), CAS 29036 (1:59), CAS 29037 (1: 31), CAS 29038 (1:51), CAS 29039 (1:60), CAS 29040 (1:60), CAS 29041 (1:65), CAS 29044 (1:67).

TRUK IS.: BPBM 7447 (3:50-54).

PONAPE IS.: USNM 201466 (1:55), UW 11369 (1:50).

KAPINGAMARANGI ATOLL: ANSP 124875 (1:72).

GILBERT IS.: AMS uncat. (2:36-52), sta. DFH 73-37, AMS uncat. (2:37-43), sta. DFH 73-49, AMS uncat. (2: 50), sta. DFH 73-51, AMS uncat. (2:33-39), sta. DFH 73-52, Apaiang Atoll.

MICRONESIA ("OCEANIA"): ANSP 124231 (1:19), 5°0′S, 169°45′E; MNHN 1973-37 (34.5, holotype of *Runula fasciata*), 2°51′S, 169°55′E.

LINE IS.: BPBM 7770 (1:46), Washington I.

MARQUESAS IS.: BPBM 10875 (1:54), Nuku Hiva I.; BPBM 11740 (2:45-48), Fatu Hiva I.

SOCIETY IS.: BPBM 14953 (1:33), Tetiaroa.

FIJI IS.: ANSP 128052 (1:65), BPBM 11360 (1:68), Viti Levu; CAS 7482 (1:61), "Fiji Is."

NEW HEBRIDES IS.: AMS I.14314 (2:33-35), "New Hebrides"; USNM 195801 (2:40-41), Espiritu Santos I.

SOLOMON IS.: AMS uncat. (1:41), Savo I., sta. GRA-50; AMS uncat. (2:54-56), Florida I., sta. GRA-52; USNM 212033 (1:42), USNM 212034 (1:56), USNM 212037 (5:32-38), Guadalcanal I.; USNM 195802 (1:38), Florida I.; USNM 195797 (1:33), USNM 195800 (2: 39-40), USNM 195803 (1:51), USNM 195804 (1:35), New Georgia Group; CSIRO uncat. (1:29), USNM uncat. (1:36), Te Vega Cr. 6, sta. 245, USNM uncat. (1:45), Te Vega Cr. 6, sta. 247, Bougainville I., vicinity Kieta.

NEW BRITAIN: CSIRO uncat. (8:32-35), DASF FO527 (1:34), Gasmata; DASF FO3722 (1:32), Cape Lambert, 4°13′S, 151°30′E; ISZZ 12309 (1:33) "New Britain"; USNM 214389 (3:39-56), Rabaul; USNM 205424 (6: 37-54), St. George Channel, Keraward I.

NEW GUINEA: CSIRO uncat. (3:33-36), SU 28227 (4: 32-40), USNM 205691 (2:39-44), USNM 205692 (4: 34-40), USNM 205693 (1:47), USNM 205694 (1:45), USNM 205695 (1:36), USNM 205696 (1:50), USNM 211112 (2:32), USNM 211113 (1:44), Madang; DASF FO3724 (1:34), "Sek Harbor"; USNM 211111 (1:35), 2°7′S, 140°10′E, stomach contents of *Thunnus albacares;* SU 28652 (1:32), Manokwari; SU 28070 (1:32), Waigeo I., 0°13′S, 130°50′E.

E. AUSTRALIA: ANSP 109674 (1:41), Queensland, Northern Escape Reef; ANSP 109675 (1:65), Qld., Big Hope I.; ANSP 109676 (1:44), Endeavour Reef; AMS I.17047 (1:56), Opal Reef off Cairns; AMS I.15682-52 (1:48), USNM uncat. (1:68), sta. VGS 66-12, Great Barrier Reef, Capricorn Group, One Tree I.

NEW CALEDONIA: AMS IB.5282 (1:51), Noumea.

LOYALTY IS.: AMS I.17467-3 (1:38), Uvea Atoll.

LORD HOWE I.: AMS uncat. (1:62), sta. 73-4, AMS uncat. (2:50-53), sta. 73-45.

Plagiotremus (*Runula*) *ewaensis* (Brock)

Specimen Figures 172-173; Tables 4, 19-25.

Petroscirtes ewaensis Brock, 1948: 125, unnumbered Fig. (type locality: Hawaiian Islands, off Ewa Beach, Oahu; holotype USNM 133821).

Description. — Dorsal fin X-XII, 34-36; total elements 45-47. Anal fin II, 31-33. Pectoral fin 12-13 (typically 12). Caudal fin: procurrent rays 6-8 + 6-7 = 12-15; segmented rays 6 + 5. Vertebrae: precaudal 14-15; caudal 36-39; total 51-53. Pleural ribs on vertebrae 3-14 or 15; epipleural ribs 15 or 16. Incisor teeth (Figs. 74 and 76), in specimens 39-88 mm, n = 27: lower jaw 33-54; upper jaw 22-40; ratio (lower/upper) 1.3-1.7, mean 1.48.

Dentary incisors with tips chisel-shaped (Fig. 72c). Snout fleshy, conical (in adults). Interorbital pores 4; supraorbital pores 2; mandibular pores 3. Preanal length 37.4-42.0% SL. Pelvic fins not sexually dimorphic 7.2-9.7% SL. Caudal fin not sexually dimorphic 18.3-20.9% SL. Outer

TABLE 25. — Comparison of regression equations and F values for slopes and heights of dentary incisors for four populations of the *Plagiotremus rhinorhynchos* species group (** = significant at p = .001; NS = not significant).[1]

Populations	N	regression equations y=	correlation coefficient
(x) Hawaii	23	28.820+0.308x	0.6493
(a) New Guinea-New Britain-Solomon Is.	18	41.196+0.178x	0.5268
(b) Palau-Yap Is.	21	37.093+0.263x	0.6653
(c) Gulf of Aqaba-Seychelles Is.	22	38.023+0.224x	0.7400

Comparisons	Covariance F Values and Probabilities			
	Slopes	P<	Heights	P<
x versus a	1.39	NS	131.97	**
x versus b	0.18	NS	60.03	**
x versus c	0.85	NS	60.54	**
a versus b	0.59	NS	0.04	NS
a versus c	0.24	NS	2.11	NS
b versus c	0.24	NS	4.81	NS

[1] Population (x) corresponds to *P. ewaensis* and populations (a-c) correspond to the three regression lines for *P. rhinorhynchos* shown in Fig. 84.

lobes of fin not elongate; innermost ray not noticeably longer than adjacent rays.

Color pattern. — The color pattern of *ewaensis* is essentially identical to that of *rhinorhynchos* with the following qualifications. The background coloration is consistently yellow or yellow-orange; and perhaps for this reason the dark borders of the pale (blue in life) body stripes are more apparent. The width of the ventral stripe (including the dark border) appears to be wider than in *rhinorhynchos.* Only in a partially banded 39 mm juvenile (USNM 203272) is there an indication of a dark blotch at the caudal-fin base, although one is usually present in *rhinorhynchos* at this size.

A good illustration of the holotype accompanied the original description, and Hobson and Chave (1972: pl. 59) present a color photograph of a young adult.

Species comparisons. — *P. ewaensis* is very closely related to *rhinorhynchos* and perhaps merits only subspecific status. There appear to be slight differences in color pattern, but more important is the fact that *ewaensis* has significantly fewer dentary incisors (see Fig. 84 and Table 25). There is little overlap in the number of teeth among the four species groups of *Plagiotremus* (Fig. 74), and only one member of each species group occurs at any locality. This suggests that differences in dentition may serve an important role in reducing interspecific

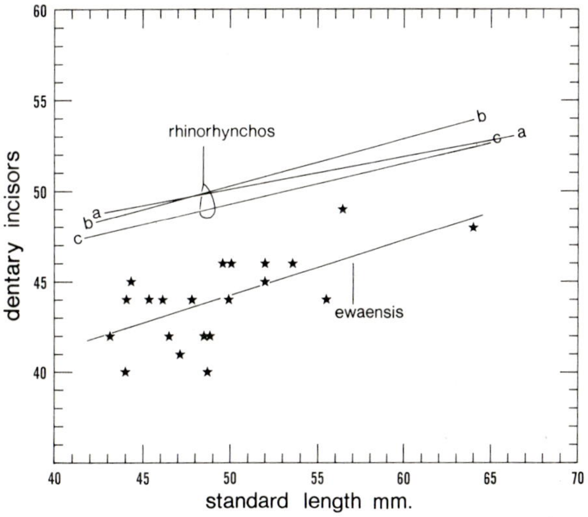

Figure 84. — Relationship between number of dentary incisors versus standard length in *Plagiotremus ewaensis* and three populations of *rhinorhynchos.* Statistics for regression lines are summarized in Table 25.

competition. Other reasons for recognition of *ewaensis* as a distinct species also apply to *P. goslinei* (refer to "species comparisons" in account of *tapeinosoma*).

Etymology. — The specific name, *ewaensis,* refers to Ewa, the district of Oahu adjacent to the type locality.

Distribution. — Endemic to the Hawaiian Islands.

The small specimen from Makua was found attached by its teeth to a surface buoy approximately 1 km offshore.

Material examined. — 28 specimens (39.4-88.3 mm SL) from 12 collections.

OAHU ISLAND: USNM 133821 (87.4 mm, holotype of *Runula ewaensis*), BPBM 12464 (1:44), off Ewa Beach; BPBM 6401 (1:43), USNM 203273 (1:88), off Lahilahi Point; BPBM 7450 (13:41-64), BPBM 9959 (2:47-54), USNM 205337 (4:47-52), Waimea Bay; BPBM 6002 (1:56), BPBM 6344 (1:49), Kahe Point; BPBM 12465 (1:56), Waikiki; USNM 111978 (1:47), "South shore"; USNM 203272 (1:39), Makua.

Plagiotremus (*Runula*) *tapeinosoma* (Bleeker)
Specimen Figures 67, 82d, 174-176;
Tables 4, 19-23, 26.

Petroskirtes tapeinosoma Bleeker, 1857a: 64 (type locality: Amboina; lectotype RMNH 26469).
Macrurrhynchus maroubrae Ogilby, 1896: 137, pl. 10, Fig. 1 (type locality: Maroubra N. S. W., Australia; holotype AMS I.3504).

Description. — Dorsal fin VII-IX, 34-39; total elements 42-47. Anal fin II, 28-33. Pectoral fin 11-13 (typically 12). Caudal fin: procurrent rays 6-8 + 6-8 = 12-16; segmented rays 6 + 5. Vertebrae: precaudal 15-16; caudal 32-38; total 48-53. Pleural ribs on vertebrae 3-15 or 16 (rarely 17); epipleural ribs 15-20. Incisor teeth (Figs. 74 and 77), in specimens 29-92 mm, n = 206: lower jaw 40-87; upper jaw 18-43; ratio (lower/upper) 1.7-2.4, mean 2.06.

Dentary incisors with tips ovoid or bluntly lanceolate (Fig. 72d). Snout fleshy, conical (in adults). Interorbital pores 2; supraorbital pores 2; mandibular pores 2. Preanal length 37.0-45.7% SL. Pelvic fins sexually dimorphic in adults (Fig. 85): males 10.1-22.5% SL, females 7.7-15.2% SL. Caudal fin not sexually dimorphic 15.2-20.8% SL. Outer lobes of fin not elongate (Fig. 86); innermost ray not noticeably longer than adjacent rays.

Color pattern. — Some populations of *P. tapeinosoma* exhibit differences in color pattern. The following description is based on adults from the Cocos-Keeling Islands; deviations from this color pattern are discussed under the section on geographic variation.

Dorsal half of head and body brown, ventral half white; the line of demarcation at about level of ventral margin of pupil. Dark stripe, approximately width of eye diameter, extends from eye to caudal fin where the stripe narrows and extends to margin of fin as a streak covering innermost caudal-fin ray. Stripe with very irregular margins, especially dorsally; the stripe largely broken into a series of vertically elongate blotches. The blotches shorten posteriorly and form oblong spots, which coalesce completely on caudal peduncle. A narrow, pale stripe (usually not apparent in preserved specimens) extends from each side of snout dorsally to dorsolateral region of caudal peduncle; pale stripe touches dorsal margin of eye and borders mid-lateral stripe. The dark area above pale stripe, which is paler than the mid-lateral stripe, narrows posteriorly and may disappear in region of caudal peduncle. A median, narrow, pale stripe (also usually not apparent in preserved specimens) extends from near anterior margin of orbits to dorsal-fin origin. The pale head stripes of *tapeinosoma* are illustrated in Figure 79d. Dorsal fin pale dusky proximally, distal third to half of fin darker with very narrow, black, submarginal stripe (indistinct posteriorly) and narrow pale margin. Spinous portion of dorsal fin paler in larger males (see discussion of sexual dichromatism below). Anal fin uniformly dusky with narrow pale margin. Caudal fin dusky except for innermost ray, which is very dark.

Juveniles (ca. 34-mm SL or smaller) are strongly bicolored like the adults, but the mid-lateral stripe is indistinct, only slightly if at all darker than the dorsum. Dorsal and anal fins hyaline except for series of dark vertical bands. Innermost ray of caudal fin usually more heavily pigmented than adjacent rays, but much lighter than in adults; remainder of caudal fin pale.

In life the dark mid-lateral stripe and associated markings are brownish-black, including the innermost ray of the caudal fin. Snout and body below mid-lateral stripe white, except for orange throat, breast and skin around mouth. Dorsum of body pale yellow-gray, that of head brownish; narrow, pale blue stripes from above eyes to caudal peduncle and a similar median predorsal stripe. Spinous portion of dorsal fin orange (males) or dusky (females) with very narrow, black margin; remainder of fin dusky distally with narrow, black, submarginal stripe

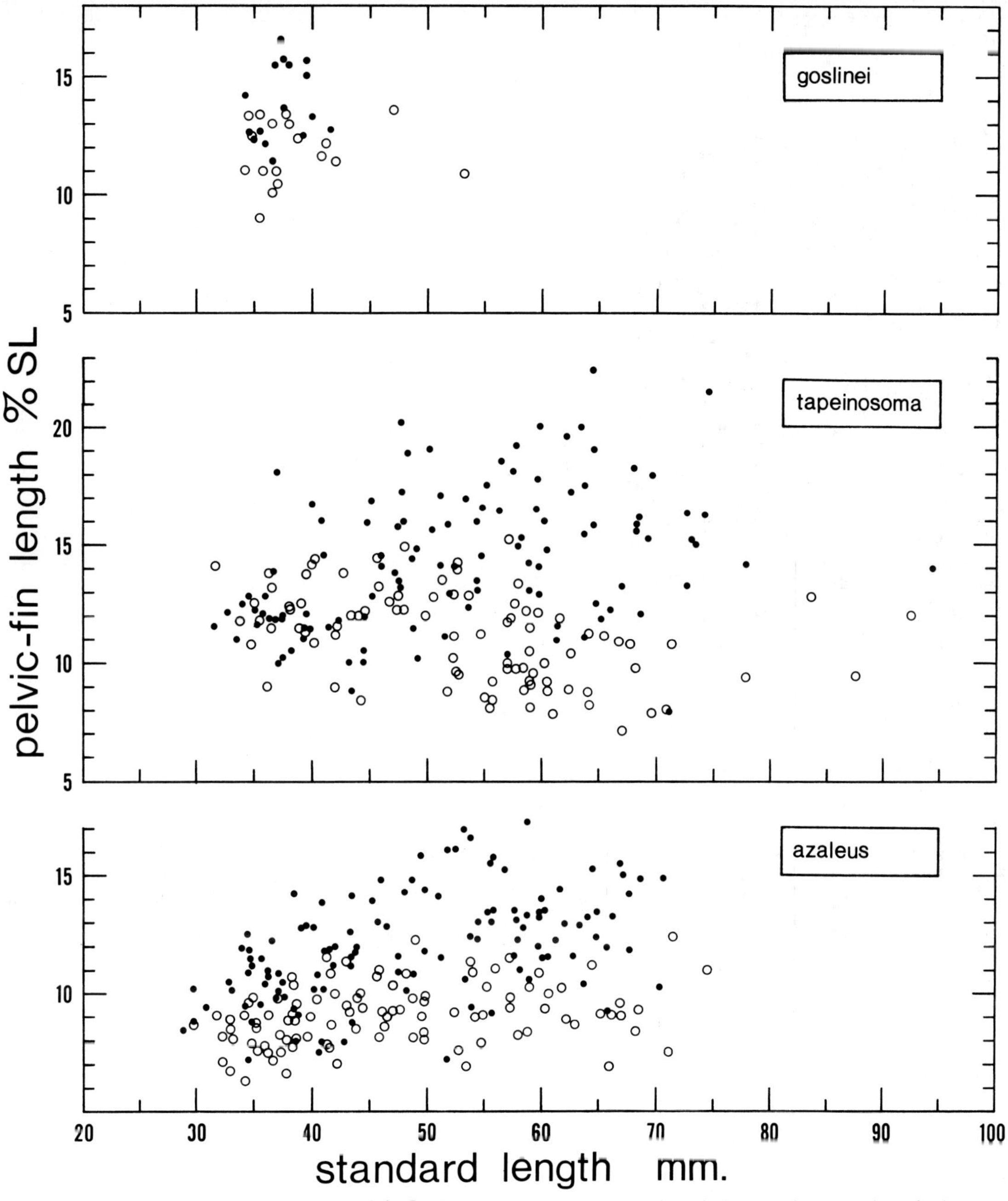

Figure 85. — Relationship between pelvic-fin length versus standard length in members of the *Plagiotremus tapeinosoma* species group. Closed symbols = males; open symbols = females.

and narrow, white margin, proximally transparent or yellowish. Anal fin dusky with narrow white margin. Caudal fin with innermost ray dark brown or black, immediately adjacent rays yellow or orange and remainder of fin dusky or paler yellow-orange. Pelvic fins white; pectoral transparent.

McCulloch (1917) gives an illustration of a banded juvenile (as *A. maroubrae*), and Tomiyama

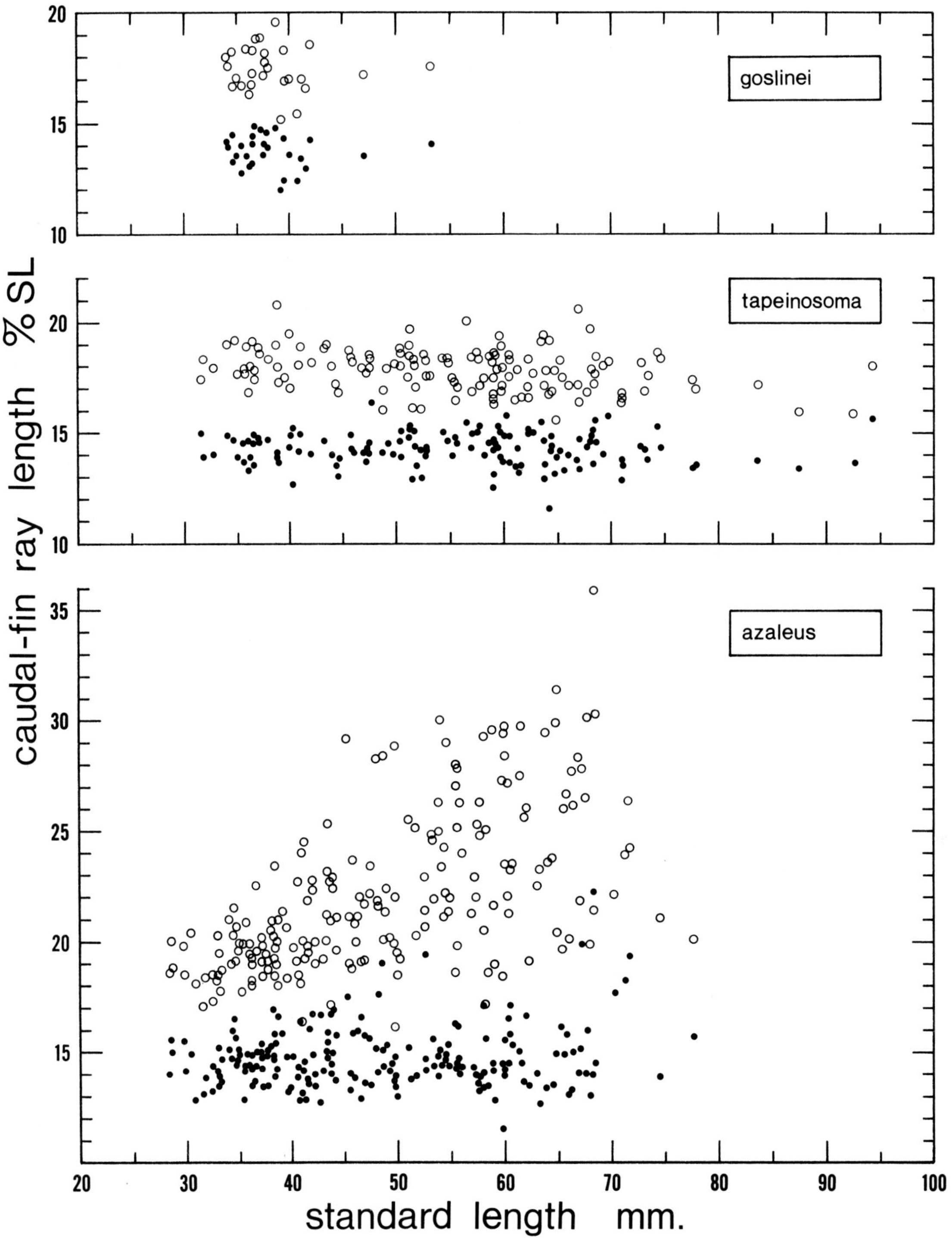

Figure 86. — Relationship between caudal-fin ray length versus standard length in members of the *Plagiotremus tapeinosoma* species group. Closed symbols = innermost segmented ray; open symbols = longest outer segmented ray.

(1956d) gives an excellent illustration of an adult from Japan. Good color photographs of adults are given in Burgess and Axelrod (1973a: 471) and Masuda *et al.* (1975: 86).

At least some populations of *P. tapeinosoma* exhibit evidence of sexual dichromatism, but more observations are needed to determine if population differences exist. Based on field observations and/or color transparencies of freshly collected specimens, the dorsal fin was bright orange anteriorly in at least some adult males from the Cocos-Keeling, Tuamotus, Gambier and Marquesas islands. The dorsal fin was dusky in at least some females from the first three islands. No data on life coloration is available for females from the Marquesas Islands, but based on a comparison of the pigmentation of the dorsal fin of preserved specimens of both sexes, the dorsal fin of the females was also orange. (In preserved specimens in which the dorsal fin is definitely known to have been orange anteriorly in life, the fin is noticeably pale anteriorly.) The dorsal fins of many well-preserved adults from other central Pacific localities, however, are not noticeably pale anteriorly and the pigmentation of the dorsal fin appears identical in both sexes. Similarly, adults of both sexes from the northern Gulf of Aqaba (where *tapeinosoma* is relatively common) that were observed on numerous occasions during August 1970 had similarly colored dorsal fins that lacked any appreciable amount of orange coloration. It would be desirable to know if these apparent differences in fin coloration merely reflect seasonal or reproductive activity (present only in nuptial individuals of one or both sexes) and if any populational differences in fin coloration exist.

Geographic variation. — Populational variation in frequencies of selected meristic characters is shown in Table 26. Specimens from the Gulf of Aqaba and northern Red Sea (specimens not available from southern Red Sea) have a well-developed dark stripe ventrally (Fig. 174) and, based on numerous observations in the Gulf of Aqaba during August 1970, do not have any bright orange coloration in the dorsal fin anteriorly (see discussion above) or on the throat and breast (the skin around the mouth may be pale orange, however). In some South African specimens the throat, at least, is bright orange (see Smith, 1949: pl. 21). No subspecific name is proposed for the Red Sea population because a broad area of intergradation seems to exist throughout most of the western Indian Ocean. Specimens from this region agree with Red Sea specimens in having high meristic values (Table 26) and a ventral stripe, although not as well defined. Only a trace of the stripe is evident in specimens from Sri Lanka; the ventral stripe is completely lacking in Cocos-Keeling and all non-Indian Ocean specimens.

Specimens from the Marquesas Islands also exhibit populational differences and perhaps warrant subspecific recognition. They have very low meristic values, at least as low as those of *goslinei* (Table 26), and also resemble *goslinei* in their diminutive size; the 23 available adults are 33-44, $\bar{x}$ 34.2 mm SL. (Adults from adjacent island groups are larger.) Unlike *goslinei,* the lateral blotches are not completely coalesced in these specimens and the dorsal fin is noticeably pale anteriorly in preservation (see discussion under color pattern above). At least some specimens from the adjacent Tuamotus and Gambier islands also have dorsal fins that are noticeably pale anteriorly. They differ, however, from the Marquesas specimens in having the posterior part of the body and the caudal fin darker. Based on color transparencies taken by John E. Randall, the posterior half of the body of a Marquesas male (excluding the lateral stripe) was white and much of the caudal fin was transparent. In a male and female from Mangareva the area above the lateral stripe on the posterior part of the body and most of the caudal fin was yellowish-brown.

Species comparisons. — *P. tapeinosoma, goslinei* and *azaleus* comprise a trio of closely related, allopatric species that can be easily distinguished collectively as a group (see Table 19). Strasburg (1956: 262) also recognized the close affinity of these three species, and suggested that the Hawaiian form (*goslinei*) was a morphological as well as geographical intermediate. I believe, however, that *tapeinosoma* and *goslinei* are much more closely related to each other than to *azaleus.*

P. tapeinosoma can be divided into at least three recognizable populations (refer to preceding discussion of geographic variation), and it could be argued that *goslinei* merely represents a fourth such population or perhaps a subspecies of *tapeinosoma.* Gos-

TABLE 26. — Geographic variation in frequencies of certain meristic characters in the *Plagiotremus tapeinosoma* species group as indicated by specimens from selected localities.

	Total vertebrae									Total dorsal-fin elements										Segmented anal-fin rays								
Species	47	48	49	50	51	52	53	N	x̄	40	41	42	43	44	45	46	47	N	x̄	27	28	29	30	31	32	33	N	x̄
tapeinosoma																												
Gulf of Aqaba				3	21	21	1	46	51.4					5	25	15	1	46	45.3				4	26	15	1	46	31.3
Seychelles					5	13	1	19	51.8						9	8	2	19	45.6					6	13		19	31.7
Aldabra					5	11	2	18	51.8						3	14	1	18	45.9					4	12	2	18	31.9
Africa					3	3	1	7	51.7						3	5	1	9	45.8					3	4	2	9	31.9
Chagos Is.					1	1	1	3	52.0							2	1	3	46.3						3		3	32.0
Cocos-Keeling Is.			1	4	2			7	50.1					4	3			7	44.4				3	3	1		7	30.7
Indonesia				3	4			7	50.6				1	3	3			7	44.3				3	3	1		7	30.7
S. China Sea				3	1			4	50.2					4				4	44.0				2	2			4	30.5
Mariana Is.			2	2	1			5	49.8				2	3				5	43.6				4	1			5	30.2
Palau-Yap Is.			8	12	3			23	49.8				6	13	4			23	43.9			5	15	3			23	29.9
Kapingamarangi			2	3				5	49.6				2	3				5	43.6			2	2	1			5	29.8
Ponape Is.			1	3	–	1		5	50.2				3	1	1			5	43.6			3	1	1			5	29.6
Marshall Is.			1	12	1			14	50.0				2	12				14	43.9			3	9	1	1		14	30.0
Gilbert Is.		1	3	8	1			13	49.7			1	4	8				13	43.6			3	9	1			13	29.8
Line Is.			16	19	2			37	49.6				19	18	1			38	43.6			19	18	1			38	29.6
Marquesas Is.		7	14	2				23	48.8			7	14	2				23	42.8		4	13	6				23	29.1
Pitcairn Group					2			2	51.0						2			2	45.0						2		2	32.0
Tuamotu-Society Is.		2	4	8	3			17	49.7			1	10	4	2			17	43.4			3	7	6			16	30.2
New Guinea				7	2			9	50.2				2	6	1			9	43.9			1	7	1			9	30.0
E. Australia				3	3	3	1	10	51.2				1	1	6	2	1	10	44.9				3	4	3		10	31.0
Lord Howe Is.				2	6	6	3	17	51.6					3	2	9	3	17	45.7				2	7	6	2	17	31.5
goslinei																												
Hawaiian Is.		5	24	6				35	49.0			8	23	5	1			37	43.0		8	17	8	2			35	29.1
azaleus																												
Gulf of Calif.	16	89	28	2				135	48.1	2	28	87	19	1				137	41.9	4	43	77	13				137	28.7
Galapagos Is.	8	27	16					51	48.2		8	34	11					53	42.1		14	33	6				53	28.8
Panama	3	10	2					15	47.9	1	5	8	1					15	41.6	2	3	8	2				15	28.7
Colombia	3	3	3					9	48.0		7	6	1					14	41.6		7	5	2				14	28.6

line and Brock (1960: 24-25) gave a thorough discussion of the taxonomic treatment of the endemic Hawaiian fishes and elected to treat as distinct species all those that could be easily and invariably distinguished. They gave two arguments for this interpretation: 1) there is no taxonomic or biological reason not to recognize them as full species (no intergrading populations exist); and 2) the Hawaiian endemics are the most distinctive offshoots of the whole Indo-West-Pacific fauna and to recognize them as subspecies would make it more difficult to deal zoogeographically with the other, relatively minor, areas of endemism within the Indo-Pacific. They emphasized that the Hawaiian endemics remain endemics regardless of what arbitrary level of taxonomic recognition is assigned to them. In the case of the two endemic species of *Plagiotremus* (*goslinei* and *ewaensis*), there is one additional reason to favor full specific recognition. It seems preferable, in the interest of stability of nomenclature, to maintain the established taxonomic rank since I have no data that unequivocally refute such an interpretation.

Nomenclature. — In the original description Ogilby did not compare *Macrurrhynchus maroubrae* with *tapeinosoma,* other than stating that both belonged in his new genus. McCulloch (1917: 92) gave a detailed redescription and illustration of Ogilby's unique type, pointing out several errors in the original description. Examination of the holotype and McCulloch's illustration clearly establishes *maroubrae* as a banded juvenile of *tapeinosoma.*

Etymology. — The specific name, derived from the Greek *tapeinos* (low, humble) and *soma* (body), refers to the slender body.

Mimicry. — Doak (1972: 105-106) described the striking similarity of *P. tapeinosoma* (as *Aspidontus maroubrae*) to an undescribed plankton feeding tripterygiid, which he called "*Tripterygion* sp. c" (= *Forsterygion* sp.). Good color photographs of both fish are given in his plate 43. He reported that on several occasions the *Plagiotremus* was observed swimming with schools of the tripterygiid, from which it attacked unsuspecting prey. Russell (1976: 328) reported that on the Great Barrier Reef *P. tapeinosoma* often swims amongst groups of the small wrasse *Thalassoma amblycephalus* (Bleeker), and in Sydney Harbor mixes with aggregations of the plesiopid *Trachinops taeniatus* Günther. In each case the resemblance is only superficial but apparently confers a real mimetic advantage; prey fishes are less likely to notice the predatory blenny.

Remarks. — At Christmas Island (Pacific Ocean) Ashmole and Ashmole (1967: 38) found *P. tapeinosoma* (15 fish in 11 of 152 samples) to be a food item of the white tern, *Gygis alba.* Although no sizes were given these fish probably were not adults, since only the banded juveniles are typically pelagic, surface dwellers.

Distribution (*Fig. 80*). — Widely distributed throughout the Indo-Pacific, but unknown from the Persian Gulf or southern Red Sea presumably due to inadequate collecting.

Material examined. — 343 specimens (29.3-121.2 mm SL) from 164 collections.

GULF OF AQABA: HUJ E65/103 (2:39), HUJ F4796 (1:40), USNM 204639 (11:34-59), USNM 204643 (3:45-54), USNM 205420 (2:49-92), USNM 214387 (3:39-51), USNM 214385 (3:35-38), USNM uncat. (3:37-41), UTAI 3819 (1:31), UTAI 3941 (2:30-32).

SINAI PENINSULA: HUJ F4752 (1:57), USNM 204638 (11:30-64), Ras Muhammad.

NORTHERN RED SEA: USNM 200338 (3:68-73), Strait of Jubal, 27°14′34″N, 33°53′55″E.

SEYCHELLES: ANSP 114808 (2:38-59), ANSP 123719 (3:45-78), RUSI uncat. (2:65-67), Mahe I.; ANSP 111017 (1:52), ANSP 111051 (11:31-73), Amirantes Is.

ALDABRA IS.: USNM 207575 (8:42-68), USNM 210624 (3:56-60), USNM uncat. (1:60), sta. HA 67-56, USNM 214386 (4:54-63), USNM uncat. (1:52), sta. HA 67-67, USNM uncat. (2:58), sta. HA 67-71.

FARQUHAR IS.: USNM 211126 (1:68), Farquhar Group, 10°7′S, 51°10′E.

AGALEGA IS.: USNM 215444 (1:48), USNM 215446 (3:34-50).

GRANDE COMORE IS.: CAS 34382 (2:43-46), CAS [illegible] (1:30).

E. AFRICA: RUSI M.202 (1:44), Mafia, RUSI uncat. (1:43), Wamizi I.; RUSI B.P.290 (1:61), RUSI B.P.494 (1:52), RUSI B.P. 679 (1:88), Pinda; RUSI 963 (1:42), Mozambique, Bazaruto I.; SAM 21422 (1:41), East London, 33°0′S, 27°54′E; SAM 22857 (1:37), False Bay, St. James, 34°10′S, 18°50′E.

MADAGASCAR: USNM 211118 (1:45), Tulear.

MAURITIUS: USNM 211119 (1:69), Mahebourg.

ST. BRANDON SHOALS (CARGADOS CARAJOS): USNM 215445 (2:49-61), USNM 215447 (3:58-62).

CHAGOS IS.: USNM 205423 (2:64-65), USNM uncat. (1:67), sta. HA 67-14, Diego Garcia Atoll.

MALDIVE IS.: FMNH 71354 (1:68), Ari Atoll.

SRI LANKA (CEYLON): USNM 207576 (1:38), Hikkaduwa.

COCOS-KEELING IS.: ANSP 128047 (1:78), ANSP 128051 (1:57), ANSP 128351 (3:39-56), ANSP 128415 (2:39-57), ANSP 128750 (1:38).

INDONESIA: RMNH 5049 (1:55), Ceram; RMNH 26469 (48.8, lectotype by present designation of *Petroskirtes tapeinosoma*), Amboina (Ambon I.); RMNH 4769 (2:51-58), originally in same jar with lectotype; RMNH 5031 (3:47-51), Banda I.; USNM 195704 (2:69-72), "Dutch East Indies-Longley", presumably Banda Sea.

W. AUSTRALIA: WAM P25050-9 (1:71), Rottnest I.

SOUTH CHINA SEA: CAS 29077 (1:53), Pulo Cecir deMir, 10°29′15″N, 108°57′30″E; CAS 29078 (2:54-65), South Vietnam, Nha Trang.

TAIWAN: USNM uncat. (1:66), sta. VGS 68-27.

RYUKYU IS.: BPBM 7262 (1:70), Okinawa, Onna Pt.

JAPAN: ANSP 128822 (1:77), ANSP 128824 (1:49), Miyake I.; USNM 148256 (1:36), No Saki, 37°8′N, 137°8′E.

MARCUS I.: BPBM 7164 (2:49-52), 26°32′N, 142°10′E.

MARIANA IS.: CAS 29059 (2:47-48), Saipan I.; ANSP 124880 (1:64), CAS 29069 (1:51), CAS 29070 (1:48), Guam I.

YAP IS.: ANSP 124855 (2:60-74), CAS 29057 (2:41-46), CAS 29073 (1:59), CAS 29074 (1:51), CAS 29061 (2:60-65), Ulithi Atoll.

PALAU IS.: ANSP 124878 (2:39-61), CAS 29054 (1:50), CAS 29055 (1:58), CAS 29056 (1:58), CAS 29058 (1:69), CAS 29060 (1:40), CAS 29067 (3:46-51), CAS 29068 (3:46-53), CAS 29071 (1:70), CAS 29072 (1:69), CAS 29075 (1:62), CAS 29076 (1:71).

IFALUK ATOLL: CAS 29048 (1:52), CAS 29049 (2:38-44), CAS 29050 (1:60).

KAPINGAMARANGI ATOLL: ANSP 124877 (1:33), CAS 29051 (1:59), CAS 29052 (2:34-43), CAS 29053 (1:55).

PONAPE IS.: ANSP 126023 (5:36-64).

WAKE I.: CAS 29079 (1:69), 19°18′N, 166°36′E.

MARSHALL IS.: BPBM 6274 (1:57), BPBM 10202 (5:48-84), Eniwetok Atoll; USNM 142225 (2:57-59), Rongelap Atoll; USNM 142227 (2:39-52), USNM 211124 (1:57), USNM 211125 (1:63), Kwajalein Atoll; USNM 211121 (1:73), Taka Atoll.

GILBERT IS.: AMS uncat. (2:41-42), sta. DFH 73-32, AMS uncat. (2:52-61), sta. DFH 73-37, AMS uncat. (2:41-53), sta. DFH 73-42, AMS uncat. (3:38-50), sta. DFH 73-51, Apaiang Atoll; USNM 167328 (3:47-52), USNM 167329 (1:52), Onotoa Atoll.

LINE IS.: ANSP 75830 (6:34-36), CAS 29046 (18:29-35), CAS 29047 (9:35-68), NMFS 134 (1:35), NMFS 1118 (1:34), USNM 207642 (3:39-58).

MARQUESAS IS.: ANSP 82161 (1:41), BPBM 10873 (1:32), BPBM 10874 (2:35-36), BPBM 10876 (11:31-34), Nuku Hiva; BPBM 10895 (1:37), Ua Pu; BPBM 10386 (1:44), Tahuata I.; BPBM 11708 (4:30-42), BPBM 11813 (2:33-40), Fatu Hiva.

PITCAIRN GROUP: BPBM 16672 (1:44), Pitcairn I.; BPBM 16514 (1:85), Oeno Atoll.

GAMBIER IS.: ANSP 124233 (1:65), 23°7′S, 134°57′W.

TUAMOTU IS.: ANSP 124236 (1:43), Rangiroa; BPBM 12467 (1:73), Duke of Gloucester Group, Anuanuraro; BPBM 12468 (2:52-62), Takaroa.

SOCIETY IS.: USNM 211116 (2:48), Bora Bora I.; USNM 211117 (3:43-48), Huahine I.; CAS 29062 (1:50), Tahiti, Teavaraa Pass; BPBM 13232 (2:50-54), CAS 29063 (2:40-45), CAS 29064 (1:47), CAS 29065 (1:53), CAS 29066 (1:64), Tahiti, Moorea.

COOK IS.: BPBM 13924 (1:54), Rarotonga I.

TOKELAU IS.: USNM 211120 (3:46-54), Nokunono Atoll.

FIJI IS.: USNM 176617 (1:94), Kandaver I., 18°35′S, 178°40′E; USNM 211127 (1:57), Great Astrolabe, 18°52′S, 178°20.5′E.

SOLOMON IS.: CAS 5652 (2:51-53), Stewart Is., Sikaiana I.; USNM 195798 (1:55), USNM 195799 (1:42), USNM 211122 (1:34), New Georgia I.

NEW GUINEA: USNM 205688 (1:35), USNM 205690 (1:55), Trobriand Is.; USNM 205687 (2:40-43), USNM 205689 (1:39), Madang Harbor, Massas I.; DASF FO1394 (1:47), Papua, Gaire Reef.

E. AUSTRALIA: ANSP 109673 (1:58), Endeavour Reef, 15°45′S, 145°42′E; QMB I.6789 (1:78), Keeper Reef, 18°45′S, 147°16′E; USNM 211111 (1:58), Great Barrier Reef, Capricorn Group, Heron I.; USNM 211123 (3:48-71), "Australia", probably Heron I.; USNM uncat. (1:45), sta. VGS 66-11, USNM uncat. (1:71), sta. VGS 66-13, Capricorn Group, One Tree I.; AMS IB. 5748 (1:59), Sydney area, Little Manly; USNM 211115 (1:45), Sydney, Port Jackson; AMS I. 3504 (43.4, holotype of *Macrurrhynchus maroubrae*), Sydney, Maroubra Bay; AMS IB. 4200 (1:40), Sydney, Collaroy Beach.

NEW CALEDONIA: ANSP 124324 (1:39).

NORFOLK I.: AMS IB. 6417-8 (2:47-61), 29°5′S, 167°59′E.

ELIZABETH REEF: NMNZ 4807 (1:52), 29°58′S, 159°13′E.

LORD HOWE I.: AMS IB. 5495-6 (2:60-64), AMS IB. 6384 (1:42), AMS uncat. (13:55-62), sta. 73-4, AMS uncat. (1:121), sta. 73-51.

NEW ZEALAND: NMNZ 5128 (1:49), Poor Knights Is.

Plagiotremus (*Runula*) *goslinei* (Strasburg)
Specimen Figure 177; Tables 4, 19-23, 26.

Runula goslinei Strasburg, 1959: 260, Fig. 4 (type locality: Hawaiian Islands, Oahu, Waimea Coast, Pupukae; holotype USNM 164202).

Description. — Dorsal fin VII-VIII, 34-37; total elements 42-45. Anal fin II, 28-31. Pectoral fin 12. Caudal fin: procurrent rays 6-8 + 6-8 = 12-16; segmented rays 6 + 5. Vertebrae: precaudal 15-16; caudal 33-35; total 48-50. Pleural ribs on vertebrae 3-15; epipleural ribs 15-17. Incisor teeth (Figs. 74 and 77), in specimens 33-53 mm, n = 36: lower jaw 37-64; upper jaw 18-33; ratio (lower/upper) 1.7-2.4, mean 1.98.

Dentary incisors with tips ovoid or bluntly lanceolate (Fig. 72d). Snout fleshy, conical (in adults). Interorbital pores 2; supraorbital pores 2; mandibular pores 2. Preanal length 42.2-45.7% SL. Pelvic fins sexually dimorphic in adults (Fig. 85): males 11.4-16.6% SL, females 9.0-13.6% SL. Caudal fin not sexually dimorphic 15.2-19.6% SL. Outer lobes of fin not elongate (Fig. 86); innermost ray not noticeably longer than adjacent rays.

Color pattern. — Adults with dorsal half of head and body brown, ventral half white; the two areas sharply separated at about level of ventral margin of pupil. Dark stripe, approximately three-fourths eye diameter in width, extends from eye to caudal fin where stripe narrows and extends to margin of fin as a streak covering innermost caudal-fin ray. Stripe either uniformly pigmented or superimposed with 10-18 pigment intensifications giving stripe a spotted appearance. A narrow pale stripe (usually not apparent in preserved specimens) extends from each side of snout (dorsally) to dorsolateral region of caudal peduncle; pale stripe touches dorsal margin of eye and dark mid-lateral stripe. The dark area above pale stripe, which is paler than the mid-lateral stripe, narrows posteriorly and may disappear in area of caudal peduncle. A median, narrow pale stripe (also usually not apparent in preserved specimens) extends from near anterior margin of orbits to dorsal-fin origin. Dorsal fin hyaline for proximal two-fifths, then a broad submarginal brown stripe, and white distal margin. Anal fin uniformly dark or with proximal fourth hyaline, then a broad submarginal brown stripe, and white distal margin. Innermost ray of caudal fin dark, other rays pale or with faint dusky outline; membranes of fin hyaline. A good illustration showing the typical adult color pattern was given in the original description.

The banded juveniles are virtually identical to those of *tapeinosoma* and *azaleus*.

In life the dark mid-lateral stripe is brownish-black, including the innermost ray of the caudal fin.

Snout and body below mid-lateral stripe white, except for orange throat, breast and skin around mouth. Median predorsal stripe and narrow stripe extending from dorsum of snout to dorsolateral region of caudal peduncle pale greenish-blue; remainder of dorsum light brown. Dorsal and anal fins transparent or yellowish proximally; followed by broad, submarginal, brown stripe and narrow, white border. Innermost ray of caudal fin dark brown, bordered by yellowish white, remainder of fin dusky or transparent. Pelvic fins white; pectorals transparent.

No orange spot, such as occurs in at least some males of *tapeinosoma* from the central Pacific, was present in the few adult males for which color observations or transparencies are available, and the fin pigmentation of the seven well-preserved adult males gives no indication that such a spot ever existed. Refer to discussion of sexual dichromatism under account of *P. tapeinosoma.*

Species comparisons. — Refer to account of *P. tapeinosoma.*

Etymology. — Named for Dr. William A. Gosline, coauthor of Handbook of Hawaiian fishes and former professor of biology (retired) at the University of Hawaii.

Distribution. — Endemic to the Hawaiian Islands.

Material examined. — 37 specimens (32.7-53.3 mm SL) from 10 collections.

OAHU I.: USNM 164202 (42.0, holotype of *Runula goslinei*), Waimea Coast, Pupukea; USNM 164203 (1:37), taken with holotype; USNM 164204 (1:37), Makapuu Beach; USNM 108503 (1:38), Waialua, reef off Mokuleia; USNM 203274 (9:35-41), Pokai Bay, ca. 1.5 km offshore, "night light and dip net"; ANSP 128931 (3:33-36), Kaneohe Bay, from stomach contents of *Coryphaena hippurus;* BPBM 7451 (2:40-42), off Pokai Bay; BPBM 9856 (1:39), off Lahilahi Point; BPBM 9960 (16:35-53), Waimea Bay.

KAUAI I.: ANSP 128930 (1:34), Nualolo.

NIHOA I.: BPBM 9165 (1:34), offshore, "night light and dip net."

Plagiotremus (*Runula*) *azaleus* (Jordan and Bollman)

Specimen Figures 178-179; Tables 4, 19-23, 26.

Runula azalea Jordan and Bollman, 1890: 171 (type locality: Galapagos Islands, Indefatigable Island; holotype USNM 44299).

Atopoclinus ringens Vaillant, 1894: 73 (type locality: Gulf of California; holotype MNHN 1894-63).

Runula albolinea Nichols, 1924: 64, Fig. 11 (type locality:

Galapagos Islands, Conway Bay, N. W. Coast, Indefatigable Island; holotype AMNH 8271).

Petroscirtes belanskei Borodin, 1928: 33, pl. 3, Fig. 3 (type locality: Canos Island, Costa Rica; holotype lost).

Petroscirtes panamensis Fowler, 1944: 291, Fig. 246 (type locality: Saboga, Perlas Islands, Panama; holotype ANSP 69995).

Description. — Dorsal fin VII-IX, 32-36; total elements 40-44. Anal fin II, 27-30. Pectoral fin 11-13 (typically 12). Caudal fin: procurrent rays 6-9 + 6-9 = 12-18; segmented rays 6 + 5. Vertebrae: precaudal 14-16; caudal 32-35; total 47-50. Pleural ribs on vertebrae 3-14 or 15; epipleural ribs 15-20. Incisor teeth (Figs. 74 and 77), in specimens 30-72 mm, n = 108: lower jaw 40-70; upper jaw 20-38; ratio (lower/upper) 1.6-2.4, mean 1.89.

Dentary incisors with tips ovoid or bluntly lanceolate (Fig. 72d). Snout fleshy, conical (in adults). Interorbital pores 2; supraorbital pores 2; mandibular pores 2. Preanal length 38.3-47.7% SL. Pelvic fins sexually dimorphic in adults (Fig. 85): males 7.2-16.2% SL, females 6.3-12.4% SL. Caudal fin also sexually dimorphic: males (50, 51.0-71.6 mm) 18.6-35.9, mean 25.8% SL; females (31, 52.3-77.8) 18.6-26.4, mean 22.2% SL, not including one female 68.4 mm with caudal fin 30.3% SL. Outer lobes of fin elongate in adults (Fig. 86); innermost ray noticeably longer than adjacent rays in large adults (both sexes).

Color pattern. — The immaculate anal fin, intense black stripe in the dorsal fin and the mid-lateral stripe that is poorly developed or absent anteriorly distinguishes adults of *azaleus* from its close relatives *tapeinosoma* and *goslinei*. Like those species, *azaleus* has narrow pale stripes on the head (Fig. 82d) (usually not apparent in preserved specimens) and the innermost caudal-fin ray heavily pigmented at least proximally. Herre (1936b: 403) described the life colors as follows:

"The color in life is clear reddish brown above and to the middle of the side, pale reddish below, the belly and under side of the head white. A pearly blue line extends from the top of the snout over the eye back to the upper part of the caudal base. From the tip of the snout a bright silver band extends back below the eye (and including its lower third) to the pectoral base. The dorsal is clear with a deep brown, marginal band, broadest on the anterior half, and a very narrow blue edge. The caudal has a central longitudinal blackish stripe and a very narrow black line on its upper and lower margins. The other fins are all colorless."

Jordan and Bollman described the juvenile holotype of *azaleus* as "reddish brown, silvery below; a black spot surrounded by paler at base of caudal; dorsal with about 6 black cross-bars; anal with four; other fins pale." A good illustration of the adult was given by Nichols (1924: 65); and Borodin (1928: pl. 3) illustrated the juvenile.

Species comparisons. — Refer to discussion under account of *P. tapeinosoma*.

Nomenclature. — Many previous workers believed that the banded juveniles and striped adults of *azaleus* were different species. This confusion was probably largely due to the fact that, as in most fishes with conspicuously dissimilar juveniles and adults, there is considerable overlap in size at transformation with relatively few specimens exhibiting transitional color patterns. (This situation has resulted in different names being proposed for the juveniles and adults of four other species of Nemophini in addition to *azaleus: Aspidontus taeniatus, A. dussumieri, Plagiotremus tapeinosoma* and *P. rhinorhynchos.*) There is at least a 10-mm overlap in size between the largest banded juvenile *azaleus* and the smallest striped "adult"; smallest striped specimen approximately 28.7 mm SL. Schultz (1950) added to the confusion when he distinguished *ringens* (adults) from *azaleus* (juveniles) on the basis of differences in the number of teeth in addition to the color pattern. He was unaware that the number of teeth changes with growth (Fig. 77).

Only Fowler compared his new species with similarly pigmented specimens of any other nominal species referable to *azaleus*. The relatively trivial differences that he emphasized were based entirely on a comparison with an illustration. *Runula azalea* and *Petroscirtes belanskei* were described from juveniles; the other nominal species were based on adults. The holotype of *P. belanskei* is apparently lost (*in litt.*, W. Fasbender, Director Vanderbilt Museum), but a good illustration accompanied the original description.

Etymology. — The specific name, according to Jordan and Evermann (1898: 2377), is derived from the Greek *azale,* meaning parched, in allusion to the brown coloration. The specific name is here spelled *azaleus* because the gender of *Plagiotremus* is masculine.

Mimicry. — Hobson (1968: 42-43; 1969) gives

an excellent discussion of the association of *P. azaleus* with the wrasse *Thalassoma lucasanum* (Gill), which it superficially resembles. He discussed several possible advantages that *azaleus* might gain from the association but avoided referring to the relationship as mimicry. In this study mimicry is defined in the broad sense and includes the interaction described by Hobson. Refer to general section on mimicry for discussion.

Distribution (*Fig. 80*). — Endemic to the eastern Pacific where it occurs from the Gulf of California to northern Peru.

Material examined. — 275 specimens (14.2-77.8 mm SL) from 52 collections.

GULF OF CALIFORNIA: MNHN 1894-63 (34.6, holotype of *Atopoclinus ringens*), LACM 1165 (4:46-55), LACM 1608 (3:35-62), SIO 59-225 (11:28-68), SIO 61-249 (10:43-67), SIO 62-55 (13:35-52), SIO 62-713 (7:56-59), SIO 62-726 (7:51-63), SIO 61-250 (6:54-66), SIO 64-60 (1:72), SIO 65-299 (27:33-75), SIO 65-351 (12:30-68), UCLA W59-249 (7:38-59), UCLA W61-31 (5:53-78), UCLA W64-84 (3:41-67), UCLA W65-77 (21:37-72), USNM 200392 (4:44-54).

MEXICO: SIO 62-66 (3:50-56), off Nayarit, Isabela I., 21°52′N, 105°54′W; SIO 62-43 (3:50-54), Jalisco, Banderas Bay, 20°33′N, 105°18′W; LACM 9056-21 (1:60), Manzanillo, 19°0′N, 104°20′W; USNM 65643 (9:28-45), Acapulco.

OFF MEXICO ("OCEANIC"): SIO 67-269 (1:14), 20°-21°N, 108°-109°W.

REVILLAGIGEDO IS.: LACM 30313-3 (1:32), Socorro I.

COSTA RICA: LACM 6894-11 (2:28-52), Playa del Coco, 10°34′N, 85°42′W.

PANAMA: SIO 70-140 (6:29-42), Gulf of Chiriqui, Secas I.; ANSP 69995 (36.5, holotype of *Petroscirtes panamensis*), Perlas Is., Saboga; ANSP 102682 (9:32-49), Pinas Bay, 7°33.5′N, 78°12′W.

MALPELO I.: USNM 207641 (1:35), 4°3′N, 81°31′W.

COLOMBIA: ANSP 102655 (13:33-44), USNM 256392-F6 (4:37-47), Gorgona I., 3°0.5′N, 78°11′W.

GALAPAGOS IS.: USNM 44299 (38.2, holotype of *Runula azalea*); AMNH 8271 (42.3, holotype of *Runula albolinea*); ANSP 85711 (3:31-34), LACM 8170 (6:35-58), SU 1848 (1:35), SU 17111 (2:38-40), UCLA W67-43 (14: 34-53), USNM 65644 (2:35-56), USNM 89776 (2:38-43), USNM 92948 (14:34-39), USNM 213476 (8:35-51), USNM 213546 (1:40), USNM 213547 (1:58), USNM 213548 (1:35), USNM 213549 (5:30-37), USNM 213550 (1:49), USNM 213551 (2:42-43), USNM 213552 (3:38-39), USNM 213553 (2:43-55), USNM 213554 (5:35-42).

PERU: IMP uncat. (1:39), Tumbes, Puerto Pizarro, 3°34′S, 80°28′W; UCLA W58-162 (3:35-39), Lobos de Afuera I., 6°57′S, 80°43′W.

LIST OF NOMINAL SPECIES OF NEMOPHINI

Nominal Species Of Tribe Nemophini, Including Some Names Originally Assigned To Nominal Genera Of The Nemophini That Were Misapplied

The following list gives in order (1) the scientific name, as it originally appeared, in alphabetical order according to the trivial name; (2) the author or authors (Cuvier and Valenciennes, 1836 is abbreviated to C. & V., 1836); (3) date of publication; (4) page references (see literature cited for complete reference); (5) present identification. Refer to text discussion under senior synonym for basis of determination for species assigned to tribe Nemophini. Refer to Springer and Gomon (1975) for discussion of nominal species assigned to *Omobranchus* or *Haptogenys* (tribe Omobranchini), and to Springer (1972) for species assigned to *Enchelychurus* (tribe Omobranchini); Springer and Smith-Vaniz (1972b) for discussion of *Petroscirtes heyligeri* (tribe Phenablenniini); and Springer (1967) for species assigned to *Hypsoblennius* (tribe Blenniini).

Species, author, publication date, and page references	Present identification
Runula albolinea Nichols, 1924: 64	*Plagiotremus azaleus* (Jordan and Bollman, 1890)
Petroscirtes altivelis Steindachner, 1863: 1191	(Omobranchini) *Omobranchus anolius* (Valenciennes, 1836)
Petroskirtes amblyrhynchos Bleeker, 1857a: 64	*Plagiotremus rhinorhynchos* (Bleeker, 1852)
Petroskirtes amboinensis Bleeker, 1852a: 144	*Petroscirtes variabilis* Cantor, 1850
Petroscirtes ancylodon Rüppell, 1838: 1	*Petroscirtes ancylodon* Rüppell, 1838

Species, author, publication date, and page references	Present identification
Petroskirtes anema Bleeker, 1852a: 273	*Meiacanthus anema* (Bleeker, 1852)
Petroscirtes annamensis Chabanaud, 1924: 363	*Petroscirtes breviceps* (Valenciennes, 1836)
Petroscirtes anolis Günther, 1861: 238 [misspelling of *anolius*]	(Omobranchini) *Omobranchus anolius* (Valenciennes, 1836)
Blennechis anolius Valenciennes, in C. & V., 1836: 288	(Omobranchini) *Omobranchus anolius* (Valenciennes, 1836)
Petroscirtes ater Günther, 1877: 199	(Omobranchini) *Enchelyurus ater* (Günther, 1877)
Petroscirtes atrodorsalis Günther, 1877: 198	*Meiacanthus atrodorsalis* (Günther, 1877)
Petroscirtes auratus Seale, 1936: 85	*Meiacanthus atrodorsalis* (Günther, 1877)
Runula azalea Jordan and Bollman, 1890: 171	*Plagiotremus azaleus* (Jordan and Bollman, 1890)
Petroscirtes azureus Jordan and Seale, 1906: 432	*Aspidontus taeniatus* Quoy and Gaimard, 1834
Petroskirtes bankanensis Bleeker, 1852b: 727	*Petroscirtes variabilis* Cantor, 1850
Petroscirtes bankieri Richardson, 1845: 136	*Petroscirtes breviceps* (Valenciennes, 1836)
Petroskirtes barbatus Peters, 1855a: 248 and 1855b: 440	*Petroscirtes mitratus* Rüppell, 1830
Petroscirtes belanskei Borodin, 1928: 33	*Plagiotremus azaleus* (Jordan and Bollman, 1890)
Petroscirtes bhattacharyae Chaudhuri, 1916: 107	(Omobranchini) *Omobranchus zebra* (Bleeker, 1868)
Blennechis biocellatus Valenciennes, in C. & V., 1836: 289	(Blenniini) *Hypsoblennius sordidus* (Bennett, 1828)
Petroscirtes bipunctatus Day, 1787: 327	(Omobranchini) *Haptogenys bipunctata* (Day, 1787)
Blennechis breviceps Valenciennes, in C. & V., 1836: 283	*Petroscirtes breviceps* (Valenciennes, 1836)
Meiacanthus (*Meiacanthus*) *bundoon* Smith-Vaniz, 1976: 107	*Meiacanthus* (*Meiacanthus*) *bundoon* Smith-Vaniz, 1976
Lembeichthys celebesensis Herre, 1936a: 283	*Plagiotremus celebesensis* (Herre, 1936)
Petroscirtes cristatus Zugmayer, 1913: 20	(Omobranchini) *Omobranchus mekranensis* (Regan, 1905)
Petroscirtes cristiceps Macleay, 1881: 9	(Omobranchini) *Omobranchus anolius* (Valenciennes, 1836)
Petroscirtes cynodon Peters, 1855a: 247; 1855b: 439	*Petroscirtes breviceps* (Valenciennes, 1836)
Blennechis cyprinoides Valenciennes, in C. & V., 1836: 286	*Petroscirtes breviceps* (Valenciennes, 1836)
Aspidontus dasson Jordan and Snyder, 1902: 456	(Omobranchini) *Omobranchus punctatus* (Valenciennes, 1836)
Petroscirtes dispar Günther, 1861: 232	(Omobranchini) *Omobranchus punctatus* (Valenciennes, 1836)
Petroscirtes dispar Fowler, 1937: 258 [junior primary homonym of *Petroscirtes dispar* Günther]	(Omobranchini) *Omobranchus elongatus* (Peters, 1855)
Meiacanthus (*Allomeiacanthus*) *ditrema* Smith-Vaniz, 1976: 87	*Meiacanthus* (*Allomeiacanthus*) *ditrema* Smith-Vaniz, 1976
Dasson duperreyi Whitley, 1945: 31	*Petroscirtes breviceps* (Valenciennes, 1836)
Blennechis dussumieri Valenciennes in C. & V., 1836: 282	*Aspidontus dussumieri* (Valenciennes, 1836)
Petroscirtes elatus Jordon and Snyder, 1902: 452	*Petroscirtes mitratus* Rüppell, 1830
Petroscirtes elegans Steindachner, 1876: 217	(Omobranchini) *Omobranchus elegans* (Steindachner, 1876)
Petroscirtes elongatus Peters, 1855a: 265	(Omobranchini) *Omobranchus elongatus* (Peters, 1855)
Petroscirtes eretes Jordan and Seale, 1907: 47	*Petroscirtes variabilis* Cantor, 1850
Petroscirtes ewaensis Brock, 1948: 125	*Plagiotremus ewaensis* (Brock, 1948)
Blennechis fasciatus Jenyns, 1842: 84	(Blenniini) *Hypsoblennius sordidus* (Bennett, 1828)
Runula fasciatus Fourmanoir, 1971: 496	*Plagiotremus rhinorhynchos* (Bleeker, 1852)
Blennechis fasciolatus Valenciennes in C. & V., 1836: 287	(Omobranchini) *Ombranchus fasciolatus* (Valenciennes, 1836)
Petroscirtes fasciolatus Macleay, 1881: 8 [secondary junior homonym of *O. fasciolatus* (Valenciennes)]	(Omobranchini) *Omobranchus rotundiceps* (Macleay, 1881)
Petroscirtes feliciana Herre, 1942: 112	(Omobranchini) *Omobranchus ferox* (Herre, 1927)
Petroscirtes ferox Herre, 1927: 277	(Omobranchini) *Omobranchus ferox* (Herre, 1927)
Blennechis filamentosus Valenciennes in C. & V., 1836: 280	*Aspidontus taeniatus* Quoy and Gaimard, 1834
Petroscirtes (*Dasson*) *fallax* Smith-Vaniz, 1976: 42	*Petroscirtes* (*Dasson*) *fallax* Smith-Vaniz, 1976
Petroscirtes fluctuans Weber, 1909: 146	*Aspidontus dussumieri* (Valenciennes, 1836)
Meiacanthus (*Meiacanthus*) *fraseri* Smith-Vaniz, 1976: 98	*Meiacanthus* (*Meiacanthus*) *fraseri* Smith-Vaniz, 1976
Lembeichthys furcocaudalis Sze-Chung, 1957: 229	*Plagiotremus spilistius* Gill, 1865
Meiacanthus (*Meiacanthus*) *geminatus* Smith-Vaniz, 1976: 95	*Meiacanthus* (*Meiacanthus*) *geminatus* Smith-Vaniz, 1976
Petroscirtes germaini Sauvage, 1883: 158	(Omobranchini) *Omobranchus germaini* (Sauvage, 1883)
Petroscirtes gorrorensis Herre, 1936a: 285	*Aspidontus dussumieri* (Valenciennes, 1836)
Runula goslinei Strasburg, 1959: 260	*Plagiotremus goslinei* (Strasburg, 1959)

Species, author, publication date, and page references	Present identification
Blennechis grammistes Valenciennes in C. & V., 1836: 284	*Meiacanthus grammistes* (Valenciennes, 1836)
Petroscirtes guttatus Macleay, 1881:9	(Omobranchini) *Omobranchus anolius* (Valenciennes, 1836)
Petroscirtes herlihyi Fowler, 1946: 183	*Meiacanthus atrodorsalis* (Günther, 1877)
Petroscirtes heyligeri Bleeker, 1859: 340	(Phenablenniini) *Phenablennius heyligeri* (Bleeker, 1859)
Petroskirtes hypselopterus Bleeker, 1855: 423	*Petroscirtes mitratus* Rüppell, 1830
Petroscirtes icelli Ogilby, 1894: 370	*Petroscirtes lupus* De Vis, 1886
Plagiotremus (*Plagiotremus*) *iosodon* Smith-Vaniz, 1976: 128	*Plagiotremus* (*Plagiotremus*) *iosodon* Smith-Vaniz, 1976
Petroscirtes japonicus Bleeker, 1869: 246	(Omobranchini) *Omobranchus punctatus* (Valenciennes, 1836)
Petroscirtes kallosoma Bleeker, 1858: 227	(Omobranchini) *Omobranchus elongatus* (Peters, 1855)
Meiacanthus kamoharai Tomiyama, 1956: 1083	*Meiacanthus kamoharai* Tomiyama, 1956
Petroscirtes kochi Weber, 1908: 263	(Omobranchini) *Omobranchus punctatus* (Valenciennes, 1836)
Petroscirtes kranjiensis Herre, 1940: 25	(Omobranchini) *Omobranchus ferox* (Herre, 1927)
Petroscirtes kraussi Klunzinger, 1871: 497	(Omobranchini) *Enchelyurus kraussi* (Klunzinger, 1871)
Petroscirtes kulambangrae Herre, 1935: 436	*Meiacanthus anema* (Bleeker, 1852)
Pescadorichthys (*Musgravius*) *laudandus* Whitley, 1961: 63	*Plagiotremus laudandus* (Whitley, 1961)
Nemophis lessoni Kaup, 1858: 168	*Xiphasia setifer* Swainson, 1839
Petroscirtes lienardi Day, 1878: 327	*Petroscirtes mitratus* Rüppell, 1830
Petroscirtes lineatus De Vis, 1884: 698	*Meiacanthus lineatus* (De Vis, 1884)
Petroscirtes lineolatus Kner, 1868a: 29; 1868b: 331	(Omobranchini) *Omobranchus lineolatus* (Kner, 1868)
Petroscirtes lineo-punctatus Sauvage, 1880: 216	(Omobranchini) *Omobranchus elegans* (Steindachner, 1876)
Petroscirtes lini Herre, 1934: 292	(Omobranchini) *Omobranchus aurosplendidus* (Richardson, 1846)
Petroscirtes longifilis Kner and Steindachner, 1867: 367	*Petroscirtes mitratus* Rüppell, 1830
Petroscirtes loxozonus Jordan and Seale, 1905: 802	(Omobranchini) *Omobranchus loxozonus* (Jordan and Seale, 1905)
Salarius lupus De Vis, 1886: 58	*Petroscirtes lupus* (De Vis, 1886)
Petroscirtes macleayi Ogilby, 1887: 38 [replacement name for *Petroscirtes fasciolatus* Macleay]	(Omobranchini) *Ombranchus rotundiceps* (Macleay, 1881)
Xiphogadus madagascariensis Playfair, 1868: 11	*Xiphasia setifer* Swainson, 1839
Petroscirtes (*Dasson*) *marginatus* Smith-Vaniz, 1976: 43	*Petroscirtes* (*Dasson*) *marginatus* Smith-Vaniz, 1976
Petroskirtes marmoratus Bleeker, 1875: 74	*Petroscirtes mitratus* Rüppell, 1830
Macrurrhynchus maroubrae Ogilby, 1896: 137	*Plagiotremus tapeinosoma* (Bleeker, 1857)
Petroscirtes masyae H. M. Smith, 1934: 316	(Omobranchini) *Omobranchus punctatus* (Valenciennes, 1836)
Xiphasia matsubarai Okada and Suzuki, 1952: 75	*Xiphasia matsubarai* Okada and Suzuki, 1952
Petroscirtes mekranensis Regan, 1905: 328	(Omobranchini) *Omobranchus mekranensis* (Regan, 1905)
Petroscirtes mitratus Rüppell, 1830: 111	*Petroscirtes mitratus* Rüppell, 1830
Meiacanthus mossambicus Smith, 1959: 233	*Meiacanthus mossambicus* Smith, 1959
Meiacanthus nigrolineatus Smith-Vaniz, 1969: 351	*Meiacanthus nigrolineatus* Smith-Vaniz, 1969
Petroscirtes obliquus Garman, 1903: 237	(Omobranchini) *Omobranchus rotundiceps* (Macleay, 1881)
Petroscirtes oualanensis Günther, 1880: 35	*Meiacanthus atrodorsalis* (Günther, 1877)
Petroscirtes panamensis Fowler, 1944: 291	*Plagiotremus azaleus* (Jordan and Bollman, 1890)
Petroskirtes paradiseus Bleeker, 1854: 495	*Aspidontus taeniatus* Quoy and Gaimard, 1834
Petroscirtes petersi Kossman and Raüber, 1877: 21	(Omobranchini) *Enchelyurus petersi* (Kossman and Raüber, 1877)
Meiacanthus (*Meiacanthus*) *phaeus* Smith-Vaniz, 1976: 107	*Meiacanthus* (*Meiacanthus*) *phaeus* Smith-Vaniz, 1976
Plagiotremus (*Musgravius*) *phenax* Smith-Vaniz, 1976: 131	*Plagiotremus* (*Musgravius*) *phenax* Smith-Vaniz, 1976
Petroscirtes pindae Smith, 1959: 231	*Petroscirtes xestus* Jordan and Seale, 1906
Blennechis polyodon Bleeker, 1851a: 254	*Petroscirtes breviceps* (Valenciennes, 1836)
Meiacanthus (*Meiacanthus*) *procne* Smith-Vaniz, 1976: 105	*Meiacanthus* (*Meiacanthus*) *procne* Smith-Vaniz, 1976
Blennechis punctatus Valenciennes in C. & V., 1836: 286	(Omobranchini) *Omobranchus punctatus* (Valenciennes, 1836)
Petroscirtes quadrimaculatus Kendall and Goldborough, 1911: 329	*Aspidontus dussumieri* (Valenciennes, 1836)

Species, author, publication date, and page references	Present identification
Meiacanthus (*Meiacanthus*) *reticulatus* Smith-Vaniz, 1976: 106	*Meiacanthus* (*Meiacanthus*) *reticulatus* Smith-Vaniz, 1976
Petroskirtes rhinorhijnchos Bleeker, 1852a: 273	*Plagiotremus rhinorhynchus* (Bleeker, 1852)
Atopoclinus ringens Vaillant, 1894: 74	*Plagiotremus azaleus* (Jordan and Bollman, 1890)
Petroskirtes rostratus Bleeker, 1857a: 62	*Aspidontus taeniatus* Quoy and Gaimard, 1834
Petroscirtes rotundiceps Macleay, 1881: 9	(Omobranchini) *Omobranchus rotundiceps* (Macleay, 1881)
Petroscirtes semilineatus Kner, 1868b: 333	(Omobranchini) *Omobranchus punctatus* (Valenciennes, 1836)
Xiphasia setifer Swainson, 1839: 259	*Xiphasia setifer* Swainson, 1839
Meiacanthus smithi Klausewitz, 1961: 17	*Meiacanthus smithi* Klausewitz, 1961
Petroskirtes solorensis Bleeker, 1853c: 81	*Petroscirtes variabilis* Cantor, 1850
Plagiotremus spilistius Gill, 1865: 140	*Plagiotremus spilistius* Gill, 1865
Petroscirtes (*Dasson*) *springeri* Smith-Vaniz, 1976: 37	*Petroscirtes* (*Dasson*) *springeri* Smith-Vaniz, 1976
Petroscirtes (*Ostreoblennius*) *steadi* Whitley, 1930: 20	*Petroscirtes lupus* (De Vis, 1886)
Petroscirtes striatus Day, 1888: 262	*Aspidontus dussumieri* (Valenciennes, 1836)
Petroscirtes striatus Jatzow and Lenz, 1898: 512 [junior primary homonym of *Petroscirtes striatus* Day]	(Omobranchini) *Omobranchus fasciolatoceps* (Richardson, 1846)
Aspidontus taeniatus Quoy and Gaimard, 1834: 719	*Aspidontus taeniatus* Quoy and Gaimard, 1834
Petroskirtes tapeinosoma Bleeker, 1857a: 64	*Plagiotremus tapeinosoma* (Bleeker, 1857)
Petroskirtes temminckii Bleeker, 1851b: 243	*Meiacanthus grammistes* (Bleeker, 1851)
Petroskirtes thepassi Bleeker, 1853b: 136	*Petroscirtes thepassi* Bleeker, 1853
Petroscirtes townsendi Regan, 1905: 328	*Plagiotremus townsendi* (Regan, 1905)
Xiphasia trachypareia Bleeker, 1864: 196; 1865: 196	*Xiphasia setifer* Swainson, 1839
Aspidontus tractus Fowler, 1903: 170	*Aspidontus taeniatus* Quoy and Gaimard, 1834
Aspidontus trossulus Jordan and Snyder, 1902: 455	*Petroscirtes breviceps* (Valenciennes, 1836)
Petroscirtes uekii Katayama, 1941: 591	(Omobranchini) *Omobranchus fasciolatoceps* (Richardson, 1846)
Petroscirtes variabilis Cantor, 1850: 1182	*Petroscirtes variabilis* Cantor, 1850
Petroscirtes vinciguerrae Borsieri, 1904: 211	(Omobranchini) *Omobranchus fasciolatus* (Valenciennes, 1836)
Salarias viperidens De Vis, 1884: 697	*Petroscirtes variabilis* Cantor, 1850
Meiacanthus (*Meiacanthus*) *vittatus* Smith-Vaniz, 1976: 93	*Meiacanthus* (*Meiacanthus*) *vittatus* Smith-Vaniz, 1976
Petroscirtes vulsus Jordan and Seale, 1907: 47	*Petroscirtes variabilis* Cantor, 1850
Aspidontus (*Escadotus*) *wamiziensis* Smith, 1959: 235	*Aspidontus dussumieri* (Valenciennes, 1836)
Petroscirtes waterousi Herre, 1942: 112	(Omobranchini) *Omobranchus ferox* (Herre, 1927)
Aspidontus woodi Gilchrist and Thompson, 1908: 105	(Omobranchini) *Omobranchus woodi* (Gilchrist and Thompson, 1908)
Petroscirtes wilsoni Macleay, 1885: 171	(Omobranchini) *Omobranchus anolius* (Valenciennes, 1836)
Petroscirtes xestus Jordan and Seale, 1906: 433	*Petroscirtes xestus* Jordan and Seale, 1906
Petroscirtes zebra Bleeker, 1868:279	(Omobranchini) *Omobranchus zebra* (Bleeker, 1868)

SPECIATION AND ZOOGEOGRAPHY

General comments on Zoogeography have already been given for each genus. The distribution of the species of Nemophini is summarized in Table 27. Figure 87 shows the geographic scheme used in the following discussion and in Table 27. The boundary between the western and central Pacific oceans is arbitrary, but hopefully represents the most natural separation of the two regions in terms of fish distributions. For example, the Palau-Yap Islands are included in the western Pacific faunal region because six of the 13 species collected there are broadly distributed in the western Pacific but have not been collected elsewhere in Oceania.

Plagiotremus azaleus, an eastern Pacific endemic, is the only species of Nemophini found outside the Indo-West Pacific Region. The greatest number of species of Nemophini occur in the western Pacific Ocean, of which slightly more than 50 percent (16 species) are restricted to that ocean. At least one

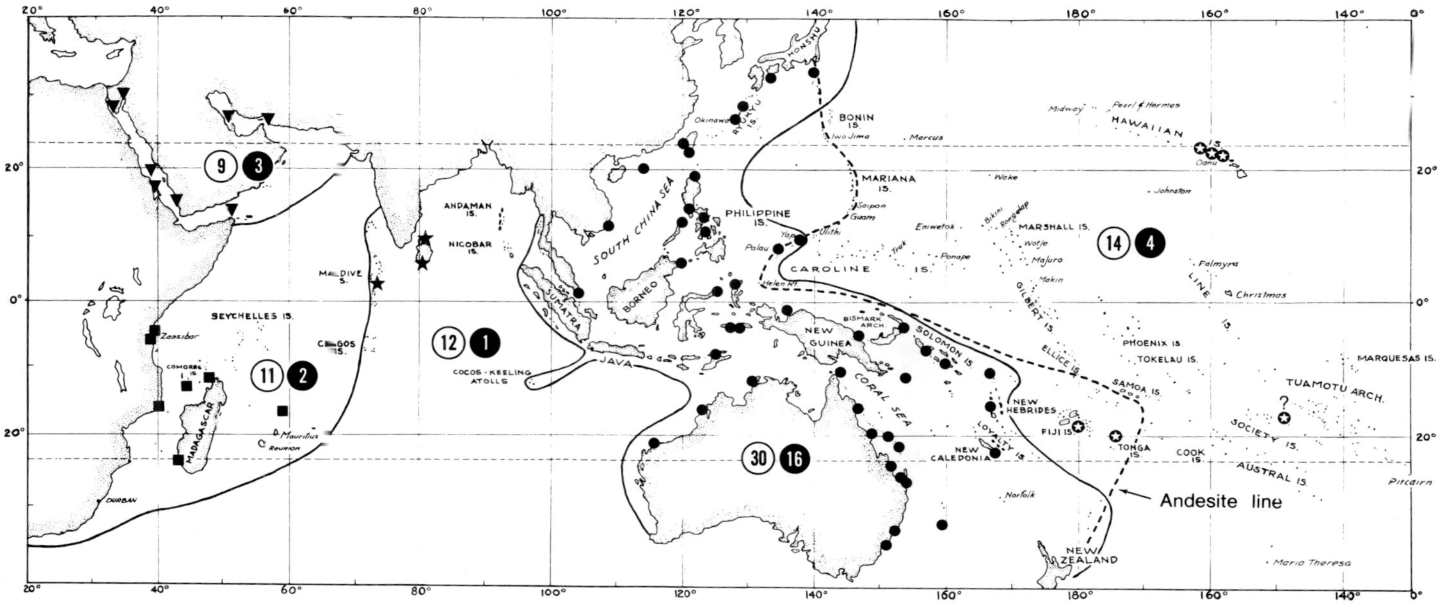

Figure 87.—Summary of the distribution of the tribe Nemophini (excluding the eastern Pacific endemic, *Plagiotremus azaleus*). Approximate boundaries of five major zoogeographic regions are indicated by bold lines. Small symbols indicate the combined distributions of all endemic species within each region. Number in solid circles is the number of endemic species in region; number in open circles indicates total number of species in region. See also Table 27.

TABLE 27.—Summary of distribution of tribe Nemophini.[1]

Genus	Petroscirtes	Aspidontus	Xiphasia	Plagiotremus	Meiacanthus
Total Species	10	2	2	11	16
Red Sea-Arabian Sea	2 (1)	2	1	3 (1)	1 (1)
W. Indian	3	2	2	2	2 (2)
E. Indian	4	2	2	3 (1)	1
W. Pacific	9 (5)	2	2	6 (3)	11 (8)
Cent. Pacific	2	2	1	5 (2)	4 (2)
E. Pacific	0	0	0	1 (1)	0

[1] Refer to Figure 87 for boundaries of regions; species endemic to a particular region are indicated by parentheses.

member of each of the major species groups within the Nemophini occurs in the western Pacific Ocean and several species groups occur only there.

Talbot (1970) noted that for most groups of tropical Indo-Pacific fishes the western Pacific Ocean (his "central area") contains the greatest number of species. He also noted that, in general, there is a progressive diminution of species the farther east or west of the area one travels. This probably is not true for the Indian Ocean (Cohen, 1973: 463), but it is certainly valid for the central Pacific Ocean. The following list of reasonably well sampled central Pacific island groups, with the total number of species of Nemophini collected at each island given in parentheses, nicely illustrates this longitudinal species diversity gradient: Palau-Yap Is. (14); Marshall Is. (7); Gilbert Is. (8); Line Is. (4); Marquesas Is. (3); Hawaiian Is. (2). Talbot (1970) concluded that the stability/time theories are the most plausible explanations for the high number of species in the tropical western Pacific Ocean and the reduced number of species elsewhere in the Indo-Pacific. He defined the two theories as follows: (1) The theory of climatic stability — "Areas of greater environmental stability allow the development of more species than areas of variable or erratic climates." (2) The time theory — "All communities tend to diversity with time, so that older communities have more species."

In a recent study, Springer and Gomon (1975: 16) noted that very few species of the blenniid tribe Omobranchini occur east of the Andesite Line (MacDonald, 1949). The Andesite Line (see Fig. 87) marks the approximate separation of the continental rocks from the oceanic rocks, which are of more recent geological age. Like Talbot, they also concluded that habitats peripheral to the central continental area are probably not species saturated. The distributional pattern of the Nemophini agrees well with this hypothesis. Whether the gradual diminution of the fish fauna from west to east in the central Pacific is due primarily to the role of distance in maintaining the geographic isolation of shore fishes with vagile larval stages, ecological conditions (Gosline, 1971) or causal geographic factors is unclear.

In discussing centers of origin, Briggs (1974b) stated that migration from the Indo-West Pacific involves competitively dominant species and that, in general, the more advanced species will be found in the center of origin. Rosenblatt *et al.* (1972: 9) attribute the success of Indo-West Pacific fishes in colonizing the eastern Pacific to their "ability to survive in the coral-poor, more variable environment of the eastern Pacific, as well as to the presence of endemic competitors." They also stated that the idea that these species are "competitively dominant species" is unwarranted, and not supported by evidence. Croizat *et al.* (1974) and Nelson (1974) reject as aprioristic all "clues" or "rules" used to resolve centers of origin and dispersal without reference to general patterns of vicariance and sympatry.

Briggs (1974a: 20) briefly discussed endemism of Red Sea fauna and suggested that the Red Sea merits recognition as a distinct zoogeographic area. Recent studies, especially of the Blenniidae, also indicate a relatively high amount of endemism in the Persian Gulf-Arabian Sea (referred to collectively as Arabian Sea in following discussion). I here treat the Red Sea and Arabian Sea as a single zoogeographic unit, but perhaps they should be kept separate. Only four species of Nemophini have been collected in the Arabian Sea. Such broadly distributed species as *Plagiotremus tapeinosoma, P. rhinorhynchos, Aspidontus taeniatus* and *A. dussumieri,* while relatively common in the Gulf of Aqaba, are unknown from the southern Red Sea and Arabian Sea. Because none of the collections available to me from the Arabian Sea was made using ichthyocides and only a few such collections have been made in the southern Red Sea, I believe that the distributional patterns and species compositions in the Red Sea and Arabian Sea are at least partially an artifact of collecting. The complete absence of *Entomacrodus,* a broadly distributed and easily collected blenniid genus, from both areas (Springer, 1971: 10)[1] suggests, however, that ecological factors may be responsible for these apparent distributional anomalies.

Cohen (1973) discussed the zoogeography of Indian Ocean fishes and concluded that the Indian Ocean is not a natural unit. He noted that while the southwestern Indian Ocean is a local center of endemism, there are few endemics in the eastern

[1] In October 1975 J. E. Randall collected specimens representing a populational variant of *Entomacrodus epalzeocheilus* (Bleeker) in the northern Red Sea (V. G. Springer, pers. comm.).

Indian Ocean and its fauna is probably best considered as part of the western Pacific fauna. Although I have maintained the eastern Indian Ocean as a distinct zoogeographic entity, the distributional patterns of the Nemophini support Cohen's conclusions. Recent collections from the Cocos-Keeling Islands contained a high percentage of fishes otherwise restricted to the western Pacific Ocean. Thus, even if the eastern Indian Ocean is maintained as a separated zoogeographic unit, I believe that this atoll and the adjacent Christmas Island are best considered peripheral "stepping-stones" of the west Pacific fauna.

PHYLOGENETIC POSITION OF THE NEMOPHINI

Previous discussions of the intrafamilial relationships of the Blenniidae appear in Springer (1968: 50-51) and Springer and Smith-Vaniz (1972b). One possible phylogeny of the six major lineages of blenniids that I recognize is shown in Figure 88. A selected list of characters found in those blenniids that have the dentaries joined anteriorly by an interdigitating (sutured) joint is given in Table 28. The sequence in which these characters are listed is not an attempt to rank them by systematic importance. Some of these characters, if considered independently, have no value in deducing the phylogeny of blenniid tribes but are included because in conjunction with other characters they facilitate interpretation of probable phyletic or directional sequence. The "Blenniini" (*sensu* Springer, 1968) is a catchbasket assemblage of approximately 60 species that share a large number of plesiomorphic characters. A monophyletic basis for recognition of this group has not been demonstrated. The following character analysis establishes the monophyletic nature of the other groups that are recognized.

Numbers given below correspond to those shown on Figure 88. For each character (or character complex) the plesiomorphic condition is given a lower case letter and the apomorphic condition a capital letter.

1. Anteroventral wall of premaxillae present and replacement teeth entering functional row through foramina in the bone (a). Anteroventral wall of premaxillae absent (except present laterally in *Hirculops*) and replacement teeth moving through a band of connective tissue (at least mesially) to obtain position in functional row (A). Smith-Vaniz and Springer (1971: 5) give a detailed description of the jaws and dentition of the speciose tribe Salariini. The nature of these structures in the Salariini seems to be correlated with a herbivorous habit, and especially with a scraping mode of feeding (Norman, 1943: 794).

2. Left and right halves of lower jaw joined anteriorly by a smooth joint; and (usually) with the ascending process of the premaxilla weakly based on the body of the premaxilla and riding over or in front of the median ethmoid (b). Left and right halves of lower jaw joined anteriorly by interdigitating knoblike processes;[1] and with the ascending process of the premaxilla moderately to strongly based on the body of the premaxilla and abutting against the ventral end of the median ethmoid (B). Character state 2B seems to comprise specializations that brace and strengthen the jaws. The interdigitation of the "knuckles" of the dentary processes make for considerable difficulty in disarticulating the two halves of the lower jaw. The "sutured" dentaries of representative species of Nemophini are shown in Figure 89. The other families of the Blenniicae, generally considered less specialized than the Blenniidae, as well as many of the perciforms usually considered more primitive, have moderately well-developed premaxillary ascending processes and dentaries that are joined by an even, nonsuturing joint (Springer, 1968: 50).

If interpretation of sutured dentaries as a derived character is correct, then on that basis *Blennius ocellaris* Linnaeus, and its close relative *B. normani* Poll, are more closely related to the Omobranchini, Phenablenniini and Nemophini than to other blenniids. *B. ocellaris* is the type species of *Blennius*, and the characters shared by *B. ocellaris* and *normani* are such that the generic name should be restricted to them alone (Springer, 1968: 47; Springer and Smith-Vaniz, 1970: 219). If the nature of the symphyseal dentary joint is the primary character I believe it to

[1] These processes and the posterior canines of both jaws are secondarily lost in females (with spent gonads) of *Omobranchus fasciolaticeps* (Richardson), possibly in conjunction with initiation of mouth brooding behavior (Springer and Gomon, 1975: 13).

be, then the tribe Blenniini also must be restricted because *Blennius* is its type-genus and only genus with sutured dentaries. See Springer (1968) for a list of the nominal genera included in the "Blenniini."

3. Base of dorsalmost pectoral-fin ray without fleshy flap (c). Short, fleshy flap extending vertically from base of dorsalmost pectoral-fin ray (C). The functional significance of this character is unknown, but since it is apparently unique within the Blenniidae it is assumed to be a specialization. Although the genus *Blennius* as here restricted (see above) seems to possess few autapomorphic characters, it is certainly monophyletic.

4. Caudal skeleton with two epural bones (d). Caudal skeleton with one epural bone (D).

5. Middle nine caudal-fin rays typically branched in adults (e). Caudal-fin rays typically unbranched in adults (E).

6. Supracleithrum articulated only with posttemporal (f). Supracleithrum articulated with epiotic and posttemporal (F).

7. Basisphenoid present (g). Basisphenoid absent (G).

8. Intercalar present (h). Intercalar absent (except subgenus *Dasson*) (H).

9. Cranial bones without ornamentation (i). Cranial bones ornamented with tiny, shallow depressions to deep pits (except weak to absent in *Xiphasia*) (I).

10. Symphyseal mandibular cirri absent (j). Symphyseal mandibular cirri present, at least in adult males (J).

11. Segmented caudal-fin rays 15-12 (typically 13) (k). Segmented caudal-fin rays 13-10 (usually 13 only in subgenus *Holomeiacanthus*, see Table 17) (K).

12. Pterosphenoid well developed (l). Pterosphenoid reduced, excluded from external surface of skull (L).

13. Interopercle without posteroventral spur (m). Interopercle with posteroventral spur (see Springer, 1968: Fig. 16) (M).

14. Pelvic-fin rays I, 3 (n). Pelvic-fin rays I, 2 (N).

15. Basisphenoid well developed and firmly attached to cranium (o). Basisphenoid greatly reduced and weakly attached to cranium (O).

16. Dorsal and ventral postcleithra well developed (except *Enchelyurus* where postcleithra consist of a few fragments of bone) (p). Dorsal postcleithrum absent; ventral postcleithrum well developed (P).

According to the preceding character anaylsis *Blennius* is the most primitive member of the four taxa with sutured dentaries, and it is of interest to note that it also has the most relict distribution. *Blennius* is restricted to the Mediterranean Sea and eastern Atlantic Ocean; the other three tribes have exclusively Indo-West Pacific distributions, except for a single eastern Pacific endemic, *Plagiotremus azaleus*.

The sister group relationship of the Omobranchini and Phenablenniini rests solely on the common possession of a reduced pterosphenoid. However, the monotypic Phenablenniini shares two additional characters with several genera of the Omobranchini which are not found in *Blennius* or the Nemophini, and thus may be indicative of close kinship. Nuptial males of *Phenablennius* have lappetlike structures present subdistal to the tips of the segmented anal-fin rays (Smith-Vaniz, 1975: Fig. 3) similar to those found in some species of *Omobranchus* and *Enchelyurus*. *Phenablennius* also has well-developed labial flaps similar to those of *Omox* and some species of *Omobranchus*.

EVOLUTIONARY TRENDS

Any meaningful attempt to establish phylogenetic relationships requires correct interpretation of evolutionary trends. The purpose of the section is to comment on the occurrence and lability of selected character states in the Nemophini, and where appropriate give a basis for interpretation of probable phyletic or directional sequence. Fusions, reductions and loss of bones invariably constitute specializations. In the absence of strong contradictory evidence, any character state of wide distribution in the other blenniid tribes, especially in those taxa with sutured dentaries (see Table 28), is assumed to be plesiomorphic for the Nemophini. The phylogenetic position of the Nemophini as outlined in the preceding section and in Figure 88 provides a foundation for the following discussion.

Frontal. — The frontal bones are fused in the adults of *Plagiotremus* (Fig. 90) and *Enchelyurus* (Omobranchini) and are separate in all other blen-

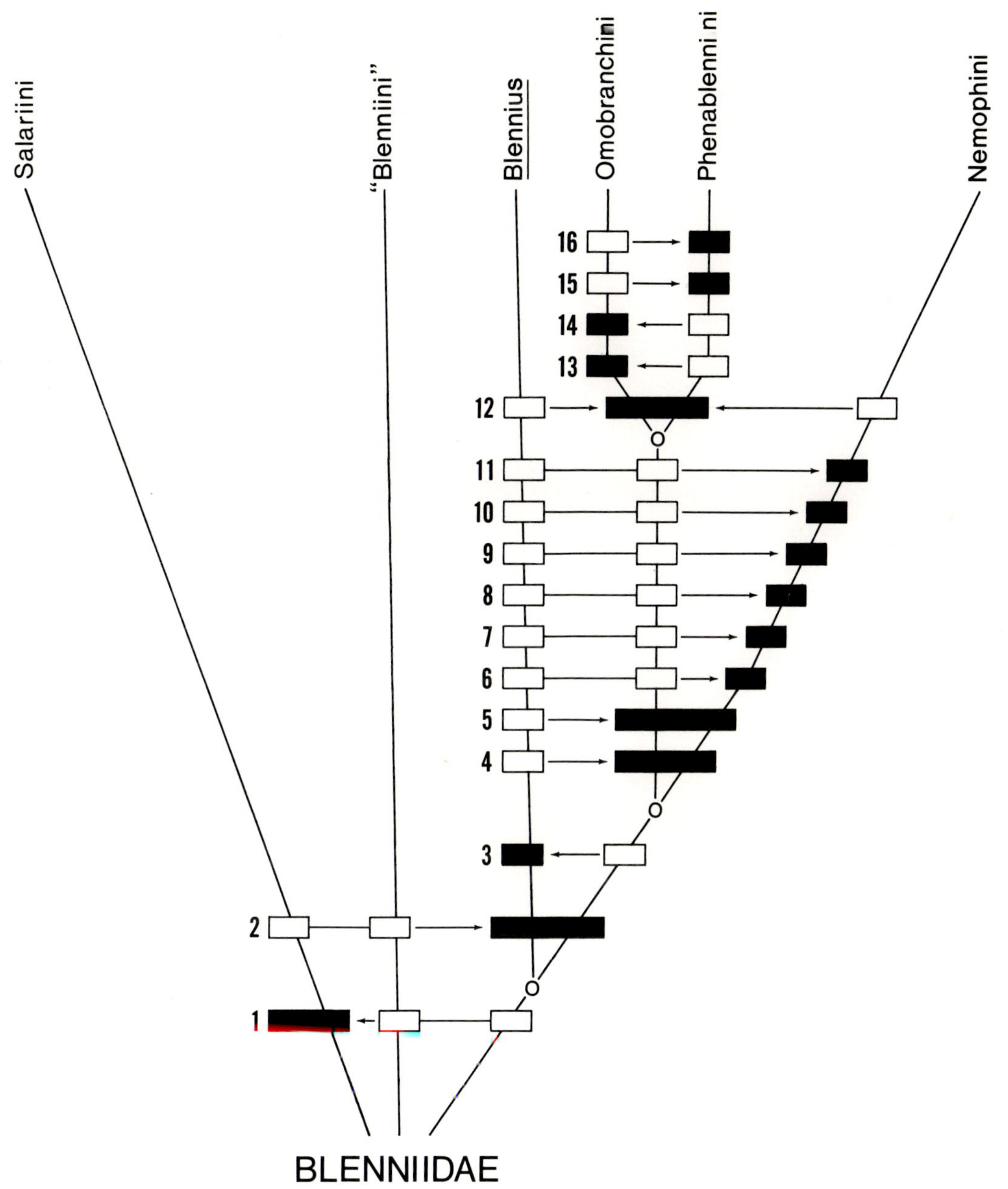

Figure 88. — A phylogenetic analysis of the six major lineages within the Blenniidae. Open squares are primitive (plesiomorph) character expressions and shaded squares are derived (apomorph) character expressions. Numbers correspond to characters discussed in text. Arrows are inferred directions of evolutionary change and do not imply ancestor-descendent relationships.

TABLE 28. — **Summary of selected characters in blenniids with dentaries joined by interdigitating (sutured) joint.**

Character	*Blennius*	Omobranchini	Phenablenniini	Nemophini
Caudal skeleton	2 epurals and hypural 5 autogenous	1 epural; hypural 5 autogenous or absent	1 epural; hypural 5 absent	1 epural (2 as rare variant); hypural 5 absent
Branched caudal-fin rays	typically middle 9 branched	none (1 or 2 as occasional variant)	none	none
Segmented caudal-fin rays	typically 13	typically 13	typically 13	11 or 10 (except *Meiacanthus*, see Table 17)
Procurrent caudal-fin rays	5 to 7	8 to 16	6	2 to 18
Basisphenoid	well developed	well developed	greatly reduced	absent
Pterosphenoid	well developed	reduced, excluded from external surface of skull	reduced, excluded from external surface of skull	well developed
Interopercle	without posteroventral spur	with posteroventral spur (Springer, 1968: Fig. 16)	without posteroventral spur	without posteroventral spur
Postcleithra	2, both well developed	2, both well developed (or vestiges *Enchelyurus* only)	ventral postcleithrum only	2, both well developed
Articulation of supracleithrum	posttemporal only	posttemporal only	posttemporal only	epiotic and posttemporal
Posttemporal	forked, with well-developed ventral arm	forked, with well-developed ventral arm	forked, with well-developed ventral arm	forked or unforked
Intercalar	well developed	well developed	well developed	absent or present (Subg. *Dasson* only)
Infraorbital bones	5	5, 4, 3	4	4 or 3 (Subg. *Meiacanthus* only)
Ornamentation of cranial bones	absent	absent	absent	moderate to well developed (except weak in *Xiphasia*)

TABLE 28. — Continued.

Neural arch of first vertebra	complete (closed dorsally)	incomplete (open dorsally)	complete (closed dorsally)	incomplete or complete (*Meiacanthus* only)
Precaudal vertebrae	10	9 to 12 (usually 10 or 11)	10	11 to 16
Dorsal-fin spines	11	6 to 14	12 or 13	3 to 14
Posteriormost dorsal-fin spine	reduced in size giving fin a distinctly notched appearance	slightly or not reduced	not reduced	not reduced
Pelvic fin	I, 3	I, 2	I, 3	I, 3 or absent (*Plagiotremus*, 2 spp. only)
Dentary canines	relatively short and robust	short and robust to relatively large	relatively short and robust	relatively large
Symphyseal mandibular cirri[1]	absent	absent	absent	present in at least mature males
Ventral restriction of gill opening in relation to pectoral-fin rays or base (in adults)[2]	unrestricted to distinctly below pectoral-fin base	entirely above to ventralmost ray or slightly below	opposite 7th to ventralmost ray or slightly below	entirely above to opposite 2nd or 3rd (except *Aspidontus*, see Table 13)
Swimbladder (in adults)[3]	absent	absent (most spp. ?) or present (*Omox*)	present	present or absent (*Xiphasia* only)

1 Refer to discussion under general account of sexual dimorphism.

2 The most plesiomorphous condition for the gill opening in blenniids is completely unrestricted laterally and united ventrally by a deep, free fold. In many blenniids with restricted gill openings, including the Nemophini, the ventral extent of the gill opening may be greater in larvae and adults than in adults.

3 A swimbladder was previously reported to be present only in the Nemophini, but see comments under methods and materials section. I have not made an exhaustive search for a swimbladder in the Omobranchini.

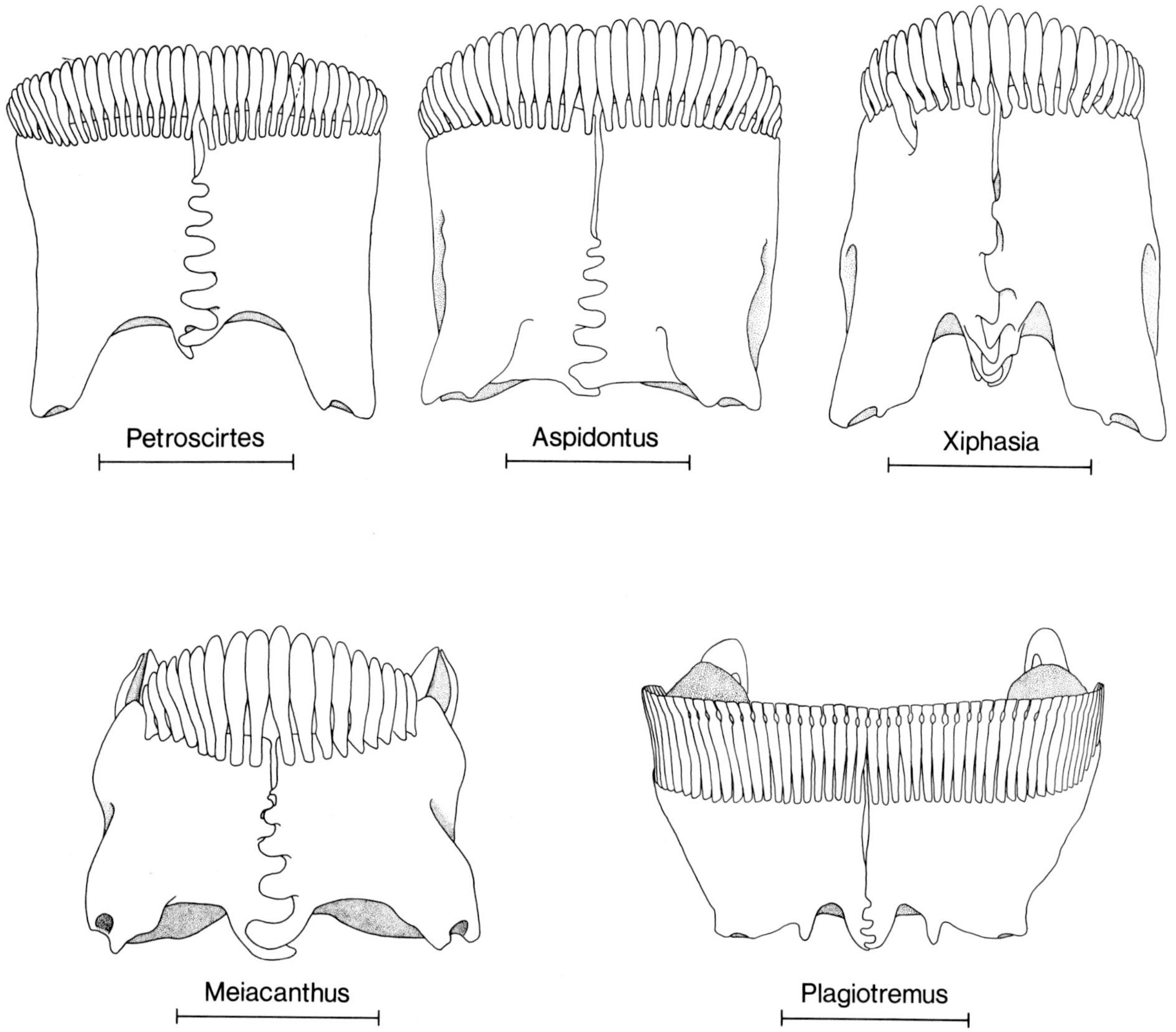

Figure 89. — Ventral view of lower jaw of selected species of Nemophini showing dentaries joined by interdigitating (sutured) joint: *Petroscirtes variabilis,* USNM 138250, 49.9 mm; *Aspidontus dussumieri,* USNM 204579, 100.3 mm; *Xiphasia setifer,* ANSP 128699, 281 mm; *Meiacanthus anema,* ANSP 124867, 29.5 mm; *Plagiotremus rhinorhynchos,* ANSP 124875, 72.4 mm.

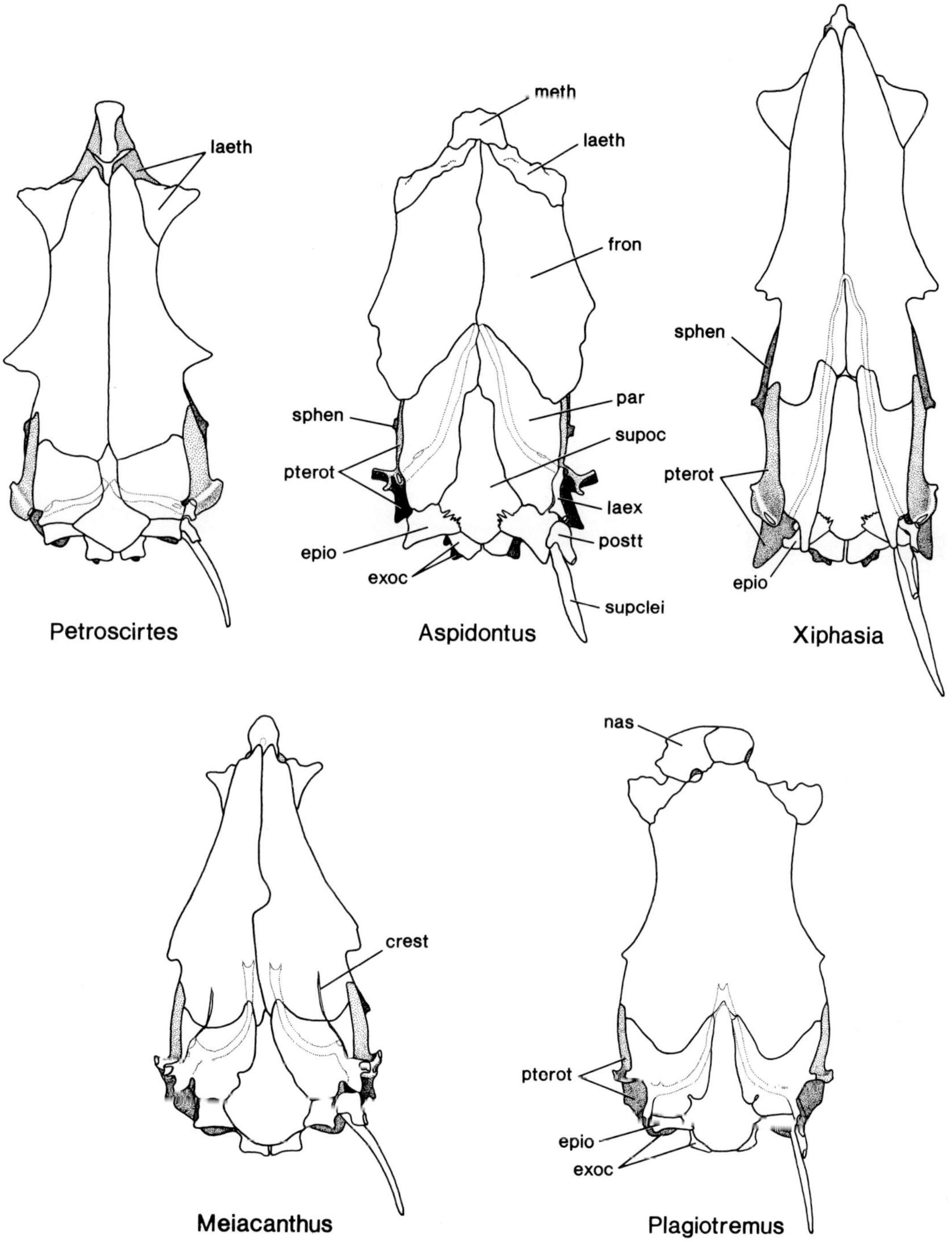

Figure 90. — Dorsal views of skulls of selected species of Nemophini: *Petroscirtes thepassi,* USNM 205296, 46.6 mm; *Aspidontus taeniatus,* USNM 142228, 90.4 mm; *Xiphasia setifer,* ANSP 128699, 267 mm; *Meiacanthus anema,* ANSP 128535, 38.8 mm; *Plagiotremus laudandus,* USNM 201413, 31.4 mm. (Ornamentation of bones not depicted.) Abbreviations: epio = epiotic; exoc = exoccipital; fron = frontal; laeth = lateral ethmoid; laex = lateral extrascapular; meth = median ethmoid; nas = nasal; par = parietal; postt = posttemporal; pterot = pterotic; sphen = sphenotic; supclei = supracleithrum; supoc = supraoccipital.

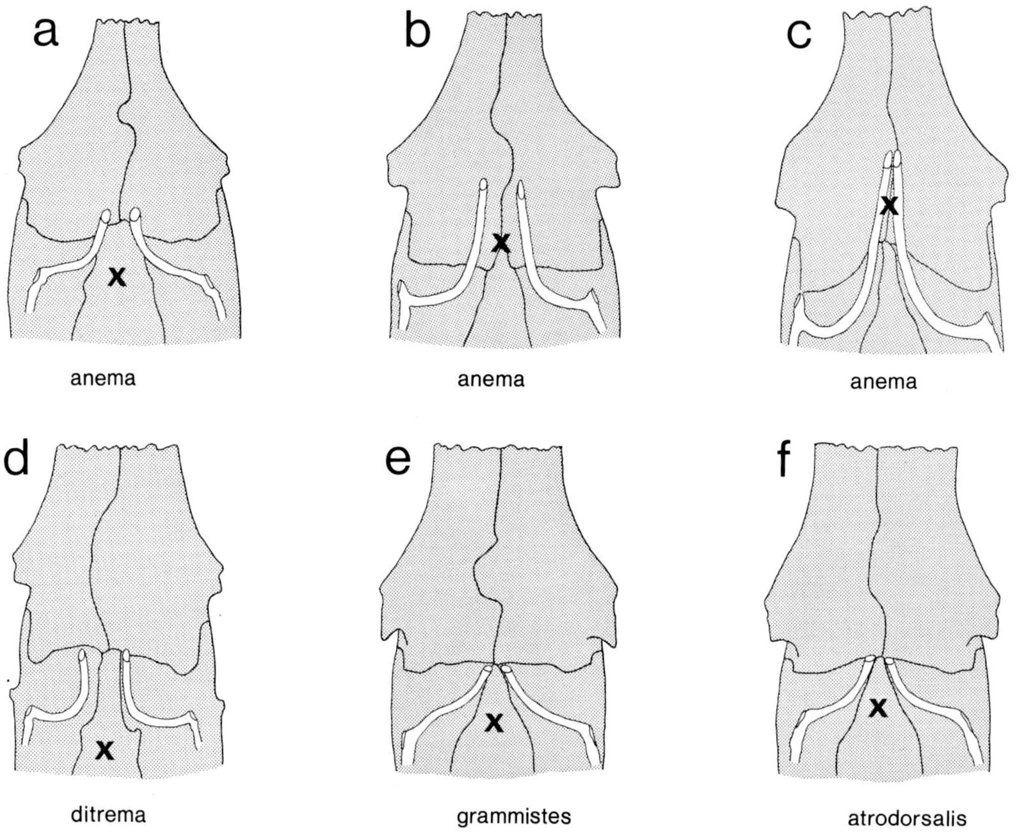

Figure 91. — Semidiagrammatic illustrations of supratemporal canal arrangement in selected species of *Meiacanthus:* a-c, *anema,* ANSP 128535, 29.0 mm, 38.3 mm, 40.6 mm, respectively; d, *ditrema,* USNM 195706, 37.1 mm; e, *grammistes,* ANSP 128536, 42.9 mm; f, *atrodorsalis,* ANSP 124873, 42.4 mm. Origin of dorsal fin indicated by "x".

niids. The frontals are relatively narrow in the interorbital region of most blenniids, but are broad in most species of Nemophini. There is also a tendency in the Nemophini for ornamentation of the skull bones, particularly the frontals (see Springer, 1968: pls. 5-8).

Supraoccipital. — *Aspidontus* and especially *Xiphasia* differ from the other genera of Nemophini in having the supraoccipital depressed medially, apparently to accommodate the pterygiophores of the dorsal-fin spines.

Parietal. — In the adults of many blenniids bony crests are formed along the lateral and posterior margins of the parietals and pterotics and at the junction of the parietals and frontals. These bony crests are present in the Nemophini only in *Petroscirtes* (Fig. 94); they are best developed in the subgenus *Petroscirtes,* but the crests are only slightly less well developed in *P. xestus.*

Intercalar and posttemporal bones. — Intercalars are absent in the Nemophini, except the subgenus *Dasson* (Fig. 18a), and present in all other blenniids. In the other blenniid tribes the posttemporal is forked and the ventral arm is attached by a ligament to the intercalar or pterotic spine. A posttemporal ligament is absent in the Nemophini. The ventral arm of the posttemporal is present or absent in the Nemophini, but is not correlated with the presence of an intercalar because the posttemporal of *Dasson* is unforked. The posttemporal of the subgenus *Allomeiacanthus* is greatly reduced and apparently unique in not encapsulating the lateral temporal sensory canal (Fig. 50b).

Supratemporal canal and mid-dorsal ST pore. —

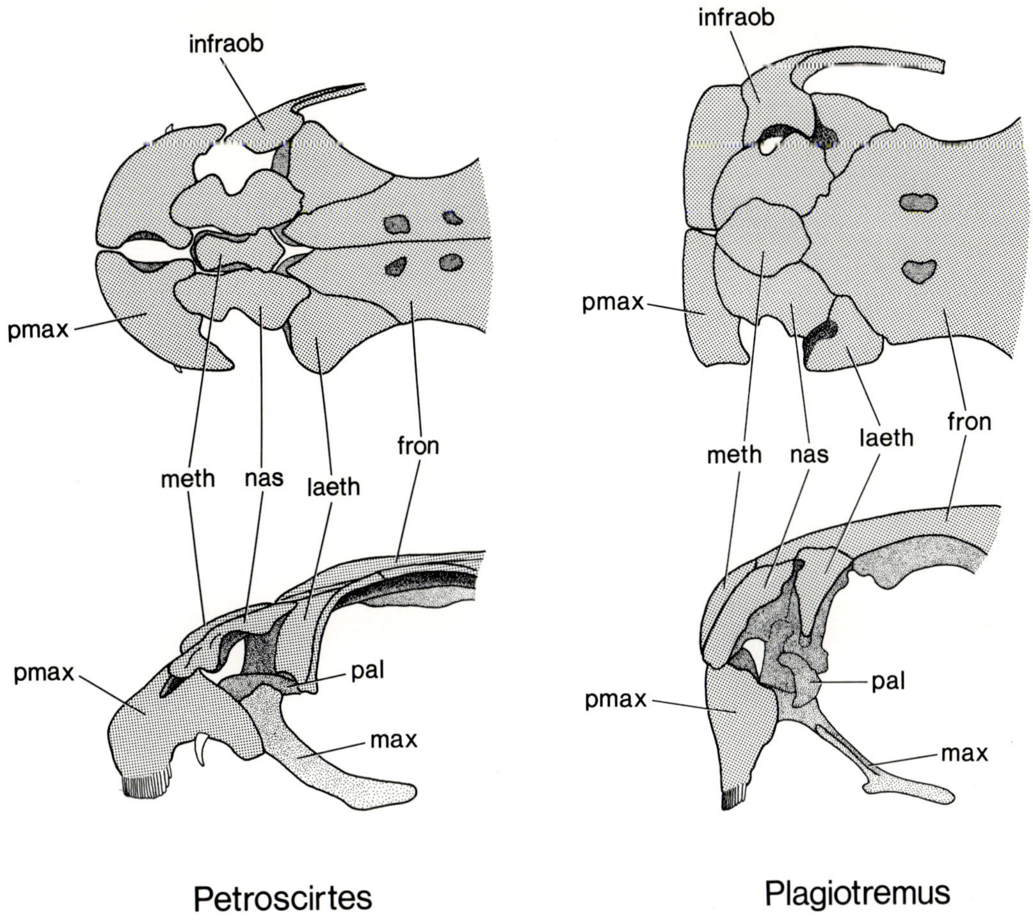

Figure 92. — Anterodorsal and lateral views of snout region of *Petroscirtes thepassi,* USNM 205296, 46.6 mm and *Plagiotremus laudandus,* USNM 201413, 31.4 mm. Abbreviations: infraob = infraorbital; pmax = premaxillary; meth = median ethmoid; laeth = lateral ethmoid; nas = nasal; fron = frontal; pal = palatine; max = maxillary.

In blenniids the two branches of the supratemporal canal typically meet just in front of the dorsal-fin origin, at which point a single pore communicates with the surface. *Xiphasia* is apparently unique in having the supratemporal canal of adults united anteriorly (Fig. 90) and terminating posterior to the dorsal-fin origin. In adults a minute pore occasionally occurs at the base of the first dorsal-fin spine (Figs. 33-34). Presumably this pore represents a rudimentary ST pore which was carried forward during the rapid ontogenetic shift in the position of the dorsal-fin origin in *Xiphasia* (see below). In *Meiacanthus anema* and *M. ditrema* the two branches of the supratemporal canal are usually widely separated anteriorly (Fig. 91) and have separate pores (refer to number 10 in discussion of intrageneric relationships).

Dorsal-fin origin. — The origin of the dorsal fin is determined, at least in part, by the number of dorsal-fin spines in proportion to the total number of dorsal-fin elements and the length of the skull. In *Plagiotremus,* the subgenus *Musgravius* has the most anteriorly positioned dorsal fin (Fig. 94i) but the subgenus *Plagiotremus* and the *rhinorhynchos* species group, which have more dorsal-fin spines, have the dorsal-fin origin more posteriorly positioned. An anteriorly positioned dorsal-fin origin is the plesiomorphic character state in *Meiacanthus* (refer to discussion of intrageneric relationships) but could have been apomorphic for the progenitor

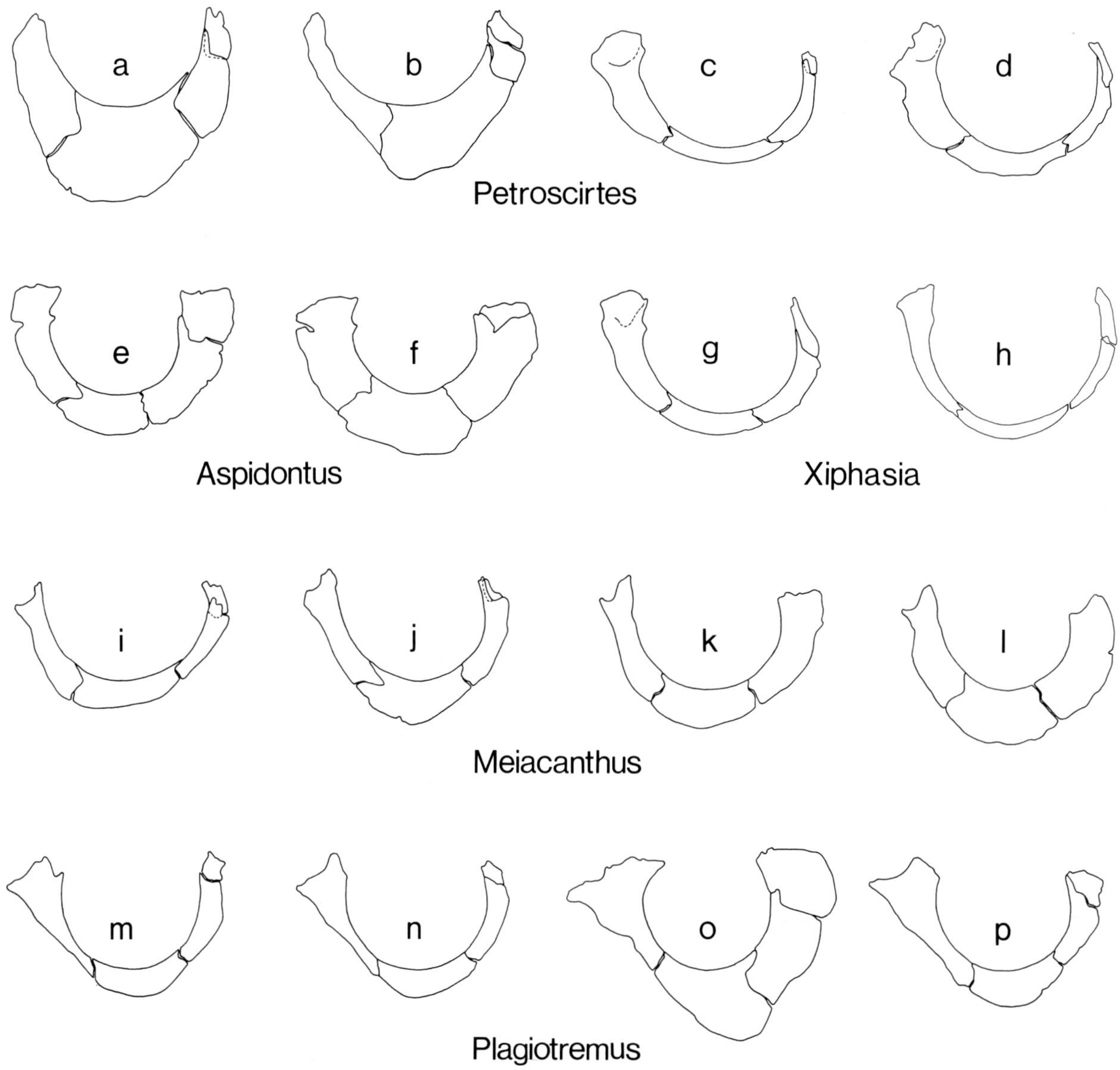

Figure 93. — Left infraorbital bones of selected species of Nemophini: *Petroscirtes,* (a) *mitratus,* USNM 200599, 51.6 mm, (b) *xestus,* ANSP 124874, 45.9 mm, (c) *thepassi,* USNM 205296, 46.6 mm, (d) *lupus,* ANSP 124405, 64.3 mm; *Aspidontus,* (e) *taeniatus,* USNM 201562, 63.9 mm, (f) *dussumieri,* USNM 204549, 100.3 mm; *Xiphasia,* (g) *setifer,* ANSP 128699, 282 mm, (h) *matsubarai,* ANSP 128537, 208 mm; *Meiacanthus,* (i) *anema,* ANSP 124867, 29.5 mm, (j) *ditrema,* USNM 195706, 37.1 mm, (k) *grammistes,* ANSP 124868, 42.9 mm, (l) *atrodorsalis,* ANSP 124872, 59.1 mm; *Plagiotremus,* (m) *celebesensis,* SU 67356, 54.9 mm, (n) *laudandus,* USNM 201413, 31.4 mm, (o) *rhinorhynchos,* ANSP 124876, 67.8 mm, (p) *tapeinosoma,* ANSP 124878, 38.9 mm.

of *Meiacanthus.* Adults of *Xiphasia* have the most anterior dorsal-fin origin of any blenniid; the more posterior position of the dorsal-fin origin in prejuveniles (Fig. 35) suggests that the progenitor of *Xiphasia* had a more posteriorly positioned dorsal fin. An ontogenetic shift in the position of the dorsal-fin origin is not known to occur in any other blenniid.

Nasals, lateral ethmoids and premaxillae. — In *Plagiotremus* the nasals and lateral ethmoids, which broadly overlap the premaxillae, are broadly expanded and firmly ankylosed to each other and to the median ethmoid (Fig. 92). This unique arrangement, together with the shape of the ascending processes of the premaxillae, allows the premaxillae to rotate anteroposteriorly (premaxillae immovable in other blenniids). Presumably the mobility of the premaxillae facilitates scale and mucus removal from prey during feeding (refer to discussion of biology).

Dentaries. — The dentary bones exhibit many differences in shape, especially in *Meiacanthus* where they are modified to accommodate the toxic buccal glands (Fig. 44). The "knuckles" of the dentary processes (Fig. 89) are also variable but exhibit few intrageneric differences. The reduced nature of the dentary processes in *Plagiotremus* probably represents a secondary specialization.

Infraorbital bones (Fig. 93). — The number of infraorbital bones varies from 5 to 2 in the Blenniidae, with 5 or 4 the usual number. Three infraorbital bones occur only in *Enchelyurus* and some species of *Parenchelyurus* and *Meiacanthus.* In the Nemophini there is a tendency for the infraorbital bones (especially the second) to become broad and deep in the adults of some species; this is particularly evident in some species of *Petroscirtes* and *Aspidontus,* but apparently does not occur in *Xiphasia.* The first infraorbital bone of *Plagiotremus* is easily distinguished by its wide dorsal end and relatively narrow ventral end.

Branchial arches. — Except for minor differences in the number and size of the pharyngobranchial teeth, no differences were noted.

Precaudal vertebrae. — There are 11-16 precaudal vertebrae in the Nemophini, with 13 or more in *Aspidontus, Xiphasia* and *Plagiotremus.* Except for rare individual variants, the number of precaudal vertebrae is 10-12 in the other blenniid tribes, with 10 the usual number. Thus, there is an obvious trend in the Nemophini to increase the number of precaudal vertebrae.

In the Nemophini the neural arches of the first vertebra are complete in *Meiacanthus,* just touch in *Xiphasia* and fail to meet dorsally in the other genera. Except for rare individual variants the neural arch of the first vertebra is complete in all other blenniids, except the Omobranchini.

"Neural arch processes" (see Methods and Materials section) occur on some of the precaudal vertebrae in adults of all blenniids except *Xiphasia.*

Caudal vertebrae. — The number of caudal vertebrae varies from 18 to 33 in the other blenniid tribes. Thus, the caudal vertebral count of 100-119 in *Xiphasia* and 43-60 in the subgenus *Plagiotremus* almost certainly constitutes a specialization.

The last few caudal vertebrae of *Plagiotremus* are unique in having the neural and haemal spines broad and bladelike (Figs. 95f-g). These specializations presumably strengthen the caudal peduncle and make it more rigid. A stiff caudal peduncle would be advantageous to a species of *Plagiotremus* during rapid-swimming, aggressive feeding attacks. The less expanded neural and haemal spines of *P. spilistius* (Fig. 95f) are probably correlated with the high number of vertebrae and more anguilliform mode of swimming in that species.

Caudal skeleton (Fig. 95). — The most specialized caudal skeleton found in blenniid fishes occurs in the Nemophini and is characterized by a single epural bone and loss of hypural 5. (The caudal skeleton of *Xiphasia* shown in Fig. 95c is exceptional in having two epurals.) The caudal skeleton of *Xiphasia* is further specialized in having the ventral hypural fused to the urostylar centrum, a condition found elsewhere in the Blenniidae only in *Enchelyurus, Medusablennius* and *Alloblennius pictus* (Lotan).

Caudal-fin rays. — Except for *Meiacanthus,* all genera of Nemophini have 11 or 10 segmented caudal-fin rays. Only the subgenus *Holomeiacanthus* typically has 13 segmented rays (Table 17). As noted by Gosline (1963), the number of principal rays tends to be reduced in elongate fishes. *Xiphasia* agrees well with that observation, having the least number of segmented (10) and procurrent (2 or 3) caudal-fin rays of any blenniid. The minimum number of procurrent rays is six in other blenniids and is the typical number in *Blennius, Phenablennius, Petroscirtes* and *Aspidontus.* In *Meiacanthus* and *Plagiotremus* the number of procurrent rays varies from 9 to 15 and 10 to 18 respectively. The labile nature of the number of procurrent rays makes this character of limited value in establishing relationships. The high number of procurrent rays in *Meiacanthus* and *Plagiotremus* is perhaps correlated

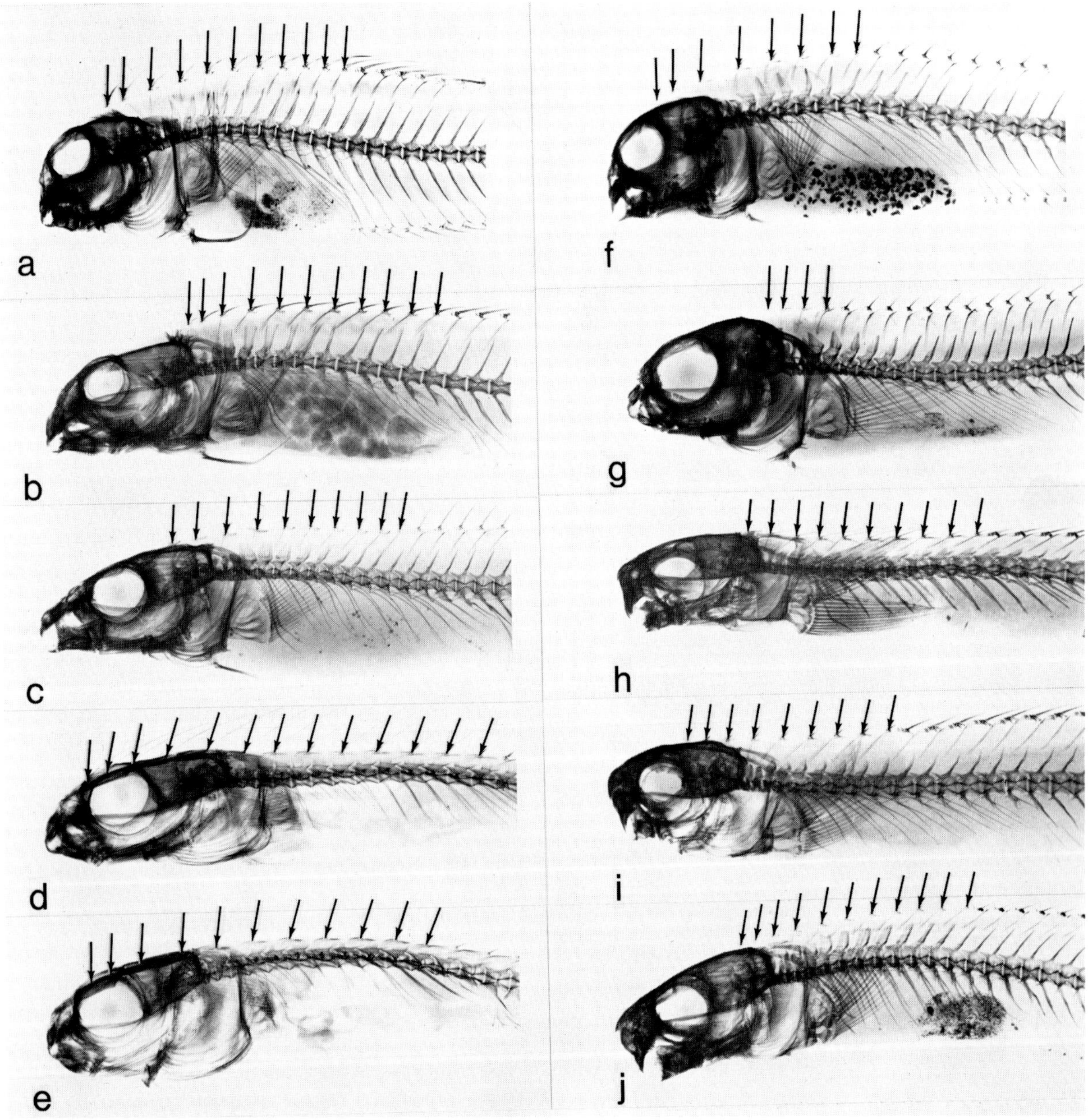

Figure 94. — Radiographs of selected species of Nemophini showing postcranial axial skeleton and positions of dorsal-fin spines (indicated by arrows): a, *Petroscirtes mitratus,* USNM 200599, 54.9 mm; b, *Petroscirtes variabilis,* RMNH 4776, 84.4 mm; c, *Aspidontus dussumieri,* RUSI 961c, 73.8 mm; d, *Xiphasia setifer,* USNM 205292, 290 mm; e, *Xiphasia matsubarai,* USNM 203276, 244 mm; f, *Meiacanthus anema,* RMNH 25232, 47.8 mm; g, *Meiacanthus grammistes,* CAS 28930, 47.0 mm; h, *Plagiotremus spilistius,* USNM 21722, 140 mm; i, *Plagiotremus laudandus,* USNM 205270, 52.0 mm; j, *Plagiotremus rhinorhynchos,* USNM 213477, 67.0 mm.

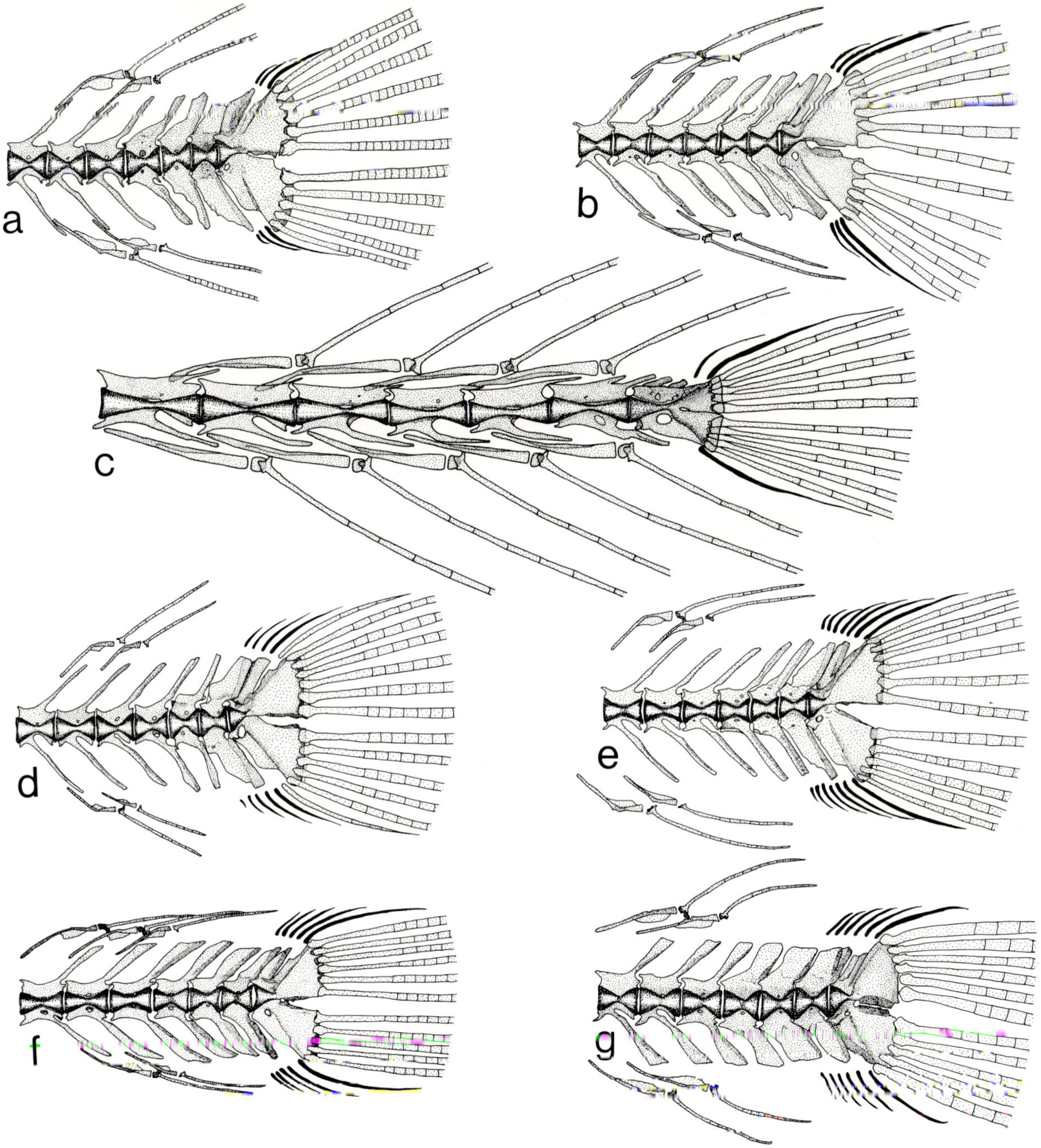

Figure 95. — Caudal skeleton and vertebrae of selected species of Nemophini: a, *Petroscirtes xestus,* USNM 115531, 46.0 mm; b, *Aspidontus taeniatus,* USNM 199432, 50.1 mm; c, *Xiphasia setifer,* ANSP 128699, 267 mm; d, *Meiacanthus anema,* ANSP 124867, 29.5 mm; e, *Meiacanthus grammistes,* ANSP 124868, 42.9 mm; f, *Plagiotremus spilistius,* MNHN 1964-628, 164.3 mm (from a radiograph); g, *Plagiotremus tapeinosoma,* ANSP 124878, 60.5 mm. Procurrent caudal-fin rays shown in black.

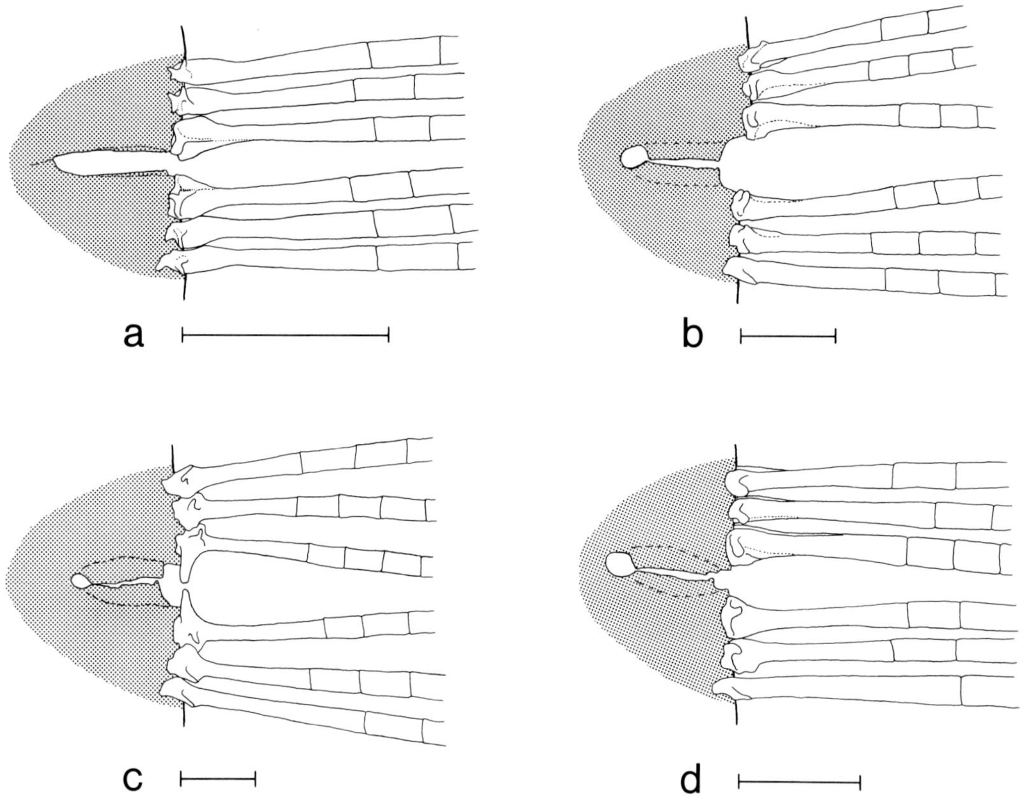

Figure 96. — Portion of caudal skeleton (shaded) showing attachment of segmented rays in selected species of *Plagiotremus:* a, *celebesensis,* SU 67356, 54.9 mm; b, *laudandus,* ANSP 126230, 49.3 mm; c, *rhinorhynchos,* ANSP 111015, 57.5 mm; d, *azaleus,* USNM 200392, 47.3 mm. Note well-developed bony spurs on innermost rays in *P. rhinorhynchos.* Scale = 1.0 mm.

with increased dependence on caudal locomotion in these active fishes, while the reduced number of procurrent rays in *Xiphasia* reflects the reverse trend. Similarly, the subgenus *Plagiotremus,* which has the highest number of caudal vertebrae in the genus, has a relatively low number of procurrent rays (Table 21).

The innermost pair of segmented caudal-fin rays of *Plagiotremus rhinorhynchos* and *ewaensis* have well-developed bony spurs proximally (Fig. 96c). This character state has not been observed in any other species of Nemophini.

Dorsal-fin elements and pterygiophores. — All dorsal-fin elements in blenniids have their own pterygiophores, except the terminal spine and first segmented ray which share the same proximal pterygiophore. *Plagiotremus* appears to be unique in having the bilaterally paired posterior extensions of the pterygiophores of the dorsal spines narrow and elongate (Fig. 97). This specialization presumably provides increased rigidity to the spinous dorsal-fin supports.

Excluding *Enchelyurus,* the number of dorsal-fin spines in the other blenniid tribes varies from 9 to 14. Thus, the relatively low number of spines in some species of *Meiacanthus* and *Plagiotremus* almost certainly represents a derived condition. Presumably the progenitor of the Nemophini had 11 or 12 dorsal-fin spines and an increase in the number of spines has occurred only in *Xiphasia.*

Usually an increase (or decrease) in the number of segmented dorsal-fin rays coincides with a similar change in the number of caudal vertebrae. In *Meiacanthus,* however, an increase in the number of segmented dorsal-fin rays coincides with a decrease in the number of dorsal spines (see Table 16).

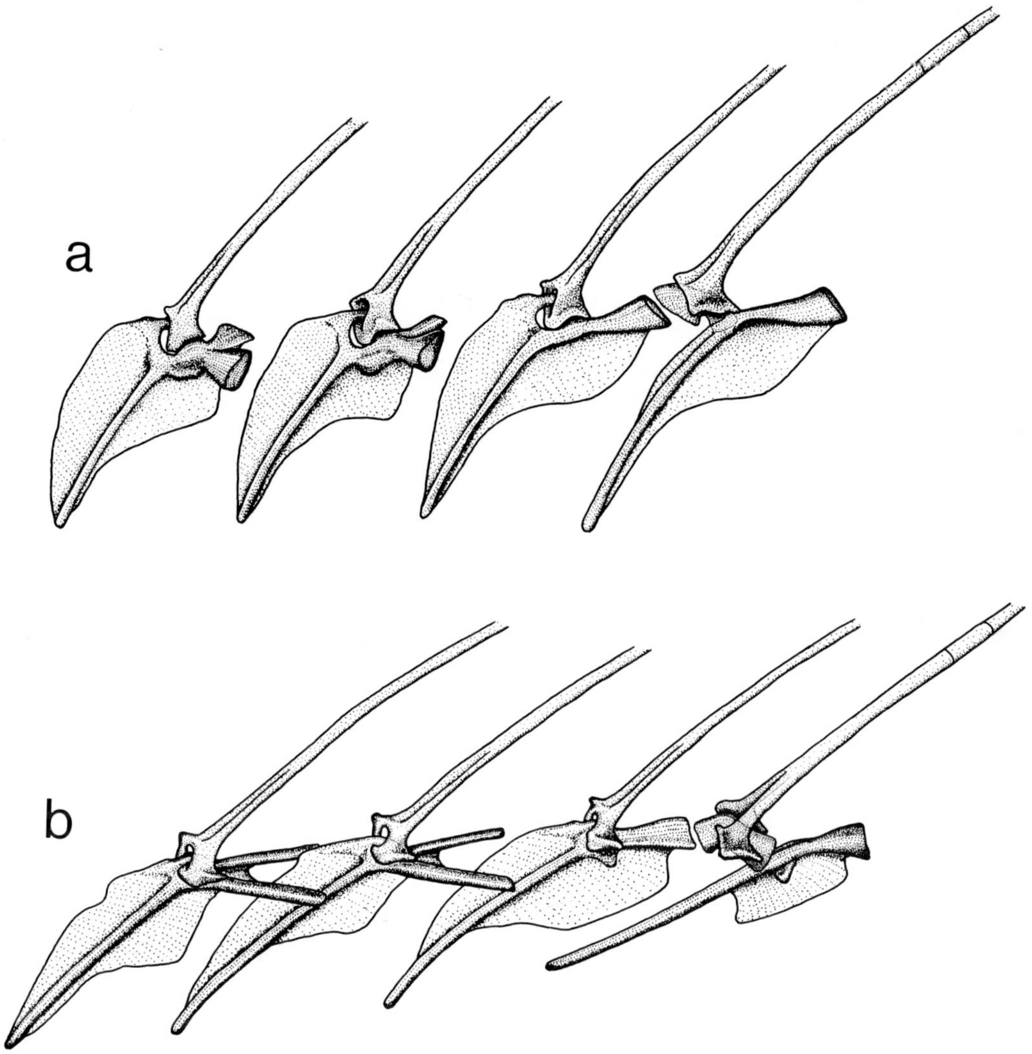

Figure 97. — Pterygiophores of last three spines and first segmented ray of dorsal fin: a, *Petroscirtes variabilis,* USNM 138250, 49.9 mm; b, *Plagiotremus tapeinosoma,* ANSP 124878, 60.5 mm.

Pelvic fin. — The subgenus *Plagiotremus* is unique within the Blenniidae in having the pelvic fins very minute or absent (absent in two of three species).

Pectoral girdle. — Numerous differences were noted in the shape of the pectoral girdle, that of *Plagiotremus* was the most dissimilar, but the structure was not studied sufficiently to draw any other conclusions.

Snout physiognomy. — The only blenniid genera with fleshy, conical snouts are *Aspidontus* and *Plagiotremus,* but the character state does not occur in all species of either genus. In blenniids as well as in at least two unrelated groups of fishes, the development of a fleshy, conical snout appears to be correlated with specialized feeding habits. Cleaner wrasses of the genus *Labroides* differ from other members of the Labridae in having well-developed fleshy snouts. Böhlke and Robins (1968: 116) recognized the *oceanops* complex as a natural unit containing the most specialized cleaning gobies of the subgenus *Elacatinus.* One of the characters that distinguishes these species from the other members of the subgenus is their well-developed snouts.

Aspidontus taeniatus is an aggressive, fin-eating

mimic of *Labroides dimidiatus,* and the similar snout shapes of the two species enhances their similarity. All species of *Plagiotremus* are obligatory scale-eaters, although only adults of the subgenus *Runula* have fleshy snouts. It does not seem likely, however, that once evolved, a fleshy, conical snout would have been secondarily lost in such a small group of closely related fishes with virtually identical feeding habits.

Cephalic sensory pores. — There is a general reductive trend in cephalic sensory pores in those blenniids that have simple pore positions. The most reduced number of pores found in the Blenniidae for those pore series listed in Table 1 occurs in the Nemophini.

Gill opening. — In most genera of Salariini and "Blenniini" the gill openings are unrestricted laterally and posteriorly. Only in *Petroscirtes, Meiacanthus* and a few species of *Omobranchus* is the gill opening restricted entirely above the level of the dorsalmost pectoral-fin ray. In many genera with partially restricted gill openings, including *Aspidontus* (Table 13) and *Xiphasia,* the gill opening is more restricted in adults than in juveniles. Thus, it seems reasonable to conclude that a restricted gill opening is a specialized character state in the Blenniidae.

The gill opening of *Xiphasia* is unique within the Blenniidae in having both a lateral and mesial fleshy flap. This arrangement undoubtedly functions as a one-way valve to prevent sand and fine sediment from entering the gill cavity.

Swimbladder. — The swimbladder of blenniids has generally been overlooked due largely to its delicate nature and absence in the adults of most species. Springer and Smith-Vaniz (1972b: 65) believed that retention of a swimbladder in the adults distinguished the Nemophini from the other blenniid tribes. I have since found a swimbladder in adults of the monotypic Phenablenniini and in one genus (*Omox*) of the Omobranchini. (I have not made an exhaustive search for a swimbladder in other genera of the Omobranchini.) The larvae of many diverse genera of blenniids are pelagic surface dwellers for a short period and it is conceivable that in most genera the swimbladder regresses, becomes nonfunctional and finally disappears before the adult stage is attained. Horn (1975) has recently documented such swimbladder regression in stromateoid fishes. Confirmation of such a process in the Blenniidae, however, would require histological studies of the larvae. Despite the apparently labile nature of

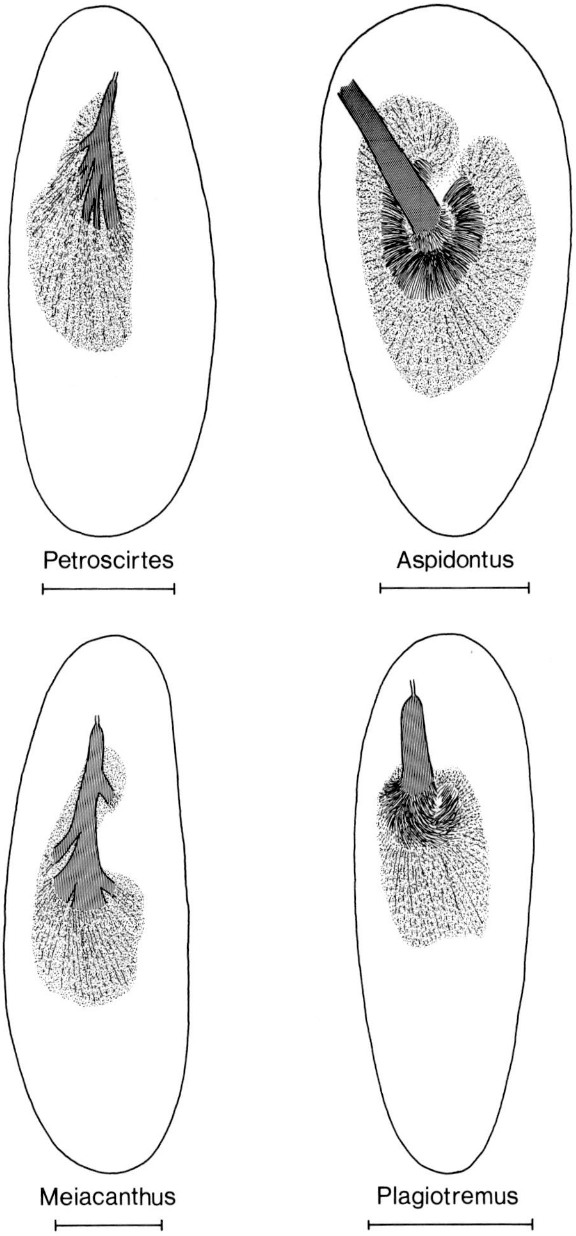

Figure 98. — Dorsal view of swimbladders of selected species of Nemophini: *Petroscirtes variabilis,* USNM 212003, 50.6 mm; *Aspidontus taeniatus,* ANSP 128048, 68.7 mm; *Meiacanthus grammistes,* USNM 211929, 50.3 mm; *Plagiotremus rhinorhynchos,* USNM 204373, 60.8 mm. Retia mirabilia striped; gas gland stippled. Scale = 2.0 mm.

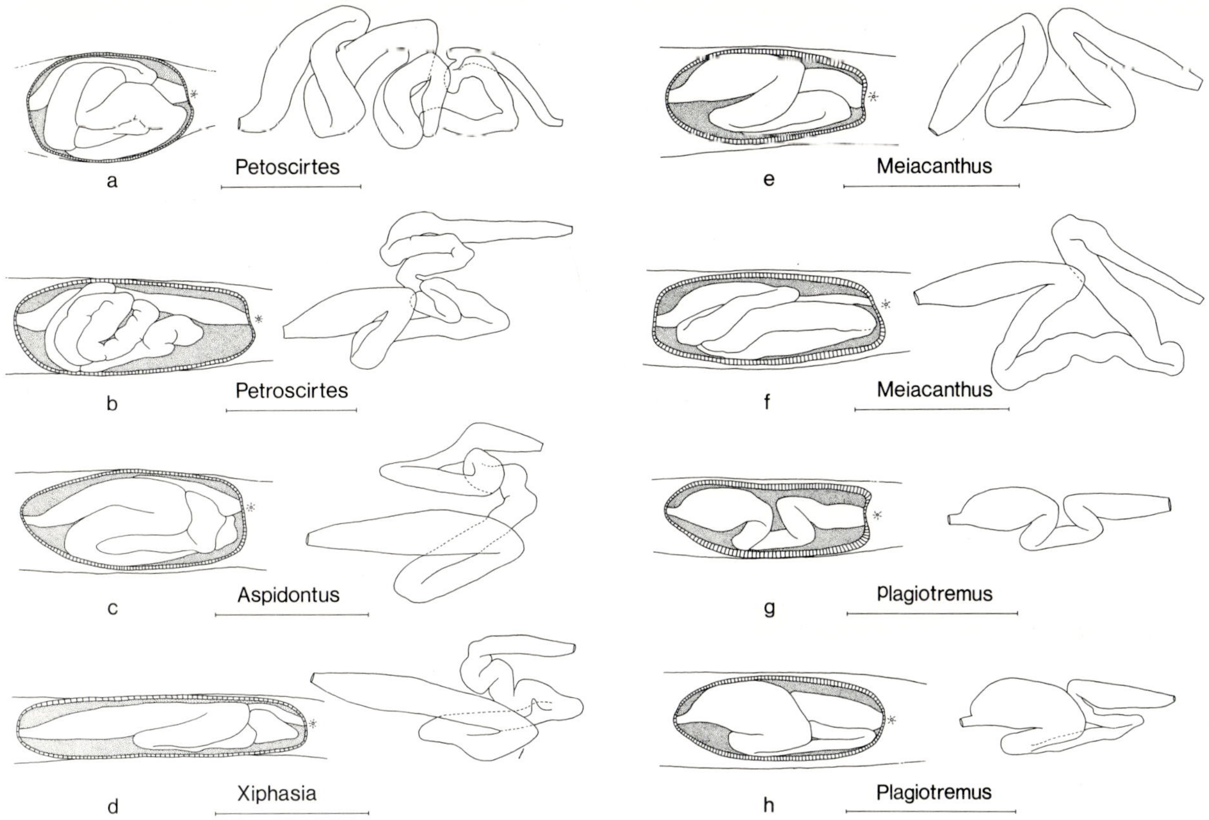

Figure 99. — Ventral views of guts in situ and unfolded of selected species of Nemophini: a, *Petroscirtes mitratus,* ANSP 124769, 43.0 mm; b, *Petroscirtes thepassi,* USNM 165710, 56.4 mm; c, *Aspidontus taeniatus,* USNM 122374, 60.2 mm; d, *Xiphasia setifer,* ANSP 124869, 296 mm; e, *Meiacanthus anema,* ANSP 124867, 42.8 mm; f, *Meiacanthus grammistes,* ANSP 124868, 47.9 mm; g, *Plagiotremus laudandus,* ANSP 126230, 49.3 mm; h, *Plagiotremus rhinorhynchos,* USNM 209845, 57.4 mm. Scale = 10.0 mm.

the structure, the presence of a well-developed swimbladder in the Phenablenniini, at least one genus of Omobranchini and in the adults of four of the five genera of Nemophini leads me to believe that its absence in *Xiphasia* is a secondary specialization.

The gross swimbladder structure of representative species of Nemophini is shown in Figure 98. I was unable to detect the presence of a diaphragm or oval, but N. B. Marshall (pers. comm.) has observed that in at least two species of *Petroscirtes* a diaphragm divides the swimbladder into anterior and posterior chambers of the type usually found in perciform fishes. The swimbladders of *Petroscirtes* and *Meiacanthus* agree in having the rete mirabile divided into several smaller bundles. A similar arrangement was also found in the Phenablenniini and Omobranchini, suggesting that this is the most generalized state. In both species of *Aspidontus* many retia are arranged in a circular pattern around the base of the central retial bundle before entering the relatively large gas gland. In *Plagiotremus* the gas gland is not as extensive and the retia are fanlike and not divided into as many discrete bundles.

Dentition. — In all species of Nemophini the number of incisors increases with increase in standard length. In most blenniids the number of premaxillary and dentary incisors is approximately equal or there are more premaxillary incisors. In most species of *Plagiotremus,* however, the dentary incisors greatly outnumber the premaxillary incisors (Table 4). The shape and articulation of the functional dentary incisors of *Plagiotremus* (Figs. 72-73) appear to be unique and facilitates their bizarre mode of feeding (refer to discussion of biology).

In most blenniids the dentary canines (if present) are short and robust, but in the Nemophini they are long and slender. *Meiacanthus* is unique among fishes in having grooved dentary canines (Figs. 43-44) associated with toxic buccal glands. *Aspidontus* is unique within the Nemophini in having the dentary canines of prejuveniles (but not adults) strongly recurved (Fig. 29). Premaxillary canines are usually present in the adults of all species of Nemophini except *Aspidontus taeniatus* (Table 12) and *Plagiotremus.* It is perhaps more than coincidental that within the Nemophini only these species are known to engage in aggressive feeding behavior.

Gut morphology. — The external gut morphology of the Nemophini (Fig. 99) appears to be correlated with feeding habits and, to a lesser extent, body shape. Feeding habits are discussed in the general section on biology for each genus. As a general rule, phytophagus fishes have long intestines, while those of zoophagus fishes are shorter. As noted by Norman (1943: 794) and confirmed by me, species of Salariini with fine movable teeth adapted for scraping algae have long, elaborately coiled intestines. Although no species of Nemophini appears to be a primary herbivore, species of *Petroscirtes* have omnivorous feeding habits and relatively elongate guts. The gut of *Aspidontus taeniatus* (Fig. 99c) is similar to that of *Petroscirtes,* although not quite as elongate, suggesting that the aggressive fin-eating behavior observed in adults of *taeniatus* probably is not the primary means of nourishment. In contrast, the short thick-walled gut of *Plagiotremus* clearly reflects its obligatory, aggressive scale-eating habit. The gut of *Plagiotremus* is unique in having an expanded, pouchlike region anteriorly. A relatively short gut, similar to that of *Meiacanthus,* is characteristic of *Blennius,* the Phenablenniini and some species of Omobranchini, and is considered to be the most generalized condition in the Nemophini.

Peritoneal pigmentation. — The peritoneal pigment (in preservation) is essentially dark brown or black in *Petroscirtes, Meiacanthus* and *Aspidontus dussumieri,* consists of a speckling of discrete melanophores in *Xiphasia* and *A. taeniatus,* and is white in *Plagiotremus.* Because the peritoneal pigment is also light or absent in *Blennius,* the Phenablenniini and a few species of Omobranchini, I believe that a heavily pigmented peritoneum is a derived character state within the Nemophini.

The peritoneal pigmentation is also quite variable in several other groups of fishes, including the freshwater Cyprinidae. Hubbs and Black (1947: 7) and Snelson (1971: 460) have noted that phytophagus cyprinids usually have a dark peritoneum and elongate intestines, while zoophagus cyprinids have silvery or speckled peritoneums and shorter intestines. A similar correlation seems to exist between feeding habits and peritoneal pigmentation in the Nemophini and other blenniids. The following hypothesis for the development of dark peritoneal pigment in cyprinids (McAllister, 1960: 42) would apply equally well to blenniids:

> "Black peritoneum in phytophagus cyprinids may be explained on the basis that light is detrimental to the digestion of plant material. The black pigment in these forms would then be preventing the penetration of external light (sunlight) to the intestine. Light may be detrimental to plant digestion either by destruction of commensal bacteria (most of which are killed or inactivated by light) used in digesting celluloses or by destruction of an enzyme such as cellulase or carbohydrase . . ."

INTERGENERIC RELATIONSHIPS

My attempts to determine the interrelationships of the five genera of Nemophini have generally resulted in frustration. Although the tribe Nemophini exhibits a remarkable array of apomorphic characters, most are autapomorphic (unique) for a given genus and provide no clues to intergeneric relationships. Because a more detailed study of the osteology than is given here holds promise of casting additional light on the phylogeny of the Nemophini, the following analysis of interrelationships is provisional. A dendogram summarizing one possible phylogeny is given below.

A sister-group relationship between *Petroscirtes* and *Meiacanthus* is suggested by two synapomorphies not found in the other genera: (1) gill opening restricted above level of dorsalmost pectoral-fin ray and (2) peritoneal lining black and gut relatively long. A dark peritoneal lining also occurs in *Aspidontus dussumieri* but not in *A. taeniatus,* so the character state in the former species is believed to be a secondary specialization (*i.e.,* the immediate ancestor of *Aspidontus* did not have a black peritoneum).

The presence of 13 or more precaudal vertebrae and relatively wide interorbital regions in all species of *Aspidontus, Xiphasia* and *Plagiotremus* are advanced features that suggest a common origin for these three genera. The alternative plesiomorphic character states (fewer precaudal vertebrae and a relatively narrow interorbital region) are found in at least one subgenus of *Petroscirtes* and *Meiacanthus.* Thus, it seems reasonable to assume that the progenitor of each genus also possessed these plesiomorphic character states.

Although I was able to identify only one shared apomorphic character (depression of the supraoccipital) that suggests a sister-group relationship between *Aspidontus* and *Xiphasia,* they share many plesiomorphic characters not present in *Plagiotremus.* The unique behavioral and morphological trophic specializations of *Plagiotremus* readily distinguish it from all other blenniids. Only the lack of a unifying character for the other genera of Nemophini prevents me from assigning *Plagiotremus* to a monotypic subtribe.

LITERATURE CITED

ADAMS, A. 1850. Preface, pp. v-viii. *In* The zoology of the voyage of H. M. S. "Samarang"; under the command of Captain Sir Edward Belcher, during the years 1843-1846. London.

ALLEN, G. R., B. C. RUSSELL, B. A. CARLSON AND W. A. STARCK, II. 1975. Mimicry in marine fishes. — *Tropical Fish Hobbyist,* 24(1): 47, 49, 52-56, 7 figs.

ASHMOLE, N. P. AND M. J. ASHMOLE. 1967. Comparative feeding ecology of sea birds of a tropical oceanic island. — *Peabody Museum of Natural History, Yale University Bulletin,* 24: 1-131, 11 figs.

BAUCHOT, M. L. 1967. Catalogue critique des types de poissons du Museum National d'Histoire Naturelle. — *Publications diverses du Museum National d'Histoire Naturelle,* 21: 1-70.

BLEEKER, P. 1851a. Over eenige nieuwe soorten van Blennioiden en Gobioiden van Indischen Archipel. — *Natuurkundig Tijdschrift voor Nederlandsch-Indië,* 1: 236-258.

————. 1851b. Bijdrage tot de kennis der ichthyologische fauna van de Banda-Eilanden. — *Natuurkundig Tijdschrift voor Nederlandsch-Indië,* 2: 225-261.

————. 1852a. Bijdrage tot de kennis der ichthyologische fauna van de Moluksche Eilanden. Visschen van Amboina en Ceram. — *Natuurkundig Tijdschrift voor Nederlandsch-Indië,* 3: 229-309.

————. 1852b. Nieuwe bijdrage tot de kennis der ichthyologische fauna van het eilang Banka. — *Natuurkundig Tijdschrift voor Nederlandsche-Indië,* 3: 715-738.

————. 1853a. Derde bijdrage tot de kennis der ichthyologische fauna van Amboina. — *Natuurkundig Tijdschrift voor Nederlandsch-Indië,* 4: 91-130.

————. 1853b. Bijdrage tot de kennis der ichthyologische fauna van Ternate. — *Natuurkundig Tijdschrift voor Nederlandsch-Indië,* 4: 131-140.

————. 1853c. Bijdrage tot de kennis der ichthyologische fauna van Solor. — *Natuurkundig Tijdschrift voor Nederlandsch-Indië,* 5: 67-96.

————. 1854. Vijfde bijdrage tot de kennis der ichthyologische fauna van Amboina. — *Natuurkundige in Nederlandsch-Indië,* 6: 455-508.

————. 1855. Zesde bijdrage tot de kennis der ichthyologische fauna van Amboina. — *Natuurkunding Tijdschrift voor Nederlandsch-Indië,* 8: 391-434.

————. 1857a. Achtste Bijdrage tot de Kennis der Vischfauna van Amboina.—*Acta Societatis Scientiarum Indo-Neerlandicae,* 2: 1-102.

————. 1857b. Bijdrage tot de Kennis der Ichthyologische Fauna van het Eiland Nias. — *Natuurkundig Tijdschrift voor Nederlandsch-Indië,* 12: 211-288.

————. 1858. Vierde Bijdrage tot de Kennis der Vischfauna van Biliton. — *Natuurkundig Tijdschrift voor Nederlandsch-Indië,* 15: 219-240.

————. 1859. Bestuursvergadering, gehouden den 14 Oktober 1858 ten huize van den heer De Lange. — *Natuurkundig Tijdschrift voor Nederlandsch-Indië,* 16: 330-342.

————. 1864. Notice sur une Nouvelle Espece de *Xiphasia.* — *Nederlandsch Tijdschrift voor de Dierkunde,* 2: 194-196.

————. 1865. Notice sur une Novelle Espece de *Xiphasia.* *Verslagen en Mededeelingen der Koinklikje van Wetenschappen, Afdeeling Natuurkunde,* series 1, 17: 194-196.

————. 1868. Description de deux Espece Nouvelles de Blennioides de l'Inde Archipelagique. — *Verslagen en Mededeelingen der Koninklikje Akademie van Wetenschappen, Letterkunde, en Schoone Kunsten te Amsterdam,* series 2, 2: 78-280.

———. 1869. Neuvieme notice sur la Fauna Ichthyologique du Japon. — *Verslagen en Mededeelingen der Koinklikje van Wetenschappen, Letterkunde, en Schoone Kunsten te Amsterdam,* series 2, 3: 237-252.

———. 1875. Poissons de Madagascar et de l'ile de la Reunion. *In* Recherches sur la Faune de Madagascar et de Ses Dependances d'apres les Decouvertes de Francois P. L. Pollen et D. C. Van Dam. Pt. 4: 1-104.

BLÜM, V. 1972. The influence of ovine follicle stimulating hormone (FSH) and luteinizing hormone (LH) on the male reproductive system and the skin of the Mediterranean blenniid fish *Blennius sphinx* (Valenciennes). — *Journal of Experimental Zoology,* 181 (2): 203-216, 9 figs.

BÖHLKE, J. E. AND C. R. ROBINS. 1968. Western Atlantic seven-spined gobies, with descriptions of ten new species and a new genus, and comments on Pacific relatives. — *Proceedings of the Academy of Natural Sciences of Philadelphia,* 120 (3): 45-174, 21 figs.

BOLIN, R. L. 1947. The evolution of the marine Cottidae of California with a discussion of the genus as a systematic category. — *Stanford Ichthyological Bulletin,* 3 (3): 153-168, 1 fig.

BORODIN, N. A. 1928. Scientific Results of the Yacht Ara Expedition during the years 1926 to 1928, while in Command of William K. Vanderbilt: Fishes. — *Bulletin of the Vanderbilt Oceanographic Museum,* 1(2): 1-37, 2 figs., 5 pls.

BORSIERI, C. 1904. Contribuzione alla Conoscenza della Fauna Ittiologica della Colonia Eritrea. — *Annali del Museo Civico di Storia Naturale di Genova,* series 3a, 1: 187-220.

BREDER, C. M. 1927. The locomotion of fishes. — *Zoologica,* 4(5): 159-297, 83 figs.

BRIGGS, J. C. 1974a. Marine zoogeography. — McGraw-Hill Book Company, New York: xiv + 475 pp.

———. 1974b. Operation of zoogeographic barriers. — *Systematic Zoology,* 23(2): 248-256.

BROCK, V. E. 1948. A new blennoid fish from Hawaii. — *Pacific Science,* 11(2): 125-127, 1 fig.

BURGESS, W. E. 1974. Salts from the seven seas. — *Tropical Fish Hobbyist,* 22 (April): 37-42.

BURGESS, W. E. AND H. R. AXELROD. 1972. Pacific marine fishes. — T. F. H. Publications, Inc., Neptune City, New Jersey, book 1: 280 pp., 489 figs.

———. 1973a. Pacific marine fishes. — book 2: 283-560, 552 figs.

———. 1973b. Pacific marine fishes. — book 3: 562-839, 343 figs.

———. 1974. Pacific marine fishes. — book 5: 1113-1381, 456 figs.

CANTOR, T. 1850. Catalogue of Malayan fishes. — *Journal of Asiatic Society of Bengal,* 1849, 18(2): i-xii + 983-1443, 14 pls.

CHABANAUD, M. P. 1924. Descriptions d'une espèce novelle et d'une forme supposée nouvelle de poissons de ner de la côte d'Annam. — *Bulletin du Museum National d'Histoire Naturelle de Paris,* 30(5): 357-363.

CHAUDHURI, B. L. 1916. Descriptions of two new fish from the Chilka Lake. — *Records of the Indian Museum,* 12(3): 105-108.

COHEN, D. M. 1973. Zoogeography of the fishes of the Indian Ocean. *In* Ecological Studies. — analysis and synthesis, Vol. 3, B. Zeitzschel, ed: 451-463, 1 fig.

CROIZAT, L., G. NELSON AND D. E. ROSEN. 1974. Centers of origin and related concepts. — *Systematic Zoology,* 23(2): 265-287, 2 figs.

CUVIER, G. L. F. D. AND A. VALENCIENNES. 1836. Histoire naturelle des Poissons. Vol. 11, 506 pp. Paris.

DAWSON, C. E. 1964. A bibliography of anomalies of fishes. — *Gulf Research Reports,* 1(6): 308-399.

DAY, F. 1878. The fishes of India . . . Burma and Ceylon, 2: i-xx + 321-778, pls. 69-195. London.

———. 1888. Observations on the fishes of India. Part I. — *Proceedings of the Scientific Meetings of the Zoological Society of London for the year 1888:* 258-265.

DE BEAUFORT, L. F. 1951. *Petroscirtes* Rüppell. *In* de Beaufort and Chapman, Fishes of the Indo-Australian Archipelago, 9: 355-381, figs. 51-53. Leiden.

DE SILVA, P. H. D. H. 1958. The occurrence of *Gonostoma elongata* Günther and *Xiphasia setifer* Swainson off Ceylon. — *Spolia Zeylanica,* 28(2): 141-144, 2 figs.

DE VIS, C. W. 1884. New fishes in the Queensland Museum. — *Proceedings of the Linnean Society of New South Wales,* 2(4): 685-698.

———. 1886. On a lizard and three species of *Salarias,* & c. — *Proceedings of the Royal Society of Queensland,* 1885, 2(2): 56-60.

DOAK, W. 1972. Fishes of the New Zealand region. — Hodder and Stoughton, Auckland: 132 pp., 48 pls.

EIBL-EIBESFELDT, I. 1959. Der Fisch *Aspidontus taeniatus* als Nachahamer des Putzers *Labroides dimidiatus.* — *Zeitschrift fur Tierpsychologie,* 16(1): 19-26.

EMILIANI, C. 1952. Nomenclature and grammar. — *Systematic Zoology,* 42(5): 137-141.

FISHELSON, L. 1974. Histology and ultrastructure of the recently found buccal toxic gland in the fish *Meiacanthus nigrolineatus* (Blenniidae).—Copeia 1974(2): 386-392, 4 figs.

———. 1975. Observations on behaviour of the fish *Meiacanthus nigrolineatus* Smith-Vaniz (Blenniidae) in nature (Red Sea) and in captivity. — *Australian Journal of Marine and Freshwater Research,* 26(3): 329-341, 7 figs.

FOURMANOIR, P. 1971. Notes ichtyologiques (IV). — *Cahiers ORSTOM, series Oceanography,* 9(4): 491-500, 8 figs.

FOWLER, H. W. 1903. Descriptions of several fishes from Zanzibar Island, two of which are new. — *Proceedings of the Academy of Natural Sciences of Philadelphia,* 4: 161-176, pls. 6-8.

———. 1928. The fishes of Oceania. — *Bernice P.*

Bishop Museum, memoir 10: iii + 540 pp., 49 pls., 82 figs.

————. 1937. Zoological results of the third de-Schauensee Siamese Expedition. Part 8. Fishes obtained in 1936. — *Proceedings of the Academy of Natural Sciences of Philadelphia,* 89: 125-264, 300 figs.

————. 1944. The fishes. *In* Results of the fifth George Vanderbilt Expedition (1941). — *Monograph of the Academy of Natural Sciences of Philadelphia,* 6: 57-530, 268 figs., 20 pls.

————. 1946. A collection of fishes obtained in the Riu Kiu Islands by Captain Ernest R. Tinkham A. U. S. — *Proceedings of the Academy of Natural Sciences of Philadelphia,* 98: 123-218, 76 figs.

FRIDMAN, D. AND D. MASRY. 1971. *Xiphasia setifer* Swainson, a blenniid fish new for the Gulf of Elat. — *Scientific Newsletter, the Hebrew University of Jerusalem Marine Biological Laboratory,* No. 1: 6.

FRYER, G., P. H. GREENWOOD AND E. TREWAVAS. 1955. Scale-eating habits of African cichlid fishes. — Nature, 175(4468): 1089-1090.

GARMAN, S. 1903. Some fishes from Australasia. — *Bulletin of the Museum of Comparative Zoology,* 39(8): 229-241.

GILCHRIST, J. D. F. AND W. W. THOMPSON. 1908. The Blenniidae of South Africa. — *Annals of the South African Museum,* 6(2): 97-143.

GILL, T. 1865. On a remarkble new type of fishes allied to *Nemophis.* — *Annals of the Lyceum of Natural History in New York,* 8: 138-141, pl. 3.

GRIFFITHS, G. C. D. 1972. The phylogenetic classification of the diptera *Cyclorrapha,* with special reference to the male postabdomen. — *The Hague, Junk, series Entomology,* Vol. 8: 340 pp.

GOSLINE, W. A. 1963. Notes on the osteology and systematic position of *Hypoptychus dybowskii* Steindachner and other elongate perciform fishes. — *Pacific Science,* 17(1): 90-101, 8 figs.

————. 1971. The zoogeographic relationships of Fanning Island inshore fishes. — *Pacific Science,* 25(2): 282-289, 2 figs.

GOSLINE, W. A. AND V. E. BROCK. 1960. Handbook of Hawaiian fishes. — The University of Hawaii Press, Honolulu: 372 pp., 277 figs.

GRANT, E. M. 1972. Guide to fishes. — Queensland Department of Primary Industries, Brisbane, 2nd ed.: 472 pp.

GÜNTHER, A. C. L. G. 1861. Catalogue of the Acanthopterygian fishes in the collection of the British Museum, 3: xxv + 586 pp.

————. 1862. Catalogue of the Acanthopterygian fishes in the collection of the British Museum, 4: xxi + 534 pp.

————. 1877. Andrew Garrett's Fische der Sudsee. IV. — *Journal des Museum Godeffroy,* 13: 169-216, pls. 101-120.

————. 1880. Report on the shore fishes procured during the voyage of H. M. S. Challenger in the years 1873-1876. — *Zoology,* 1(6):1-82.

HENNIG, W. 1966. Phylogenetic systematics. University of Illinois Press, Urbana: 269 pp.

HERRE, A. W. C. T. 1926. Four rare Philippine fishes. — *The Philippine Journal of Science,* 31(3): 217-225, 1 fig., 1 pl.

————. 1927. Four new fishes from Lake Taal (Bombon).—*The Philippine Journal of Science,* 34(3): 273-279, 3 pls.

————. 1931. A check list of fishes from the Solomon Islands. — *Journal of the Pan-Pacific Research Institution,* 6(4): 4-9.

————. 1934. Notes on new or little known fishes from southeastern China. — *Lingnan Science Journal,* 13(2): 285-296.

————. 1935. New fishes obtained by the Crane Pacific Expedition. — *Field Museum of Natural History, Zoological series,* 335, 18(12): 383-455, figs. 31-33.

————. 1936a. Fishes in the zoological museum of Stanford University. III. New Genera and Species of Gobies and Blennies and a new *Myxus,* from the Pelew Islands and Celebes. — *The Philippine Journal of Science,* 59(2): 275-287, 1 pl.

————. 1936b. Fishes of the Crane Pacific Expedition. — *Field Museum of Natural History, Zoological series,* 353, 21: 472 pp., 50 figs.

————. 1939. The Philippine blennies. — *The Philippine Journal of Science,* 70(4): 315-373, 5 pls.

————. 1940. New species of fishes from the Malay Peninsula and Borneo. — *Bulletin of the Raffles Museum,* 16: 5-26.

————. 1942. Contributions from the zoological museum of Stanford University, California, IX. Two new species of *Petroscirtes* and a key to the Philippine species. — *Copeia,* 1942(2): 111-116.

HERRE, A. W. C. T. AND E. S. HERALD. 1951. Noteworthy additions to the Philippine fish fauna with descriptions of a new genus and species. — *The Philippine Journal of Science,* 79(3): 309-340, 12 figs.

HOBSON, E. S. 1968. Predatory behavior of some shore fishes in the Gulf of California. — *United States Department of the Interior, Bureau of Sport Fisheries and Wildlife, Research Report* 73: 1-92, 26 figs.

————. 1969. Possible advantages to the blenny *Runula azalea* in aggregating with the wrasse *Thalassoma lucasanum* in the Tropical Eastern Pacific. *Copeia,* 1969(1): 191-193, 1 fig.

————. 1974. Feeding relationships of teleostean fishes on coral reefs in Kona, Hawaii. — *Fishery Bulletin,* 72(4): 915-1031, 42 figs.

HOBSON, E. S. AND E. H. CHAVE. 1972. Hawaiian reef animals. The University Press of Hawaii, Honolulu: i-xiv + 135 pp., 86 pls.

HOESE, H. D. 1966. Ectoparasitism by juvenile sea catfish, *Galeichthys felis*. — *Copeia*, 1966(4): 880-881.

HORN, M. H. 1975. Swim-bladder state and structure in relation to behavior and mode of life in stromateoid fishes. — *Fishery Bulletin*, 73(1): 95-109, 10 figs.

HUBBS, C. L. AND J. D. BLACK. 1947. Revision of *Ceratichthys*, a genus of American cyprinid fishes. — *Miscellaneous Publications of the Museum of Zoology, University of Michigan*, No. 66: 1-56, 1 fig., 2 pls.

HUBBS, C. L., L. P. SCHULTZ, V. G. SPRINGER AND J. E. RANDALL. 1963. Comments on the proposed rejection of the type designations of Jordan and Evermann, 1896-1900 and 1896. Z. N. (S.) 1279. — *Bulletin of Zoological Nomenclature*, 20(4): 257-259.

JATZOW, R. AND H. LENZ. 1898. Fische von Ost-Afrika, Madagaskar und Aldabra. — *Abhandlungen der Senckenbergischen Naturforschenden Gesellschaft*, 21(3): 497-531.

JONES, S. AND M. KUMARAN. 1965. New records of fishes from the seas around India. Part II. — *Journal of the Marine Biological Association of India*, 7(1): 108-123, 18 figs.

JORDAN, D. S. AND C. H. BOLLMAN. 1890. Scientific results of explorations by the U. S. Fish Commission steamer Albatross. No. IV. — Descriptions of new species of fishes collected at the Galapagos Islands and along the coast of the United States of Colombia, 1887-'88. — *Proceedings of the United States National Museum*, 12(770): 149-183.

JORDAN, D. S. AND B. W. EVERMANN. 1898. The fishes of North and Middle America. — *United States National Museum Bulletin*, 47(3): i-xxiv + 2183a-3136.

JORDAN, D. S. AND C. L. HUBBS. 1925. Record of fishes obtained by David Starr Jordan in Japan, 1922. — *Memoirs of the Carnegie Museum*, 10(2): 93-346, pls. 5-12.

JORDAN, D. S. AND R. E. RICHARDSON. 1908. Fishes from islands of the Philippine Archipelago. — *Bulletin of the Bureau of Fisheries*, 1907, 27: 235-287, 12 figs.

JORDAN, D. S. AND A. SEALE. 1905. List of fishes collected by Bashford Dean on the island of Negros, Philippines. — *Proceedings of the United States National Museum*, 28 (1407): 769-803, 20 figs.

————. 1906. The fishes of Samoa. — *Bulletin of the Bureau of Fisheries*, 1905, 25: 173-455, 110 figs., pls. 38-53.

————. 1907. Fishes of the islands of Luzon and Panay. — *Bulletin of the Bureau of Fisheries*, 1906, 26: 1-48, 20 figs.

JORDAN, D. S. AND J. O. SNYDER. 1902. A review of the blennoid fishes of Japan. — *Proceedings of the United States National Museum*, 25 (1293): 441-504, 28 figs.

KATAYAMA, M. 1941. A new blennoid fish from Toyama Bay.—*The Zoological Magazine* [*Dobutsugaku Zasshi*], 53(12): 591-592.

KAUP, J. 1858. On *Nemophis*, a new genus of riband-shaped fishes. — *Proceedings of the Zoological Society of London*, 26: 168-169, 1 fig.

KAVANAUGH, D. H. 1972. Hennig's principles and methods of phylogenetic systematics. — *The Biologist*, 54(3): 115-127, 1 fig.

KENDALL, W. C. AND E. L. GOLDSBOROUGH. 1911. The shore fishes. XIII. *In* Reports on the scientific results of the expedition to the tropical Pacific, in charge of Alexander Agassiz, by the U. S. Fish Commission steamer "Albatross", from August, 1899, to March, 1900, Commander Jefferson F. Moser, U. S. N., commanding. — *Museum of Comparative Zoology, Harvard, Memoir* 26(7): 239-344, 7 pls.

KLAUSEWITZ, W. 1962. *Meiacanthus smithi* n.sp. aus dem Indischen Ozean (Pisces, Percomorphi, Blenniidae). — *Senckenbergiana Biologica*, 43(1): 17-19, 1 fig.

KLUNZINGER, C. B. 1871. Synopsis der Fische des Rothen Meeres. Part 2. — *Verhandlungen der Kaiser-Königlichen Zoologisch-botanischen Gesellschaft in Wien*, 21: 441-688.

KNER, R. 1868a. Uber neue Fische aus dem Museum der Herren Johann Casär Godeffroy und Sohn in Hamburg. — *Sitzungsberichte der Kaiserlichen Akademie der Wissenschaften in Wien*, 58(1-2), 1(7): 26-31.

————. 1868b. Folge neurer Fischen aus dem Museum der Herren Joh. Cas. Godeffroy and Sohn in Hamburg. — *Sitzungsberichte der Kaiserlichen Akademie der Wissenschaften in Wien*, 58(3-4), 1(8): 293-356.

KNER, R. AND F. STEINDACHNER. 1866. Neue Fische aus dem Museum der Herren Joh. C. Godeffroy und Sohn in Hamburg. — *Sitzungsberichte der Kaiserlichen Akademie der Wissenschaften in Wien*, 54(9): 356-395, 5 pls.

KOSSMAN, R. AND H. RAÜBER. 1877. Pisces. Pages 1-34. *In* R. Kossman, Zoologische Ergebnisse einer in Auftrage der Königlichen Academie der Wissenschaften zu Berlin ausgefuhrten Reise in die Küstengebiete des Rothen Meeres. [Each article individually paginated.]

KUTHALINGAM, M. D. K. AND K. K. P. MENON. 1965. A note on the occurrence of *Xiphasia setifer* (Swainson) off Mangalore, west coast of India. — *Journal of the Marine Biological Association of India*, 7(1): 214-215.

LIEM, K. F. AND D. J. STEWART. 1976. Evolution of the scale-eating cichlid fishes of Lake Tanganyika: A generic revision with a description of a new species. — *Bulletin of the Museum of Comparative Zoology*, 147 (7): 319-350, 16 figs., 2 pls.

LIENARD, E. 1839. Zoologie. *In* Rapport Annel sur les Travaux. — *Societe Royale des Arts et des Sciences de l'lile Maurice*, 10: 27-49.

LOSEY, G. S. 1972. Predation protection in the poison-fang blenny, *Meiacanthus atrodorsalis*, and its mimics *Ecsenius bicolor* and *Runula laudandus* (Blenniidae). — *Pacific Science*, 26(2): 129-139, 2 figs.

————. 1974. *Aspidontus taeniatus:* effects of increased abundance on cleaning symbiosis with notes on pelagic dispersion and *A. filamentosus* (pisces, Blenniidae). — *Zeitschrift für Tierpsychologie,* 34: 430-435.

MACLEAY, W. 1881. Descriptive catalogue of the fishes of Australia. Part 3. — *Proceedings of the Linnean Society of New South Wales,* 6(1): 138 pp., 2 pls.

————. 1885. Notices on new fishes. — *Proceedings of the Linnean Society of New South Wales,* 9(1): 170-172.

MACDONALD, G. A. 1949. Hawaiian petrographic providence. — *Bulletin of the Geographical Society of America,* 60: 1541-1596.

MAJOR, P. F. 1973. Scale feeding behavior of the leatherjacket, *Scomberoides lysan* and two species of the genus *Oligoplites* (Pisces: Carangidae). — *Copeia,* 1973(1): 151-154.

MASUDA, H., C. ARAGA AND T. YOSHINO. 1975. Coastal fishes of southern Japan. — Tokai University Press, Tokyo: 379 pp., Illus.

MENON, A. G. K. AND P. C. VARMA. 1967. On a rare blennid fish *Xiphasia setifer* Swainson from the Madras Coast. — *Journal of the Bombay Natural History Society,* 63(3): 752-755, 2 figs.

MOHAN, R. S. L. 1968. On a collection of blennies from Gujarat Coast with some new records. — *Journal of the Marine Biological Association of India,* 10(1): 118-125, 6 figs.

MUNRO, I. S. R. 1955. Eggs and larvae of the sabre-toothed oyster blenny, *Dasson steadi* (Whitley) (Blenniidae).—*Australian Journal of Marine and Freshwater Research,* 6(1): 30-34, 2 figs.

————. 1958. The fishes of the New Guinea Region. — *Papua and New Guinea Agricultural Journal,* 10(4): 97-369.

————. 1967. The fishes of New Guinea. — Department of Agriculture, Stock and Fisheries, Port Moresby, New Guinea: 650 pp., 84 pls.

MCALLISTER, D. E. 1960. A collection of oceanic fishes from off British Columbia with a discussion of the evolution of black peritoneum. — *Bulletin of the National Museum of Canada,* No. 172: 39-43.

MCCULLOCH, A. R. 1917. Ichthyological notes. — *Australian Zoologist,* 1(4): 89-93, 1 pl.

NELLEN, W. 1973. Fischlarven des Indischen Ozeans. Ergebnisse der Fischbrutuntersuchungen wahrend der ersten Expedition des Forschungsschiffes "Meteor" in den Indischen Ozean und den Persischen Golf, Oktober 1964 bis April 1965. — *"Meteor" Forschungsergebnisse-Ergebnisse, series D,* No. 14: 1-66, 47 figs.

NELSON, G. 1974. Historical biogeography: an alternative formalization. — *Systematic Zoology,* 23(4): 555-558, 2 figs.

NICHOLS, J. T. 1924. A contribution to the ichthyology of the Galapagos. — *Zoologica,* 5(4): 63-65, 1 fig.

NORMAN, J. R. 1943. Notes on the blennioid fishes. I. A provisional synopsis of the genera of the family Blenniidae. — *Annals and Magazine of Natural History, series 11,* 10(72): 793-812.

OGILBY, J. D. 1887. Catalogue of the fishes of New South Wales, with their principal synonyms. — Appendix A of Report of the Commissioners of Fisheries up to 31 December 1886, Department of Fisheries, New South Wales: 67 pp.

————. 1894. Description of five new fishes from the Australasian Region. — *Proceedings of the Linnean Society of New South Wales,* 9(2): 367-374.

————. 1896. Descriptions of two new genera and species of Australian fishes. — *Proceedings of the Linnean Society of New South Wales,* 1896(2): 136-142.

————. 1910. On new or insufficiently described fishes. — *Proceedings of the Royal Society of Queensland,* 23: 1-55.

OKADA, Y. AND K. SUZUKI. 1952. A new blennoid fish from the Sea of Japan. — *Report of Faculty of Fisheries, Prefectural University of Mie,* 1(2): 75-77, 1 fig.

PENRITH, M. J. AND M. L. PENRITH. 1972. The Blenniidae of western southern Africa. — *Cimbebasia, series A,* 2(5): 66-89, 28 figs., 2 pls.

PETERS, W. 1855a. Uebersicht der in Mossambique Beobachten Fische. — *Archiv für Naturgeschichte,* 1: 234-282.

————. 1855b. Übersicht der in Mossambique Beobachten Seefische. — *Monatsberichte der Koniglich Preussischen Akademie der Wissenschaften zu Berlin,* 1855: 428-466.

PLAYFAIR, P. L. 1868. On a collection of fishes from Madagascar. — *Proceedings of the Zoological Society of London,* 1868: 9-12.

PUTNAM, F. W. 1874. Remarks on the family Nemophidae. — *Proceedings of the Boston Society of Natural History,* 16: 366-374, 1 fig.

QUOY, J. R. N. AND P. GAIMARD. 1834. Voyage decouvertes de L'Astrolabe, Execute par Ordre du Roi, Pendant les annees 1826-1827-1828-1829, sous de commandement de M. J. Dumont d'Urville. — *Zoologie,* 3:647-729.

RANDALL, J. E. 1955. A revision of the surgeon fish genera *Zebrasoma* and *Paracanthurus.* — *Pacific Science,* 9(4): 396-412, 8 figs.

RANDALL, J. E. AND H. A. RANDALL. 1960. Examples of mimicry and protective resemblance in tropical marine fishes. — *Bulletin of Marine Science of the Gulf and Caribbean,* 10(4): 444-480, 15 figs.

READ, K. R. H. 1970. From Catalina to Cabo San Lucas with Calypso. — *Oceans,* 3(4): 24-32.

REGAN, C. T. 1905. On fishes from the Persian Gulf, the Sea of Oman and Karachi collected by Mr. F. W. Townsend. — *Journal of the Bombay Natural History Society,* 16(2): 318-333, 3 pls.

————. 1912. The classification of the blennioid fishes. — *The Annals and Magazine of Natural History, 8th series,* 10(36): 265-280, 4 figs.

RICHARDSON, J. 1845. Ichthyology. *In* the zoology of the voyage of H. M. S. Sulpher, under the command of Sir

Edward Belcher, R. N., C. B., F. R. G. S., etc., during the years 1836-42, 1: 50-150, pls. 35-64.

Roberts, T. R. 1970. Scale-eating American characoid fishes, with special reference to *Probolodus heterostomus.* — *Proceedings of the California Academy of Sciences, 4th series,* 38(20): 383-390.

————. 1972. Ecology of fishes in the Amazon and Congo Basins. — *Bulletin of the Museum of Comparative Zoology,* 143(2): 117-147.

Rosenblatt, R. H., J. E. McCosker and I. Rubinoff. 1972. Indo-west Pacific fishes from the Gulf of Chiriqui, Panama. — *Contributions in Science, Los Angeles County, Natural History Museum,* No. 234: 1-18, 3 figs.

Rüppell, E. 1830. Atlas zu der Reise im Nördlichen Afrika. Zoologie. — *Fische des Rothen Meeres,* 16: 95-118.

————. 1838. Fisches des Rothen Meeres. *In* Neu Wirbelthiere zu der Fauna von Abyssinien Gehörig, 4: 1-28, pls. 1-7.

Russell, B. C. 1976. Disguise, defense and aggression. — Australian Natural History, 18 (19): 324-329, 13 figs.

Russell, B. C., G. R. Allen and H. R. Lubbock. (in press). New cases of mimicry in marine fishes. — *Journal of Zoology, London.*

Ryan, M. 1973. How to catch a canary. — *The Marine Aquarist,* 4(3): 43-47.

Sauvage, H. E. 1880. Description de Quelques Blennioides de la Collection de ·Museum d'Histoire Naturelle. — *Bulletin de la Societe Philomatique de Paris, series 7,* 4(1879-1880): 215-220.

Sawyer, F. C. 1952. The dates of publications of Wilhelm Peter Eduard Simon Rüppell's [1794-1884] "Neu Wirbelthiere zu der Fauna von Abyssinien Gehorig" (fol., Frankfurt a.M., 1835-1840). — *The Journal of the Society for the Bibliography of Natural History,* 2(9): 407.

Schoener, T. W. 1967. The ecological significance of sexual dimorphism in size in the lizard *Anolis conspersus.* — *Science,* 155: 474-477.

Schultz, L. P. 1950. Ichthyology. Notes on the blennioid fish genera *Runula* (subfamily Petroscirtinae) and *Tripterygion* and *Helcogramma* (family Clinidae), of the American tropical Pacific. — *Journal of the Washington Academy of Sciences,* 40(8): 266-268.

Seale, A. 1936. A new member of the blenny family. — *Proceedings of the California Academy of Sciences,* 23(6): 85-86.

Selander, R. K. 1966. Sexual dimorphism and differential niche utilization in birds. — *Condor,* 68(2): 113-151, 8 figs.

Shen, S. C. and W. H. Ting. 1972. Ecological and morphological study on fish-fauna from the waters around Taiwan and its adjacent islands. No. 2. Notes on some rare continental shelf fishes and description of two new species. — *Bulletin of the Institute of Zoology, Academia Sinica,* 11(1): 13-31, 7 figs.

Smith, C. L. 1971. A revision of the American groupers: *Epinephelus* and allied genera. — *Bulletin of the American Museum of Natural History,* 146(2): 67-242, 41 figs.

Smith, H. M. 1934. Contributions to the ichthyology of Siam. XVIII. The blenniid fishes of Siam, with descriptions of new species. — *Journal of the Siam Society, Natural History Supplement,* 9(3): 315-323, pls. 13-14.

Smith, J. L. B. 1949. The sea fishes of South Africa. Central News Agency Limited, Cape Town, 550 pp., 103 pls., 1232 text-figs. (Also, 1953, 1961 and 1965 editions).

————. 1959. Fishes of the families Blenniidae and Salariidae of the western Indian Ocean. — *Rhodes University, Department of Ichthyology, Ichthyological Bulletin,* 14: 229-252, 16 figs., pls. 14-17.

Smith, R. L. 1974. On the biology of *Blennius cristatus* with special reference to anal fin morphology. — *Bulletin of Marine Science,* 24(3): 595-605, 5 figs.

Smith-Vaniz, W. F. 1969. A new species of *Meiacanthus* (Pisces: Blenniidae: Nemophidinae) from the Red Sea, with a review of the Indian Ocean species. — *Proceedings of the Biological Society of Washington,* 82: 349-354, 1 fig.

————. 1975. Supplemental description of rare blenniid fish *Phenablennius heyligeri* (Bleeker). — *Proceedings of the Academy of Natural Sciences of Philadelphia,* 127(6): 53-55, 3 figs.

Smith-Vaniz, W. F. and J. E. Randall. 1973. *Blennechis* filamentosus Valenciennes, the prejuvenile of *Aspidontus taeniatus* Quoy and Gaimard (Pisces: Blenniidae). — *Notulae Naturae,* Academy of Natural Sciences of Philadelphia, No. 448: 1-11, 3 figs.

Smith-Vaniz, W. F. and V. G. Springer. 1971. Synopsis of the tribe Salariini, with description of five new genera and three new species (Pisces: Blenniidae). — *Smithsonian Contributions to Zoology,* No. 73: 1-72, 51 figs.

Smith-Vaniz, W. F. and J. C. Staiger. 1973. Comparative revision of *Scomberoides, Oligoplites, Parona,* and *Hypacanthus* with comments on the phylogenetic position of *Campogramma* (Pisces: Carangidae). — *Proceedings of the California Academy of Sciences,* 39(13): 185-256, 26 figs.

Snelson, F. F., Jr. 1971. *Notropis mekistocholas,* a new herbivorous cyprinid fish endemic to the Cape Fear River Basin, North Carolina. — *Copeia,* 1971(3): 449-462, 2 figs.

Springer, V. G. 1967. The Pacific South American blenniid fish, *Hypsoblennius sordidus.* — *Copeia,* 1967(2): 461-465, 1 fig.

————. 1968. Osteology and classification of the fishes of the family Blenniidae. — *United States National Museum Bulletin,* 284: 1-85, 16 figs., 11 pls.

————. 1970. The blennies. — *Tropical Fish Hobbyist,* 19(2): 54-66, 11 figs.

————. 1971. Revision of the fish genus *Ecsenius* (Blenniidae, Blenniinae, Salariini). — *Smithsonian Contributions to Zoology,* No. 72: 1-74, 36 figs.

————. 1972. Synopsis of the tribe Omobranchini with descriptions of three new genera and two new species (Pisces: Blenniidae). — *Smithsonian Contributions to Zoology,* No. 130: 1-31, 16 figs.

SPRINGER, V. G. AND M. F. GOMON. 1975. Revision of the blenniid fish genus *Omobranchus,* with descriptions of three new species and notes on other species of the tribe Omobranchini. — *Smithsonian Contributions to Zoology,* No. 177: 1-135, 52 figs.

SPRINGER, V. G. AND W. F. SMITH-VANIZ. 1970. *Blennius antholops,* new deep-water fish, from the Gulf of Guinea, with comments on the bathymetric distribution of the family Blenniidae. — *Proceedings of the Biological Society of Washington,* 83(20): 215-220, 2 figs.

————. 1972a. Mimetic relationships involving fishes of the family Blenniidae. — *Smithsonian Contributions to Zoology,* No. 112: 1-36, 4 figs., 7 pls.

————. 1972b. A new tribe (Phenablenniini) and genus (*Phenablennius*) of blenniid fishes based on *Petroscirtes heyligeri* Bleeker. — *Copeia,* 1972(1): 64-71, 5 figs.

SPRINGER, V. G. AND K. D. WOODBURN. 1960. An ecological study of the fishes of the Tampa Bay Area. — *Professional papers of the Florida Board of Conservation,* 1: 1-104.

STRASBURG, D. W. 1959. Notes on the blennioid fishes of Hawaii with descriptions of two new species. — *Pacific Science,* 10(3): 241-267, 4 figs.

STRASBURG, D. W., E. C. JONES AND R. T. B. IVERSEN. 1968. Use of a small submarine for biological and oceanographic research. — *Journal du Conseil. Conseil Permanent International pour l'Exploration de la Mer,* 31(3): 410-426, 7 figs.

STEINDACHNER, F. 1863. Ichthyologische Mittheilungen (VI). — *Verhandlungen der Zoologisch-Botanischen Gesellschaft in Wien,* 13: 1189-1192.

————. 1876. Ichthyologische Beitrage (V). — *Sitzungsberichte der Kaiserlichen Akademie der Wissenschaften in Wien,* 74(1-2, 1): 49-240.

SWAINSON, W. 1839. The natural history of fishes, amphibians, + reptiles, or monocardian animals. Vol. 2: vi + 448 pp., 135 figs. London.

SZE-CHUNG, L. 1957. On a new fish, *Lembeichthys furcocaudalis* sp. nov. (Blenniidae) from South China Sea. — *Acta Zoologica Sinica,* 9(3): 227-232, 4 figs. [In Chinese and English.]

TALBOT, F. H. 1970. The South East Asian Area as a center of marine speciation: an ecological analysis of causes. *In* Symposium on Marine Sciences. — *Reports of the Australian Academy of Science,* No. 12: 43-50, 4 figs.

TANAKA, S. 1908. Notes on some Japanese fishes, with descriptions of fourteen new species. — *Journal of the College of Science, Imperial University,* 23(7): 1-54, 4 pls.

TAYLOR, W. R. 1967a. An enzyme method for cleaning and staining small invertebrates. — *Proceedings of the United States National Museum,* 122(3596): 1-17.

————. 1967b. Outline of a method of clearing tissues with pancreatic enzymes and staining bones of small invertebrates. — *Turtox News,* 45(12): 308-309, 1 fig.

THORNE, R. F. 1963. Biotic distributional patterns in the tropical Pacific. — *In* Pacific Basin Biogeography. J. L. Gressitt, ed. Bishop Museum, Honolulu: 311-350.

TOMIYAMA, I. 1951. On a Japanese blennioid fish, *Dasson trossulus* (Jordan et Snyder). — *The Zoological Magazine (Dobutsugaka Zasshi),* 60(8): 159-161, 7 figs. (In Japanese with English summary).

————. 1955. *Aspidontus fluctuans* (Weber) (Blenniidae). — *In* Tomiyama and Abe, Figures and Descriptions of the Fishes of Japan, 51(301): 1040-1043, 1 pl.

————. 1956a. *Aspidontus filamentosus* (Cuvier and Valenciennes) (Blenniidae). — *Ibid.,* 52(307): 1073-1076, 1 pl.

————. 1956b. *Meiacanthus kamoharai* new species (Blenniidae). — *Ibid.,* 53(309): 1083-1086, 2 pls.

————. 1956c. *Aspidontus rhinorhynchos* (Bleeker) (Blenniidae). — *Ibid.,* 55(316): 1128-1131, 1 pl.

————. 1956d. *Aspidontus tapeinosoma* (Bleeker) (Blenniidae). — *Ibid.,* 55(317): 1132-1134, 1 pl.

TREWAVAS, E. 1947. An example of 'mimicry' in fishes. — *Nature,* 160(4056): 120.

TYLER, J. C. 1968. A monograph on plectognath fishes of the superfamily Triacanthoidea. — *Monograph of the Academy of Natural Sciences of Philadelphia,* 16: 1-364, 209 figs.

VAILLANT, L. 1894. Sur une Collection de Poissons Recueillie en Basse-Californie et dans le Golfe par M. Leon Diguel. — *Extrait du Bulletin de la Societe Philomathique de Paris, series 3,* 6: 69-75.

WAITE, E. R. 1904. Additions to the fish fauna of Lord Howe Island, No. 4. — *Records of the Australian Museum,* 5(3): 135-186, pls. 17-24.

————. 1910. A list of the known fishes of Kermadec and Norfolk Islands, and a comparison with those of Lord Howe Island. — *Transactions of the New Zealand Institute,* 42(47): 370-391, pls. 37-38.

WEBER, M. 1908. Süsswasserfische von Neu-Guinea; ein Beitrag zur Frage nach dem Früheren Zusammenhang von Neu-Guinea und Australien. Pages 201-267. *In* Nova Guinea. *Resultats de l'Expedition Scientifique Neerlandaise a la Nouvelle-Guinee en 1903 sous les auspices de Arthur Wichmann,* 5(Zoology), part 2.

————. 1909. Note IV. Diagnosen Neuer Fischer der Siboga-Expedition. — *Notes from the Leyden Museum,* 31: 143-169.

WHITEHEAD, P. J. P., M. BOESEMAN AND A. C. WHEELER. 1966. The types of Bleeker's Indo-Pacific elopoid and clupeoid fishes. — *Zoologische Verhandelingen,* Leiden No. 84: 1-152, 1 fig., 19 pls.

WHITLEY, G. P. 1926. The biology of North-West Islet, Capricorn Group. — *Australian Zoologist,* 4(4): 227-236, 1 fig., pls. 33-34.

————. 1930. Ichthyological miscellanea. — *Memoirs of the Queensland Museum,* 10(1): 8-31, 1 fig., 1 pl.

————. 1932. Fishes. Great Barrier Reef Expedition 1928-29. — *Scientific Reports,* 4(9): 267-316, 5 figs., 4 pls.

————. 1945. New sharks and fishes from Western Australia. Part 2. — *Australian Zoologist,* 11(1): 1-45, 15 figs., 1 pl.

————. 1961. Fishes from New Caledonia. — *Proceedings of the Royal Zoological Society of New South Wales,* 1958-1959: 60-65, 2 figs.

————. 1968. A check-list of the fishes recorded from the New Zealand Region. — *The Australian Zoologist,* 15(1): 1-102.

WICKLER, W. 1960. Aquarienbeobachtungen an *Aspidontus,* einer ektoparasitischen Fisch. — *Zeitschrift für Tierpsychologie,* 17(3): 277-292.

————. 1961. Über das Verhalten der Blenniiden *Runula* und *Aspidontus* (Pisces, Blenniidae). — *Zeitschrift für Tierpsychologie,* 18(4): 421-440.

————. 1962. A sabre-toothed blenny. — *Tropical Fish Hobbyist,* 10(10): 11-24, 12 figs.

————. 1963. Zum Problem der Signalbildung, am Beispiel der Verhaltens-Mimikry zwischen *Aspidontus* und *Labroides* (Pisces, Acanthopterygii). — *Zeitschrift für Tierpsychologie,* 20(6): 657-679.

————. 1963a. *Aspidontus taeniatus* (Blenniidae). Nahrungsaufnahme.—*Encyclopaedia Cinematographica (Gottingen)* E 140/1962: 1-7.

————. 1968. Mimicry in plants and animals. — McGraw-Hill Book Co., New York. 256 pp., Illus.

WILLIAMS, G. C. 1954. Differential vertical distribution of the sexes in *Gibbonsia elegans* with remarks on the two nominal subspecies of this fish. — *Copeia,* 1954(4): 267-273, 1 fig.

ZUGMAYER, E. 1913. Die Fische von Balutschistan. — *Abhandlungen der Königlich Bayerischen Akademie der Wissenschaften, Mathematisch-Physikalische Klasse,* 26(6): 1-35.

100

101

102

103

Petroscirtes: mitratus (**Fig. 100**), USNM 199433, male, 56.7 mm SL, Madgascar, Tulear; *mitratus* (**Fig. 101**), ANSP 124769, female, 41.9 mm, Australia, Queensland, Little Hope Island; *mitratus* (**Fig. 102**), HUJ F4372, female, 34.3 mm, Red Sea, Sinai Peninsula, Ras Muhammad; *xestus* (**Fig. 103**), NFIS 9765, male, 56.8 mm, Nicobar Islands.

104

105

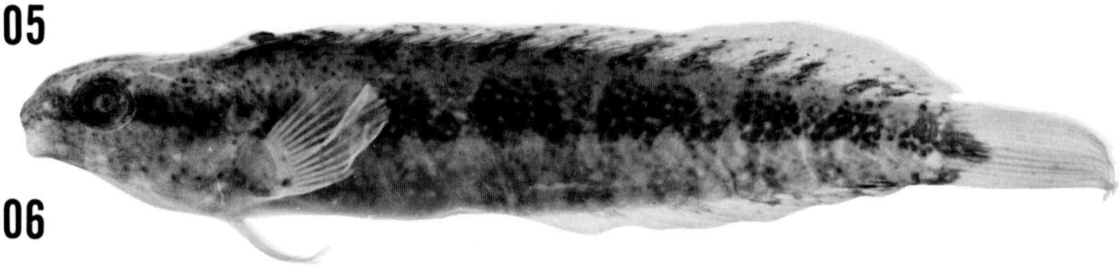

106

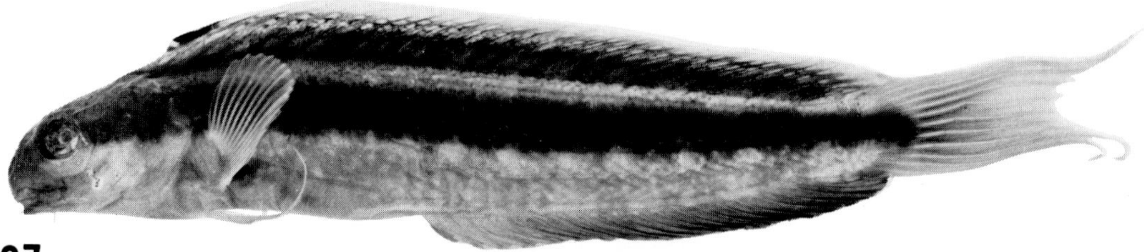

107

108

Petroscirtes: springeri (**Fig. 104**), USNM 203279, holotype, male, 49.7 mm SL, Taiwan (drawn by Mary H. Fuges); *springeri* (**Fig. 105**), data same as preceding; *breviceps* (**Fig. 106**), WAM P.5381, male, 73.1 mm, western Australia, Shark Bay; *breviceps* (**Fig. 107**), WAM P.5516, male, 95.7 mm, western Australia, Woodmans Point; *breviceps* (**Fig. 108**), ANSP 123735, male, 60.8 mm, Hong Kong.

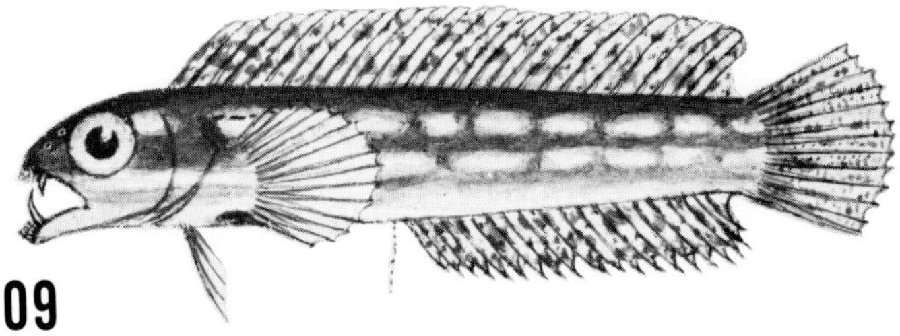

109

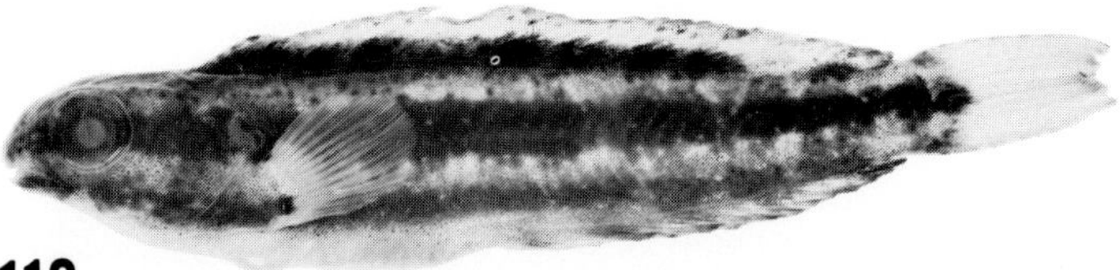

110

111

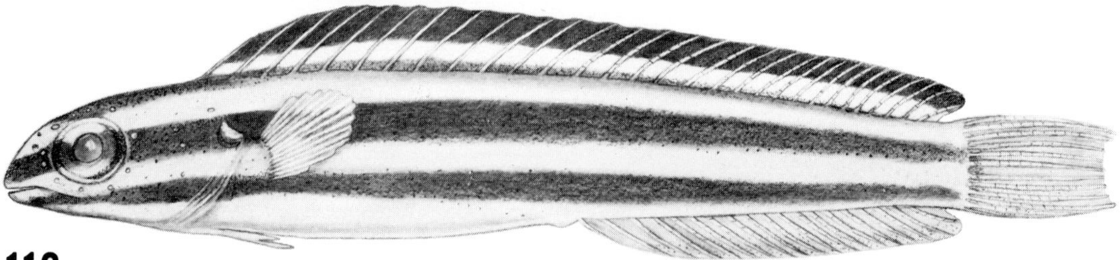

112

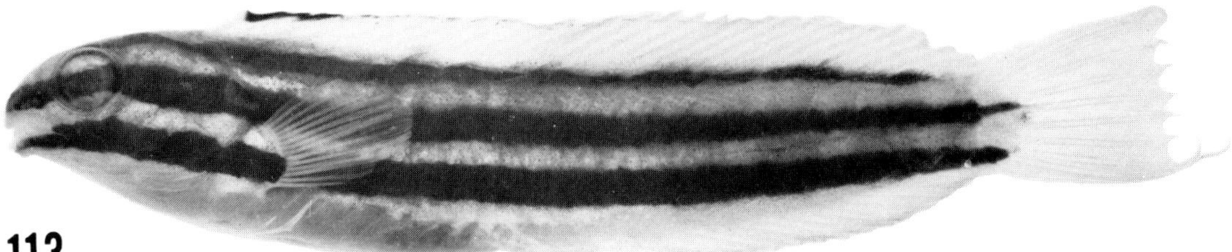

113

Petroscirtes: solorensis (**Fig. 109**), photograph of color illustration of holotype among plates of Bleeker's unpublished atlas of fishes in the Leiden Museum; *breviceps* (**Fig. 110**), USNM 211964, female, 34.1 mm SL, Celebes, Buton (Boeton) Island; *breviceps* (**Fig. 111**), NMV 71761, female, 74.5 mm, Borneo, Salang; *marginatus* (**Fig. 112**), BMNH 1848.3.16.151, holotype, female, 39.2 mm, "China" (Drawn by Mary H. Fuges.); fallax (**Fig. 113**), USNM 201799, male, 54.8 mm, Australia, Queensland, One Tree Island.

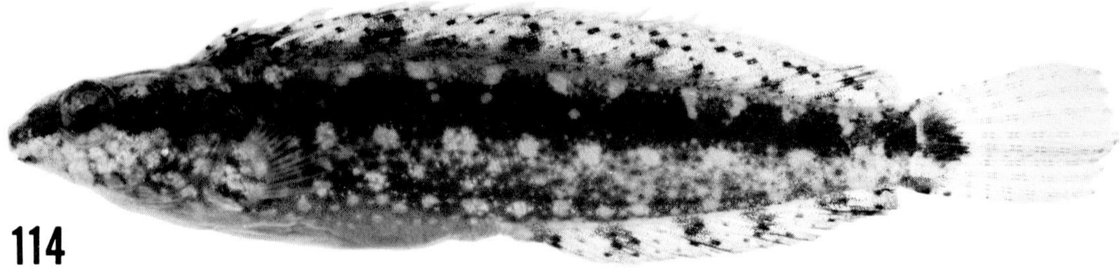

114

115

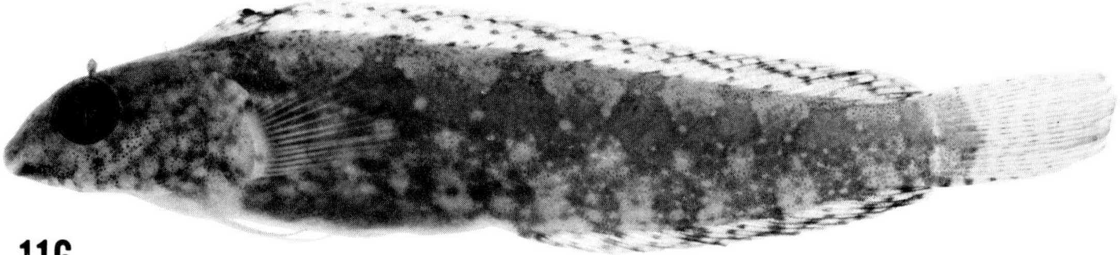

116

117

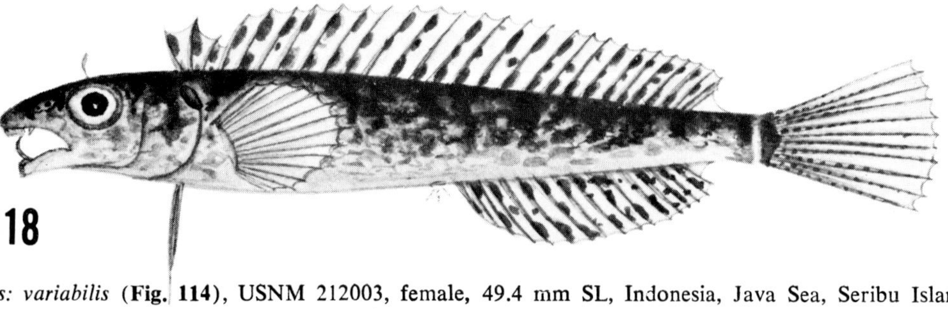

118

Petroscirtes: variabilis (**Fig. 114**), USNM 212003, female, 49.4 mm SL, Indonesia, Java Sea, Seribu Island; *variabilis* (**Fig. 115**), SU 67355, neotype, male, 65.8 mm, Singapore; *variabilis* (**Fig. 116**), ANSP 109685, male, 43.0 mm, Australia, Queensland, Little Hope Island; *thepassi* (**Fig. 117**), USNM 205296, male, 57.3 mm, Palau Islands; *thepassi* (**Fig. 118**), photograph of color illustration of holotype among plates of Bleeker's unpublished atlas of fishes in the Leiden Museum.

119

120

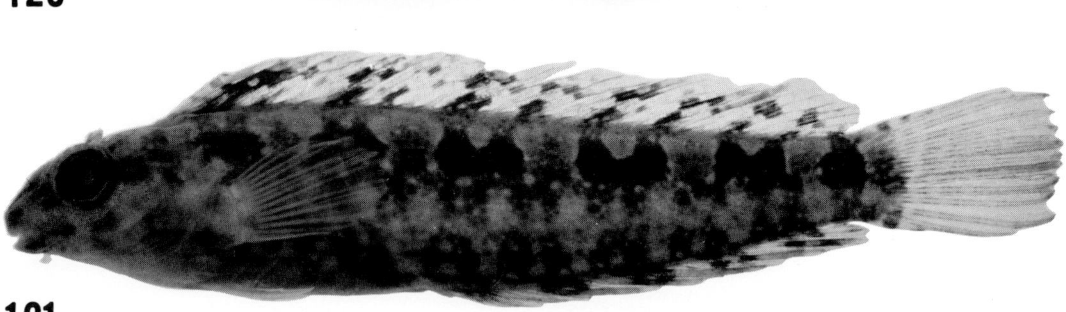

121

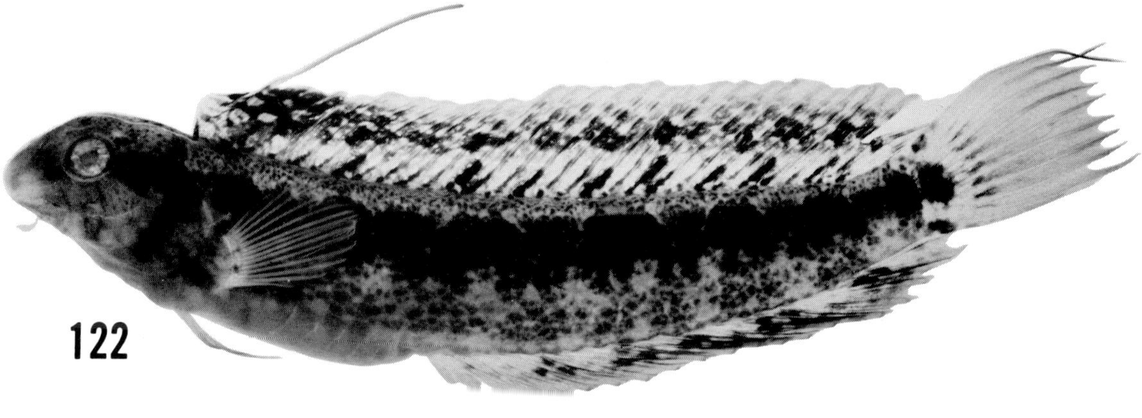

122

Petroscirtes: ancylodon (**Fig. 119**), USNM 147974, female, 57.2 mm SL, Persian Gulf, Tarut Bay, Zoal Island; *ancylodon* (**Fig. 120**), USNM 167027, female, 74.4 mm, Red Sea, Egypt, Hurghada; *lupus* (**Fig. 121**), SU 47023, female, 52.2 mm, New Caledonia, Noumea; *lupus* (**Fig. 122**), AMS uncatalogued, sta. LHI 73-67, male, 104.3 mm, Lord Howe Island.

123

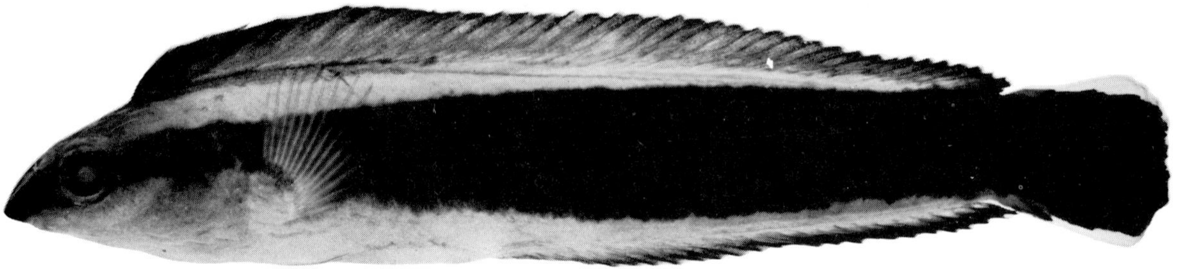

124

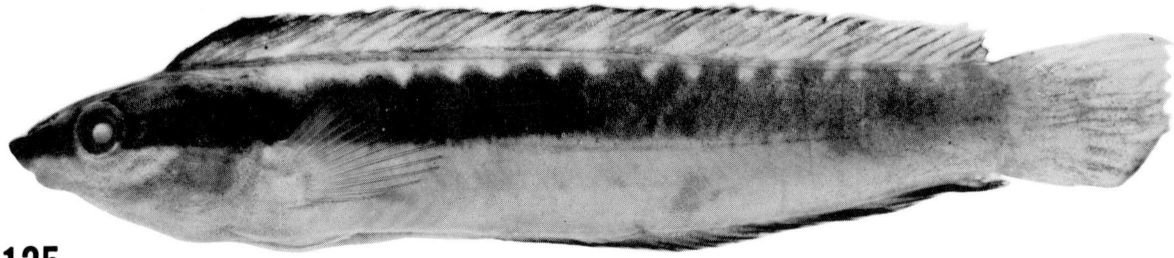

125

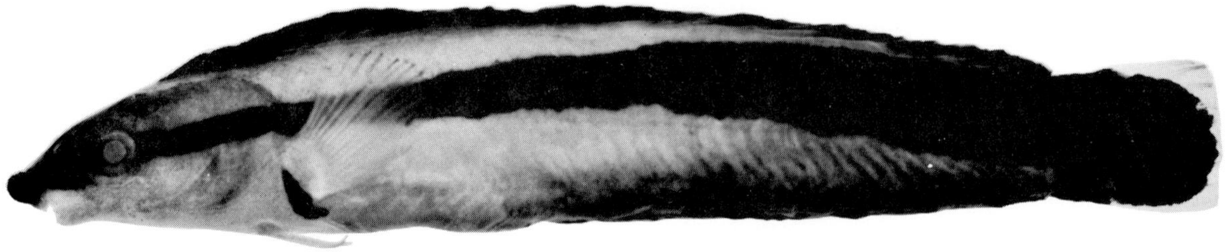

126

Aspidontus taeniatus: t. taeniatus (**Fig. 123**), ANSP 124233, male, 46.8 mm SL, Tuamotu Islands, Rangiroa Atoll; *t. taeniatus* (**Fig. 124**), BPBM 10881, male, 66.1 mm, Marquesas Islands; *t. taeniatus* (**Fig. 125**), BPBM 10871, male, 59.2 mm, Marquesas Islands; *t. tractus* (**Fig. 126**), USNM 204552, female, 61.5 mm, Red Sea, Sinai Peninsula, Ras Muhammad.

127

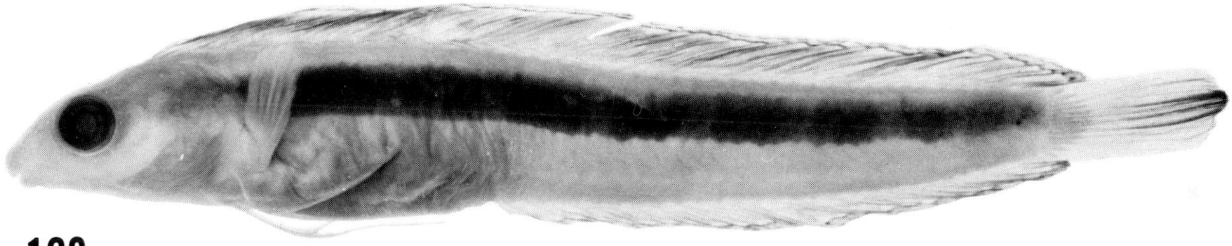

128

129

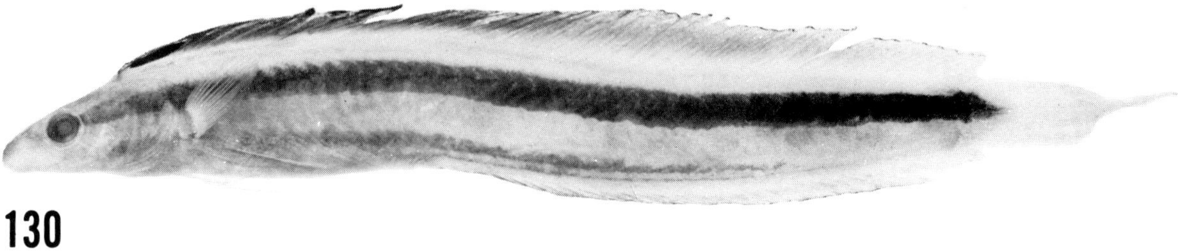

130

Aspidontus dussumieri: (**Fig. 127**), CAS 28880, female, 39.2 mm SL, South China Sea, off South Vietnam; (**Fig. 128**), ANSP 109806, male, 62.7 mm, Amirantes Islands, D'Arros Island; (**Fig. 129**), USNM 204549, male, 98.3 mm, Gulf of Aqaba, Sinai Peninsula, El Himeira; (**Fig. 130**), USNM 204549, female, 100.3 mm, data same as preceding.

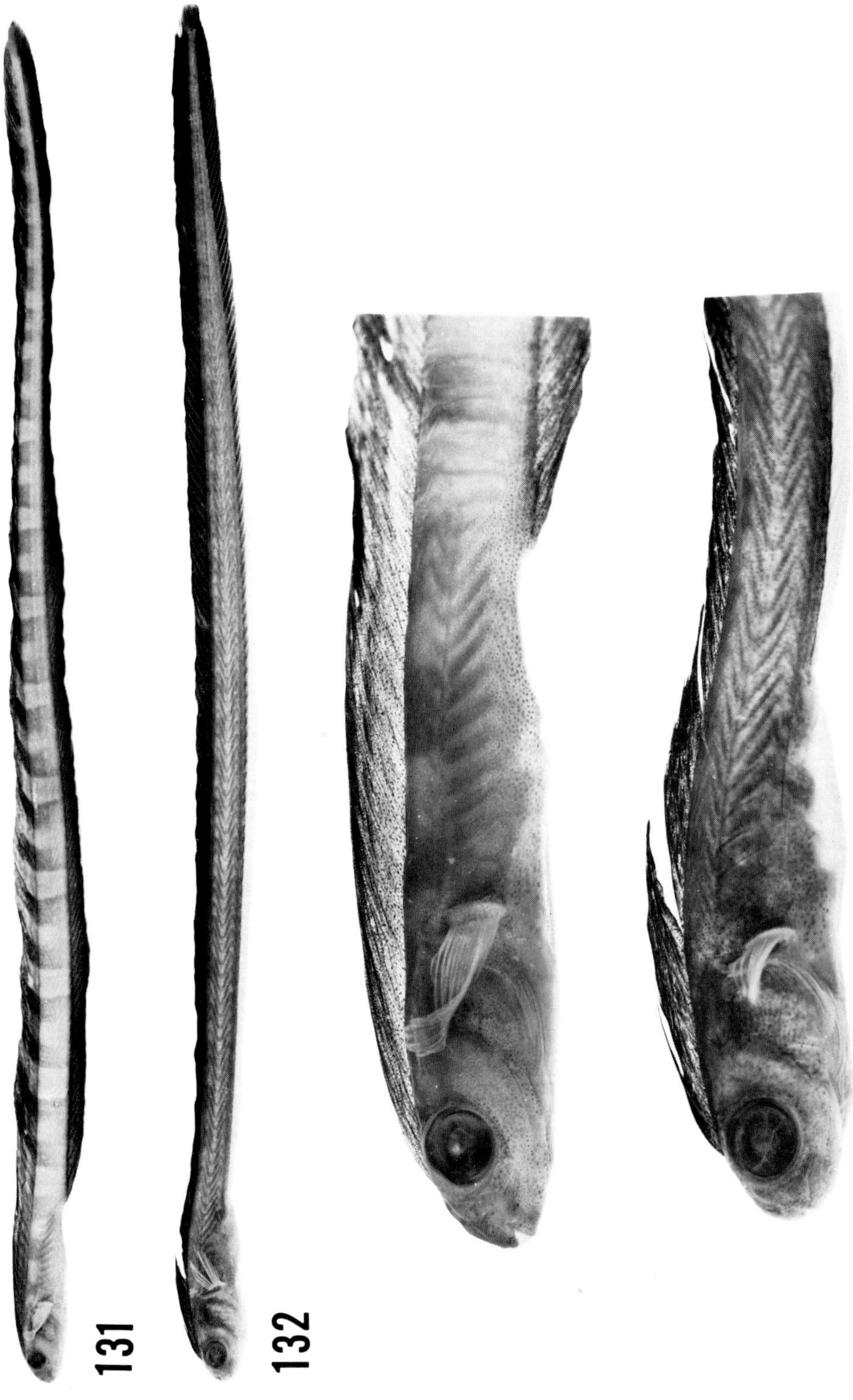

131

132

Xiphasia: setifer (**Fig. 131**), USNM 205292, females, 290 mm SL, Gulf of Thailand (enlarged view below); *matsubarai* (**Fig. 132**), USNM 203276, female, 244 mm, Solomon Islands, Rennell Island (enlarged view below).

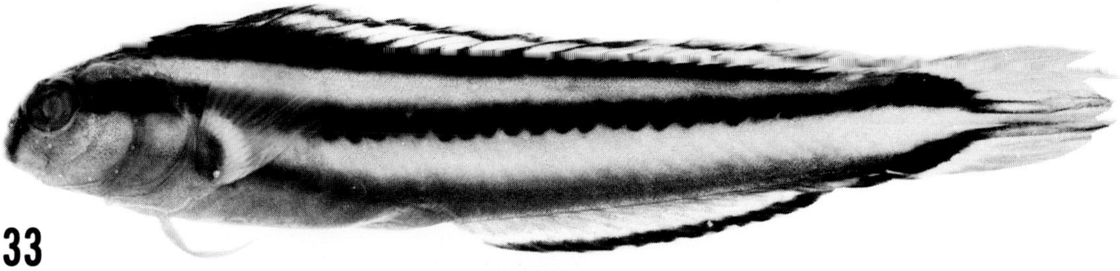

133

134

135

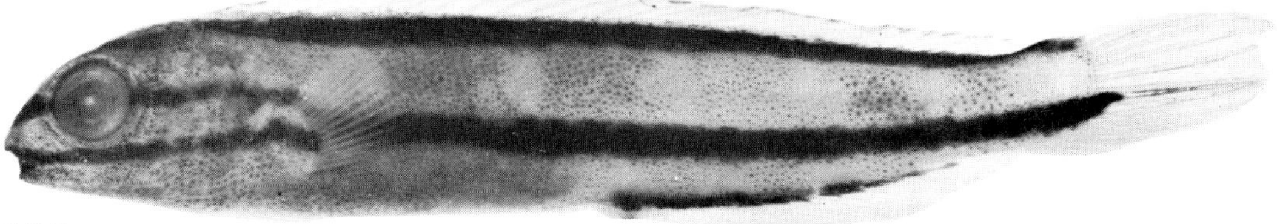

136

Meiacanthus: anema (**Fig. 133**), USNM 211956, male, 40.4 mm SL, Amboina, Wailella River; *anema* (**Fig. 134**), USNM 211956, female, 30.0 mm, same data as preceding (below ventral view); *ditrema* (**Fig. 135**), BPBM 7453, male, 39.6 mm, Ryukyu Islands; *ditrema* (**Fig. 136**), USNM 205294, female, 27.2 mm, Palau Islands.

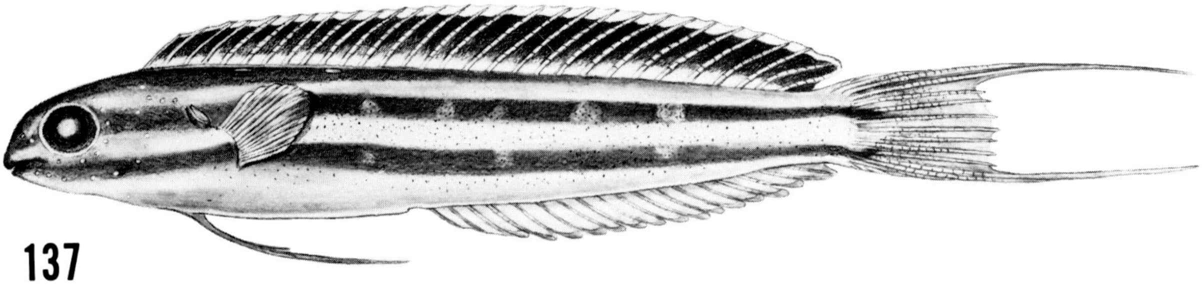

137

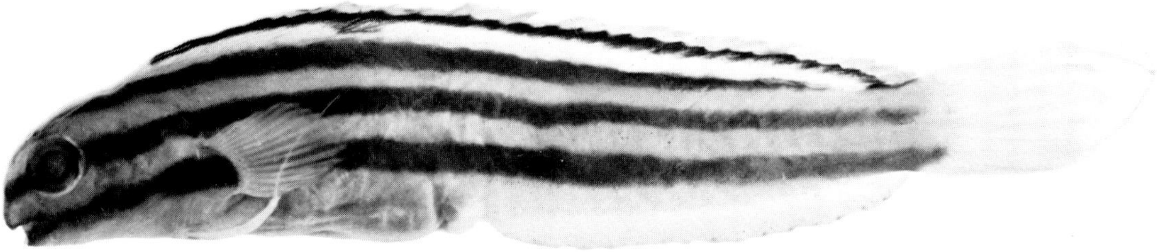

138

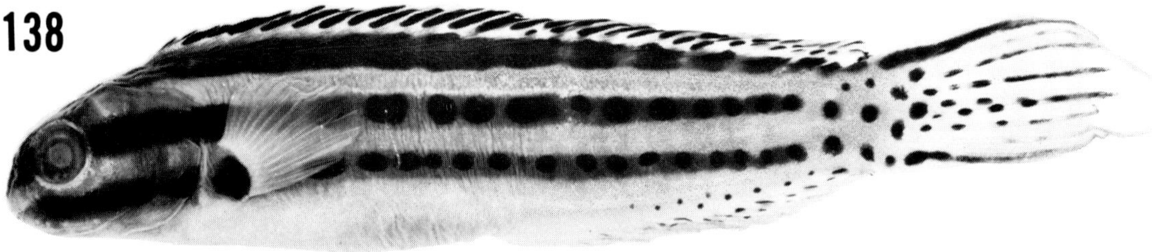

139

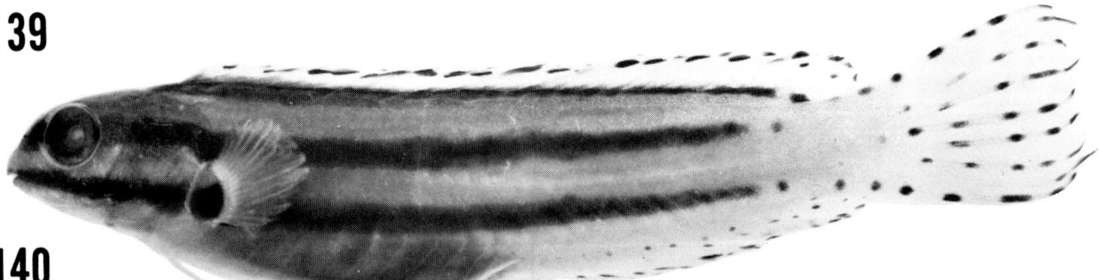

140

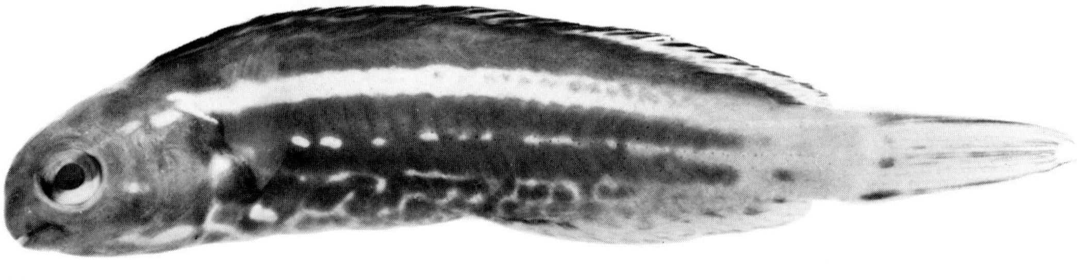

141

Meiacanthus: lineatus (**Fig. 137**), USNM 99370, male, 48.3 mm SL, Sulu Archipelago, Sirun Island (drawn by Mary H. Fuges); *lineatus* (**Fig. 138**), USNM 201363, male, 62.8 mm, Australia, Queensland, One Tree Island; *grammistes* (**Fig. 139**), USNM 211929, male, 55.5 mm, Banda Sea, Banda Island; *grammistes* (**Fig. 140**), ANSP 126026, female, 58.9 mm, Santa Cruz Islands; *kamoharai* (**Fig. 141**), USNM 132705, female, 54.2 mm, Okinawa.

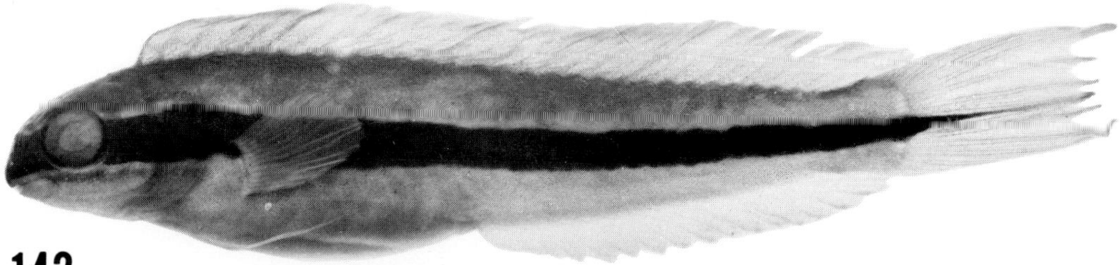

142

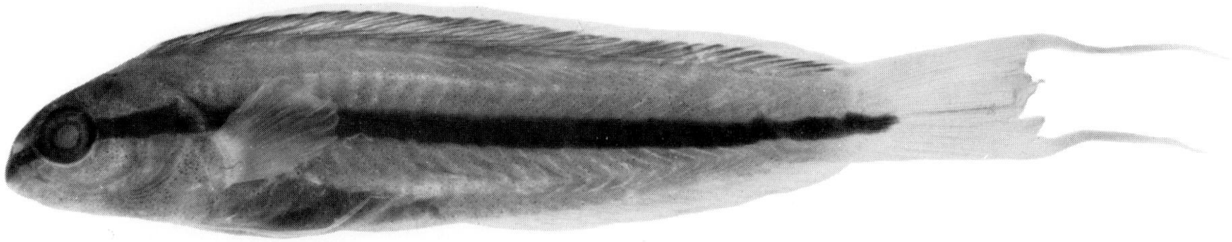

143

144

145

146

Meiacanthus: vittatus (**Fig. 142**), USNM 201453, holotype, male, 45.5 mm SL, New Guinea, Madang; *geminatus* (**Fig. 143**), USNM 205286, holotype, male, 44.0 mm, Borneo; *nigrolineatus* (**Fig. 144**), USNM 205378, male, 45.6 mm, Gulf of Aqaba, near Eilat; *nigrolineatus* (**Fig. 145**), USNM 204486, male, 44.2 mm, off Ethiopia, Sheikh el Abu; *nigrolineatus* (**Fig. 146**), USNM 205378, juvenile, 16.9 mm, Gulf of Aqaba, near Eilat.

147

148

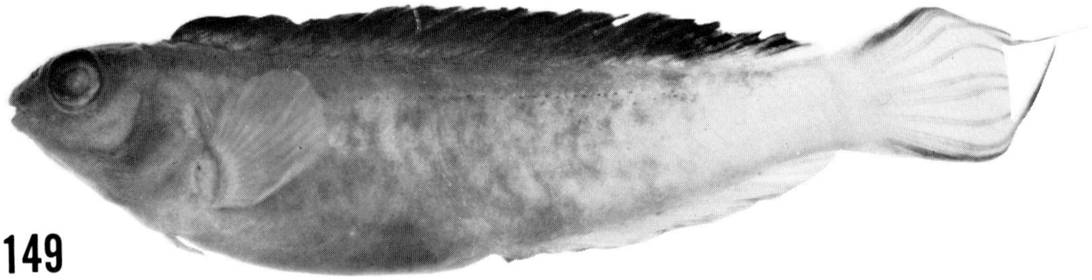

149

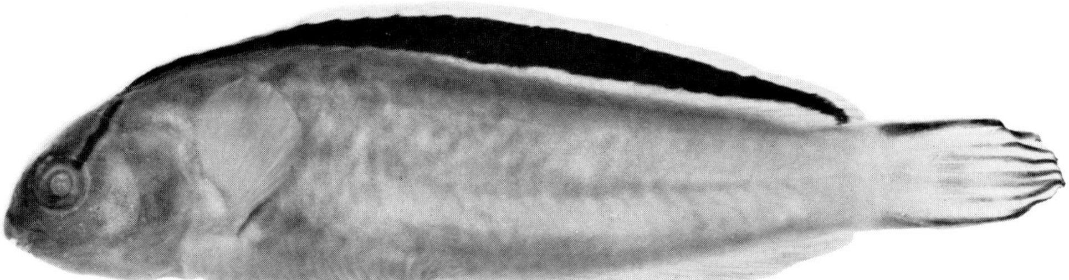

150

Meiacanthus: mossambicus (**Fig. 147**), RUSI Z.516, juvenile, 17.5 mm SL, Zanzibar; *mossambicus* (**Fig. 148**), ANSP 122784, male, 62.5 mm, Mozambique, Pinda; *fraseri* (**Fig. 149**), RUSI 74-309, female, 61.0 mm, St. Brandon Shoals (Cargados Carajos); *smithi* (**Fig. 150**), USNM 205284, male, 54.2 mm, Sri Lanka, Trincomalee (below dorsal view).

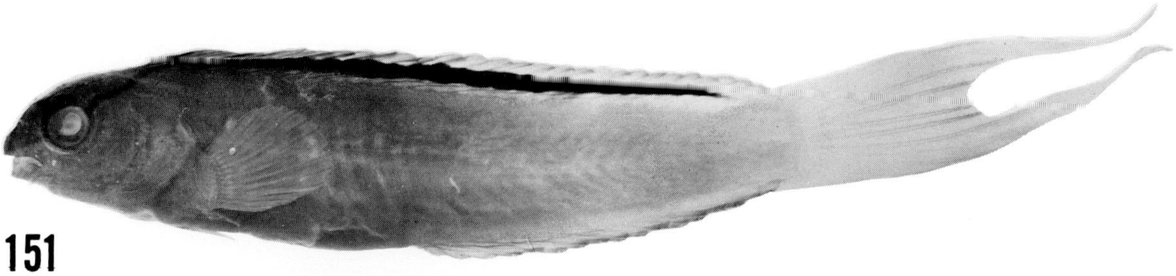

151

152

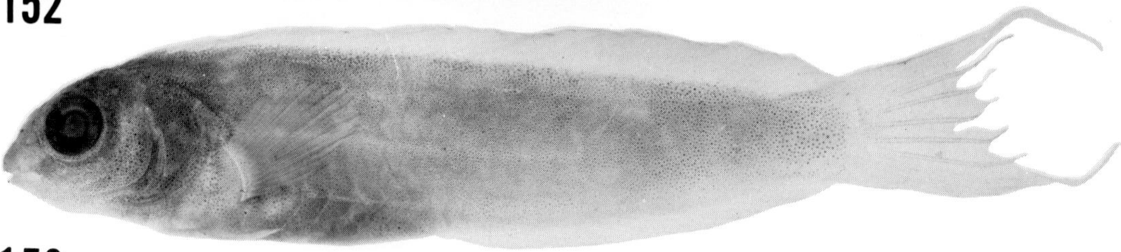

153

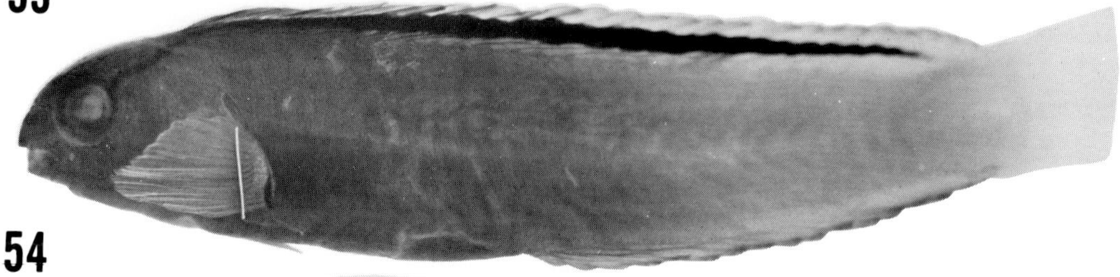

154

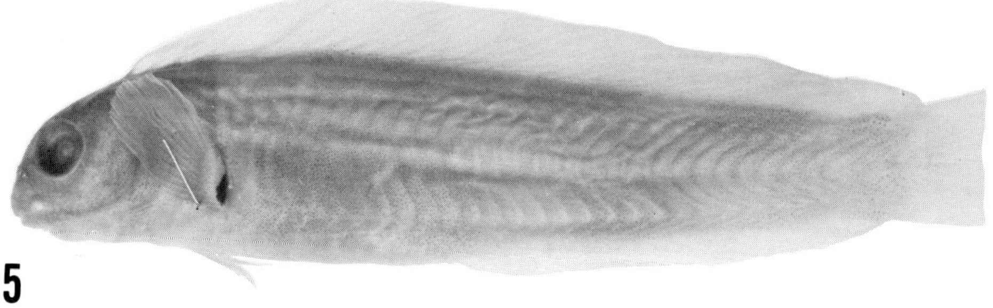

155

Meiacanthus atrodorsalis: a. atrodorsalis (**Fig. 151**), USNM 201593, male, 50.2 mm SL, Australia, Queensland, One Tree Island; *a. atrodorsalis* (**Fig. 152**), USNM 211912, female, 31.7 mm, Celebes, Kabaena Island; *a. ovalauensis* (**Fig. 153**), ANSP 128238, male, 43.7 mm, Fiji Islands, Mbengga Island; *a. atrodorsalis* (**Fig. 154**), same data as Fig. 151; *a. ovalauensis* (**Fig. 155**), USNM 201455, male, 58.5 mm, Fiji Islands, Great Astrolabe.

156

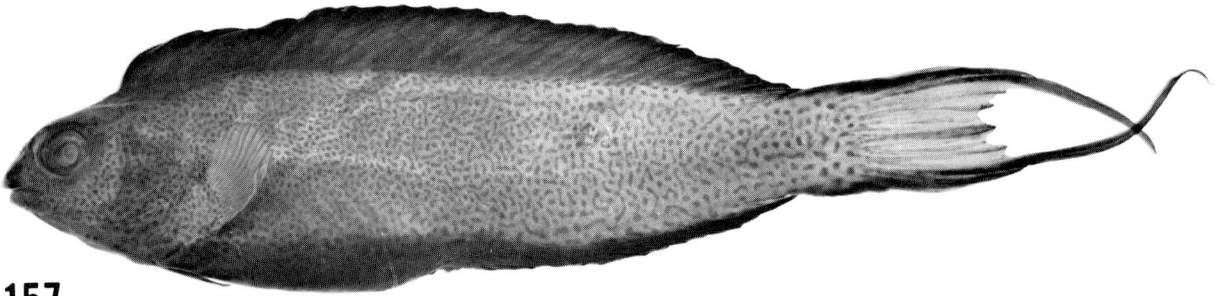

157

158

159

Meiacanthus: procne (**Fig. 156**), USNM 203275, holotype, male, 44.0 mm SL, Tonga Islands, Neiafu Vavau; *reticulatus* (**Fig. 157**), BPBM 13648, holotype, female, 70.5 mm, Solomon Sea, Egum Atoll; *bundoon* (**Fig. 158**), USNM 212257, holotype, male, 63.0 mm, Fiji Islands, Lau Group, Moala Island; *phaeus* (**Fig. 159**), AMS I.17466-004, holotype, female, 55.8 mm, New Caledonia, Puetege Reef.

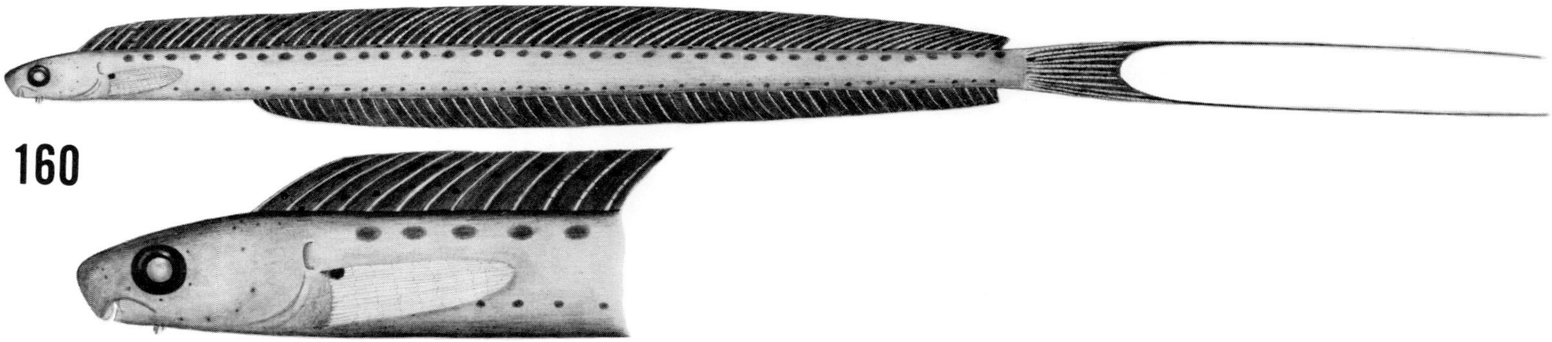

160

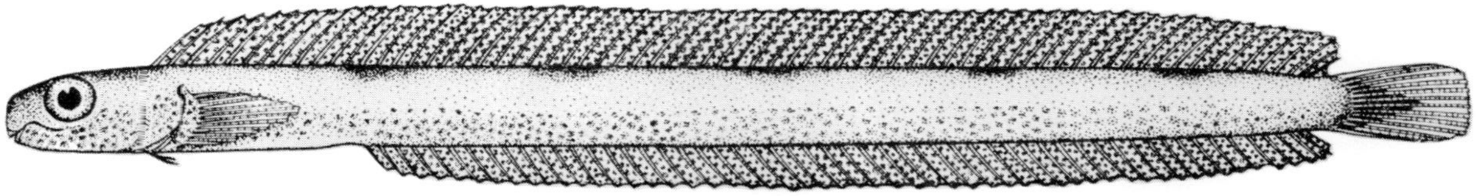

161

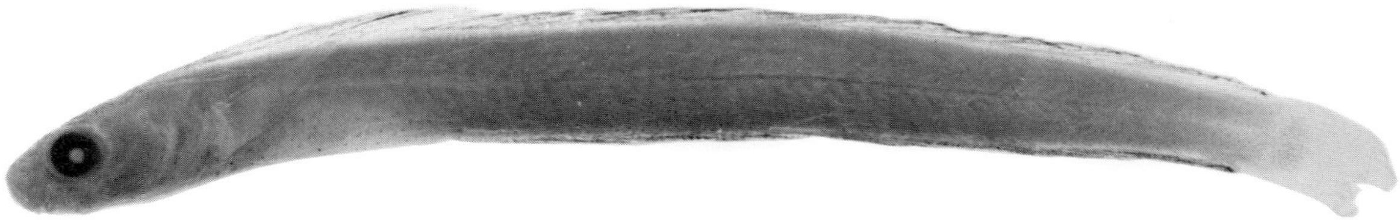

162

Plagiotremus: spilistius (**Fig. 160**), MNHN 1964-628, male, 164.3 mm SL, South Vietnam (enlarged view below); *celebesensis* (**Fig. 161**), SU 29081, holotype 54.5 mm, after Herre (1936a); *iosodon* (**Fig. 162**), USNM 99392, holotype, male, 38.8 mm, Sulu Archipelago, Jolo Island.

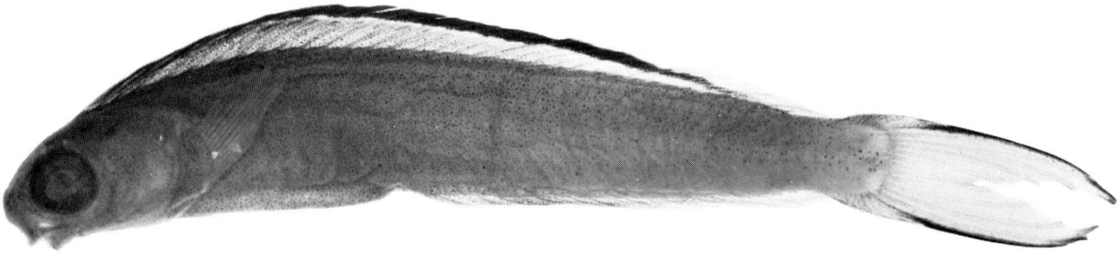

163

164

165

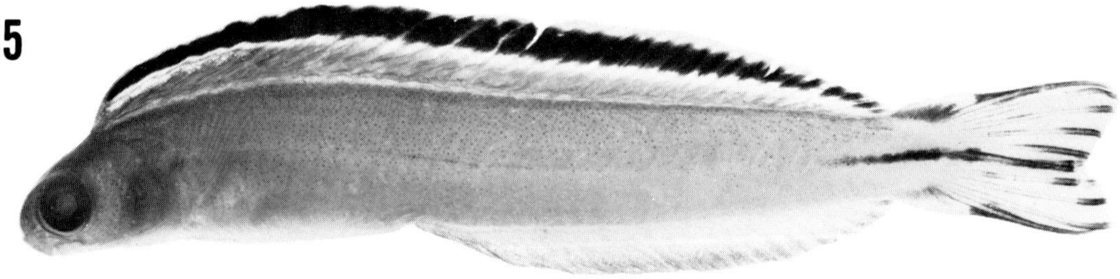

166

167

Plagiotremus: l. laudandus (**Fig. 163**), USNM 201413, male, 28.0 mm SL, Santa Cruz Islands, Vanikoro Island; *l. laudandus* (**Fig. 164**), USNM 209744, male, 40.6 mm, Moluccas, Amboina; *laudandus flavus* (**Fig. 165**), ANSP 128024, female, 44.4 mm, Fiji Islands, Viti Levu; *phenax* (**Fig. 166**), USNM 204699, holotype, male, 34.1 mm, Sri Lanka, Trincomalee; *townsendi* (**Fig. 167**), USNM 205379, male, 34.6 mm, Gulf of Aqaba, near Eilat.

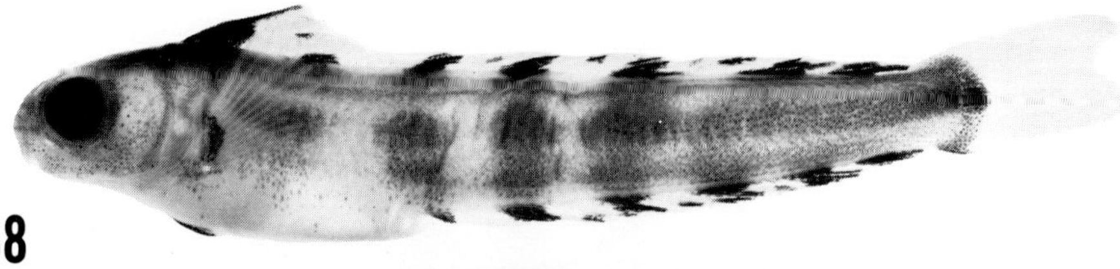

168

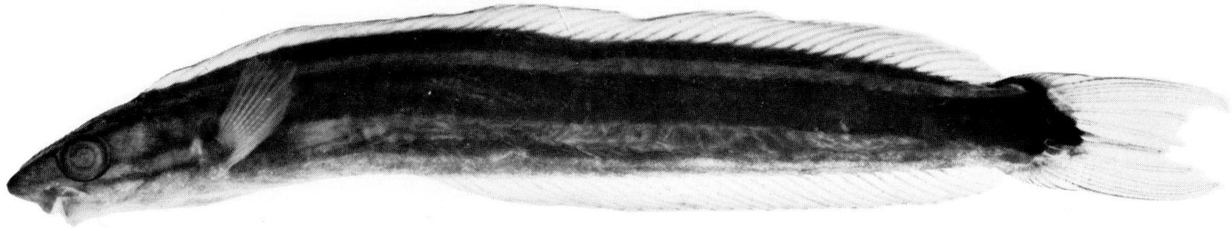

169

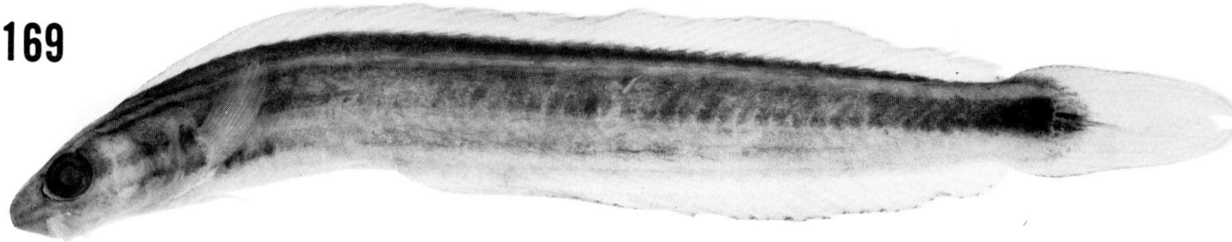

170

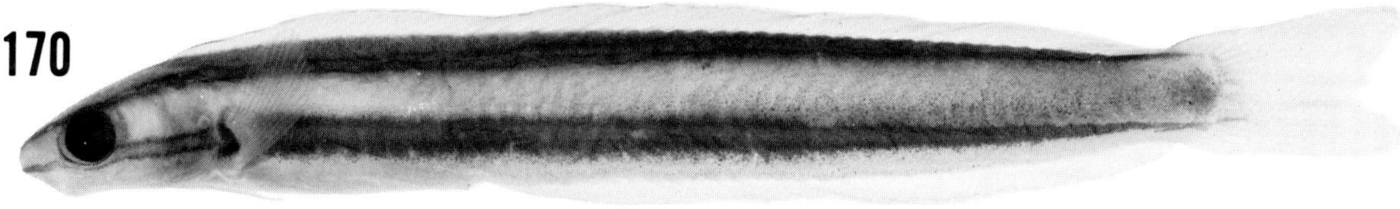

171

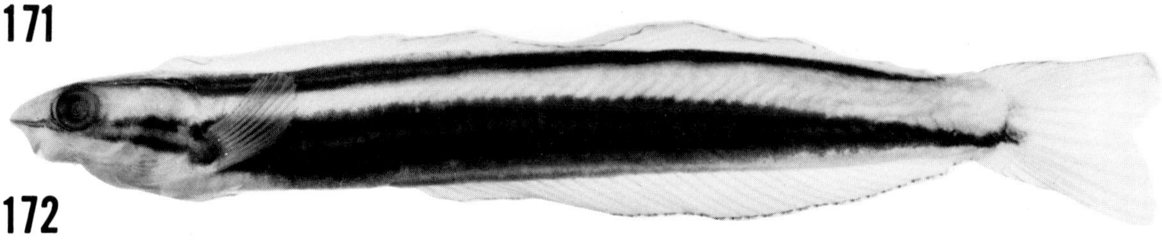

172

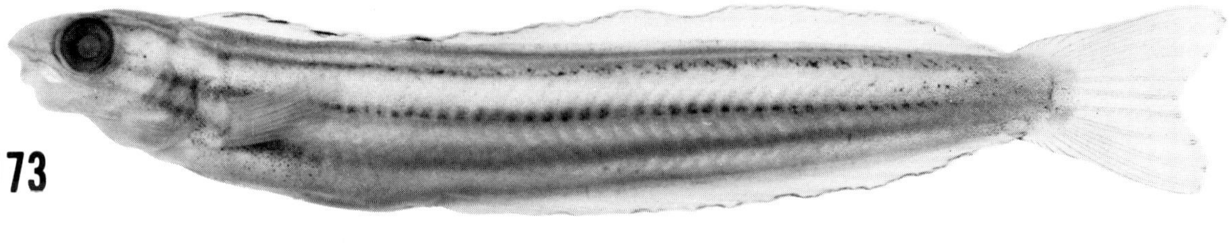

173

Plagiotremus: rhinorhynchos (**Fig. 168**), MNHN 1973-37 (holotype of *Runula fasciata*), juvenile, 34.5 mm SL, central Pacific Ocean, 2°15′S, 169°55′E; *rhinorhynchos* (**Fig. 169**), USNM 204573, male, 67.3 mm, Red Sea, Strait of Jubal; *rhinorhynchos* (**Fig. 170**), USNM 211971, male, 62.0 mm, Celebes, Buton (Boeton) Island; *rhinorhynchos* (**Fig. 171**), ANSP 109805, male, 62.2 mm, Amirantes Islands, D'Arros Island; *ewaensis* (**Fig. 172**), BPBM 7450, male, 64.0 mm, Hawaiian Islands, Oahu; *ewaensis* (**Fig. 173**), BPBM 7450, female, 48.7 mm, Hawaiian Islands, Oahu.

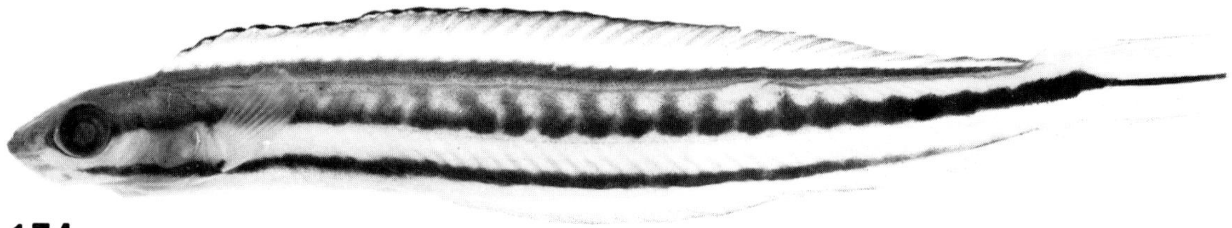

174

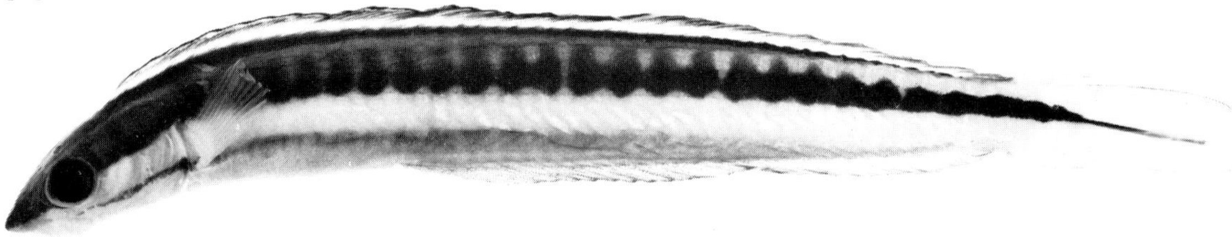

175

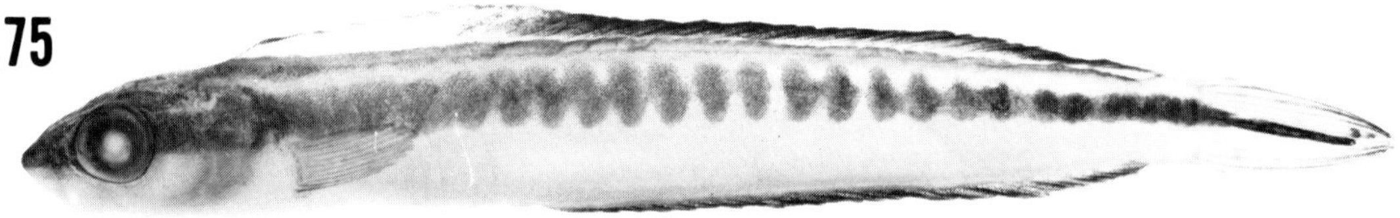

176

177

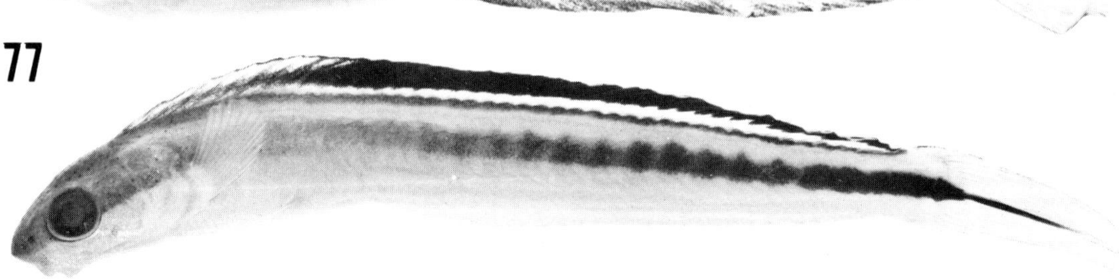

178

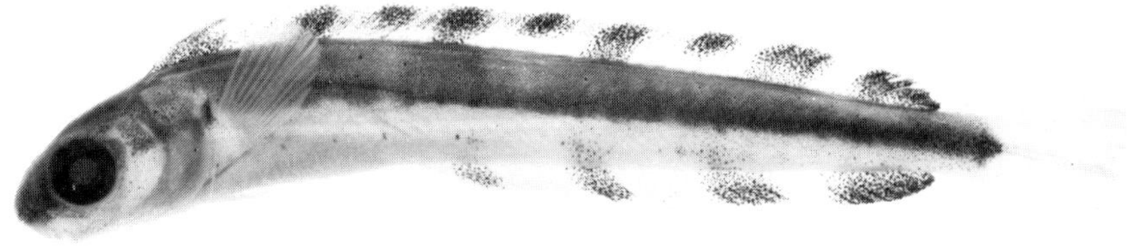

179

Plagiotremus: tapeinosoma (**Fig. 174**), USNM 204643, male, 49.1 mm SL, Gulf of Aqaba, Ras Burqa; *tapeinosoma* (**Fig. 175**), USNM 205423, male, 63.7 mm, Diego Garcia Atoll; *tapeinosoma* (**Fig. 176**), BPBM 10874, male, 34.7 mm, Marquesas Islands; *goslinei* (**Fig. 177**), BPBM 9960, female, 37.7 mm, Hawaiian Islands, Oahu; *azaleus* (**Fig. 178**), USNM 213476, male, 48.0 mm, Galapagos Islands; *azaleus* (**Fig. 179**), LACM 30313-3, juvenile, 32.0 mm, Revillagigedo Islands, Socorro Island.